Rainwater Harvesting for Agriculture and Water Supply

Qiang Zhu · John Gould · Yuanhong Li
Chengxiang Ma
Editors

Rainwater Harvesting for Agriculture and Water Supply

Science Press
Beijing

Springer

Editors
Qiang Zhu
Gansu Research Institute for Water Conservancy
Lanzhou, Gansu
China

Yuanhong Li
Gansu Research Institute for Water Conservancy
Lanzhou, Gansu
China

John Gould
Lincoln University
Canterbury
New Zealand

Chengxiang Ma
Gansu Research Institute for Water Conservancy
Lanzhou, Gansu
China

Additional material to this book can be downloaded from http://extras.springer.com.

ISBN 978-981-13-5734-3 ISBN 978-981-287-964-6 (eBook)
DOI 10.1007/978-981-287-964-6

Jointly published with Science Press, Beijing
ISBN: 978-7-03-045705-9 Science Press, Beijing

Springer Singapore Heidelberg New York Dordrecht London

Softcover re-print of the Hardcover 1st edition 2015

Printed on acid-free paper

Springer Science+Business Media Singapore Pte Ltd. is part of Springer Science+Business Media (www.springer.com)

Preface

China in common with most other countries around the world is facing unprecedented challenges due to the growing demands on its water resources. The problem faced is how to manage these sustainably so they meet both the needs of a rapidly developing nation today, while also providing for generations to come. One major part of this challenge involves how to carefully manage ground and surface water resources to protect them from overexploitation and pollution through promoting water conservation, recycling, and reuse of urban and industrial water supplies. An equally important part is how to provide improved water supplies to meet the increasing demands from agriculture and the needs and growing aspirations of rural populations. A failure to meet the needs of the rural populations in developing nations will simply lead to millions more remaining in impoverished conditions or joining the flood of poor rural migrants moving into new megacities. This will put even greater pressure on their water supplies, wastewater systems, and other infrastructure if the urbanization is not managed properly.

While most developing countries are struggling with these problems, in the realm of rural development China has made great strides. Since the 1980s, hundreds of millions of rural people have been lifted out of extreme poverty and had their livelihoods improved. This achievement has of course been the result of numerous different projects and interventions. One particularly successful intervention, which is the focus of this book is the promotion and widespread implementation of rainwater harvesting (RWH) across the loess plateau in Gansu and from there to other provinces across China. The new improved rainwater supplies, initially provided a lifeline to communities which were desperately short of water, but over time the careful utilization of the stored rainwater, in conjunction with low rate irrigation (LORI) methods and use of greenhouses, led to significant improvements in the livelihoods of individuals and the rural economy as a whole. In addition to providing food and water security, using RWH and LORI systems farmers have been able to grow a far greater range of produce which has led to greatly improved diet and opportunities in the market. By 2007, at least 22 million people across China were using rainwater supplies as their main water source for domestic use. In addition, 2.8 million hectare of farmland was using RWH for supplementary irrigation, leading to crop yield increases averaging 40 %. RWH has also played an important role in soil and water conservation in semi-arid regions along with the construction of terraces, contour bunds, and microcatchments. This has assisted in the re-vegetation and ecological restoration of large swathes of previously degraded land and helped reduce soil erosion and flood hazards. In China, RWH has become an important alternative to conventional water resources and a key to fostering sustainable development and environmental restoration especially in areas facing water scarcity.

Internationally, over the past few decades there has been a growing interest in RWH for improving domestic water supplies, for small-scale irrigation and for environmental conservation. In some countries, RWH has had a significant impact on rural food and water security, but with over a billion people across the developing world still lacking access to either safe convenient water supply or adequate food, much still needs to be done.

In order to share China's positive experiences and to help promote RWH worldwide, since 2003 the China Ministry of Commerce has been supporting an International Training Course on RWH and Utilization for Developing Countries for the professional development of participants working in the field of water resources and agriculture in developing countries. Course sponsors have also included the Ministry of Water Resources, the Gansu Provincial Government, and UNESCO. The Gansu Bureau of Water Resources (GBWR) and the Gansu Research Institute for Water Conservancy (GRIWAC) are in the responsibility of organization of the course. They have provided experts who in cooperation with a number of international RWH specialists with links to the International Rainwater Catchment System Association (IRCSA) and other agencies have been delivering the course. To date, over 848 International participants from over 81 countries have taken part in the course. This book has evolved out of the material used for the training course and is intended to provide a practical background reader for practitioners and others with an interest in RWH.

The purpose of this book is to present some of the key resource material developed for the training course with a particular focus on the RWH program in China, and international experience relating to RWH for water supply and agriculture with an emphasis on rural areas from across the developing world. The first six chapters focus on the development of RWH in the arid province of Gansu, China and cover the design, construction, operation, and management of systems developed over the past three decades for both domestic use and irrigation. The final five chapters include a global overview of RWH with chapters covering topics including roof water harvesting in the tropics, rainwater quality issues, and runoff farming. While most of the content deals with RWH for arid and semi-arid contexts, some of the material in these chapters is universally relevant and will be of use to those in more humid areas.

A great debt of gratitude is afforded to all the sponsors, lecturers, and participants for their contribution to the training course and a special acknowledgment afforded to all those who have contributed to this publication. It should be noted that the ideas and views expressed in this book represent those of the authors and not necessarily those of the editors or the organizations they represent. Thanks are also due to Ms. Xiaojuan Tang of GRIWAC, who has checked the style and format of the manuscript. Special thanks are due to Ms. Becky Zhao, Ms. Abbey Xiaojin Huang and Jenna Mengyuan Zhou of Springer Beijing and Ms. Xiuwei Li of China Science Press for their constructive comments on improving the manuscript and kind assistance in enabling this book to be published.

The editors would appreciate any feedback and comments to the material presented in this book, please send these to GRIWAC (gssk@163.com, machengxiang@hotmail.com).

July 2015

John Gould
Qiang Zhu

Contents

Contributors

Zhijun Chen FAO Regional Office for Asia and the Pacific, Bangkok, Thailand

Shiming Gao Gansu Academy of Agricultural Sciences, Lanzhou, China

John Gould Canterbury, New Zealand

Yuanhong Li Gansu Research Institute for Water Conservancy, Lanzhou, China

Andrew Guangfei Lo Chinese Cultural University, Taipei, Taiwan

Chengxiang Ma Gansu Research Institute for Water Conservancy, Lanzhou, China

Xiaojuan Tang Gansu Research Institute for Water Conservancy, Lanzhou, China

Terry Thomas DTU, School of Engineering, University of Warwick, Coventry, UK

Fengke Yang Gansu Academy of Agricultural Sciences, Lanzhou, China

Qiang Zhu Gansu Research Institute for Water Conservancy, Lanzhou, China

Part I
Rainwater Harvesting Experiences from China

Chapter 1
Why Harvesting Rainwater—China's Experiences

Qiang Zhu, Yuanhong Li and Xiaojuan Tang

Keywords Water scarcity · Loess plateau and hilly area · Rainwater harvesting · Poverty alleviation

1.1 General Information on the Water Resources Globally and in China

1.1.1 Global Water Resources

Water is one of the most important natural resources for human beings. With no water, there will be no life. All human productive activities require water. The relationship between water and agriculture is even closer. Crops need water from germination to harvesting. Water is an irreplaceable resource on our planet.

Water is a limited resource. The total amount of all kinds (liquid, solid, or gaseous state) of water (H_2O) in the hydrosphere of our planet is about 1.39 billion cubic kilometer (Shiklomanov et al. 1999). Of this, water in the oceans amounts to 97.5 %. World's fresh water only accounts for 35.2 million cubic kilometer, 2.5 % of the total water amount. Most of the fresh water (68.7 %) exists in the form of

Q. Zhu (✉) · Y. Li · X. Tang
Gansu Research Institute for Water Conservancy, Lanzhou, China
e-mail: zhuq70@163.com

Y. Li
e-mail: gssk@163.com

X. Tang
e-mail: tangxiaojuanrwh@163.com

Q. Zhu et al. (eds.), *Rainwater Harvesting for Agriculture and Water Supply*,
DOI 10.1007/978-981-287-964-6_1

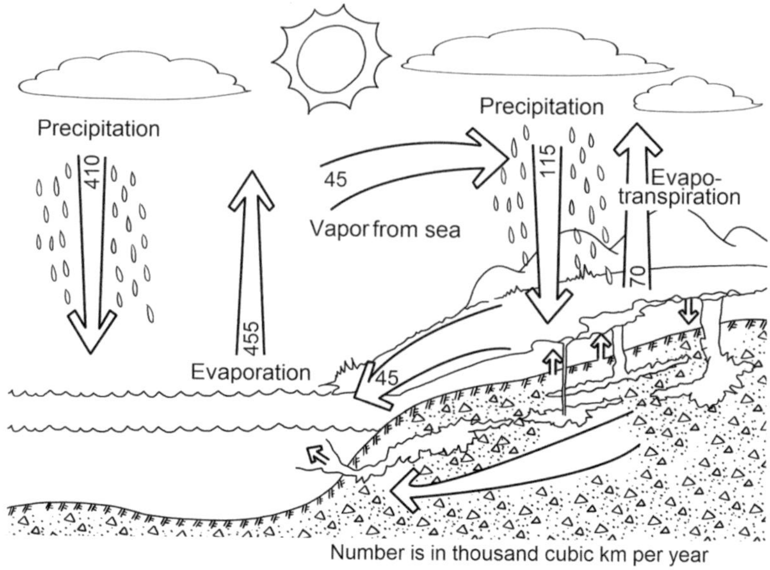

Fig. 1.1 Global hydrological cycle

ice and permanent snow cover in the Antarctic, Arctic, and mountainous regions. Next 30.1 % are fresh groundwater, and 0.8 % is in the permafrost. The remaining 0.4 % of the fresh water is the surface and atmospheric water. This portion is divided as follows: among freshwater lakes (67.4 %), soil moisture (12.2 %), wetlands (8.5 %), rivers (1.5 %), plants and animals (0.8 %), and atmospheric water (9.5 %) (Quoted from Water a share responsibility—The United Nations World Water Development Report 2, Shiklomanov and Rodda 2003). Figure 1.1 illustrates the global hydrological cycle.

When considering the sustainable use of fresh water, the amount of renewable water available is of most concern. The precipitation falling on the continents annually is from 113,500 to 120,000 km^3 (Quoted from Water a share responsibility—The United Nations World Water Development Report 2, Shiklomanov and Rodda 2003; FAO-AQUASTAT 2003), which is the only renewable water source on the land. This (rain and snow) produces some 42,600 km^3 of runoff in river systems and 2200 km^3 of underground flow (Shiklomanov et al. 1999). These water resources denote the sum of the surface runoff and the non-repeated groundwater flow. So the world water resources are in the order of 45,000 km^3. If this amount is divided by the global population (7.149 billion—in July 2013 as estimated by the United States Census Bureau), then each person would have a share of about 6300 m^3 of water annually. Although with population growth, this figure will reduce over time.

Moreover, these limited water resources are unevenly distributed. According to the data of 176 countries by the World Bank for 2011 (data.worldbank.org/

indicator), 44 countries have annual internal renewable water resources (AIRWR) per capita less than 1000 m^3, while 14 countries have AIRWR/head greater than 50,000 m^3. The lowest is only 0–3 m^3, while the highest is 532,892 m^3.

A United Nations' country-by-country assessment, published in 2000 jointly by WHO, UNICEF, and WSSCC, indicated that when entering the new century, the right of humanity's basic need of safe water is denied to nearly 1.1 billion persons who lack access to improved source of water. The Ministerial Declaration made at the Second World Water forum in Hague in March 2000 reiterated the need "to recognize that access to safe and sufficient water and sanitation are basic needs and are essential to health and well-being, through a participatory process of water management." It urged the global community to ensure that by 2025, the vision of water, sanitation, and hygiene for all should be a reality.

The 2000 UN General Assembly established the Millennium Goals, with 2015 as the horizon, which has been extended and further agreed by 2005 World Summit. Many targets such as to halve the proportion of people living in extreme poverty and suffered from hunger, to halve the proportion of people without access to safe drinking water and basic sanitation, to ensure environmental sustainability, etc., are directly related to water resources development. Water is also a pre-condition for reaching the goals for improving health condition. Reaching the MGDs applies great challenges to the water resources management worldwide.

According to the *Millennium Development Goals Report 2014* (http://www.un.org/en/development/desa/publications/mdg-report-2014.html), there has been important progress across all goals with some targets already having been met well ahead of the 2015 deadline. Global estimates indicate that goals such as reduction of extreme poverty, access to safe drinking water, improved lives for slum dwellers, and gender parity in education have been met. However, the remaining MDGs have been lagging, and the regional progress towards the MDGs is more diverse. Proportion of people suffered from hunger has continued to decline, but major efforts are needed to achieve the hunger target globally by 2015. In 2011–2013, 842 million people, or about one in eight people in the world, were estimated to be suffering from chronic hunger. On the sanitation target, 1 billion people still resort to open defecation, and 0.8 billion people still use unimproved toilet. On the goal for environment sustainability, there were still around 13 million hectares of forest lost worldwide each year between 2000 and 2010. Global emissions of carbon dioxide (CO_2) have continued their upward trend, increasing by 2.6 % between 2010 and 2011 and had a 48.9 % rise above their 1990 level.

Even in those targets already being met, the post 2015 jobs are still huge. Globally, in 2010, 1.2 billion people still live below the poverty line, and 76 % of those living in extreme poverty are in rural area. There are 748 million people still lack access to clean water accessible supplies. Additionally, those populations using an improved drinking water source may not necessarily have safe water. Many improved facilities are microbiologically contaminated, and water is not easily accessible to many households. The above facts indicate even greater challenges would be faced by the water resources management in the future.

1.1.2 Water Resources in China and the Challenges

China ranks sixth largest in the world in terms of the total amount of water resources. The mean annual precipitation (1956–1979) falling on the continental amounts to 619 billion cubic meter. The annual internal renewable water resources, composed of river runoff and non-repeated groundwater, are about 280 billion cubic meters (Hydrological Bureau, China Ministry of Water Resources and Electric Power 1987). However, owing to the huge population, the AIRWR is only 2060 m^3 per capita (population in the end of 2013 1.36 Billion), less than 1/3 of the global average. So despite having significant total water resources, China is a country facing some serious water shortages. The situation is further exacerbated by the uneven water distribution both spatially and temporally. In the southwest of the country, the annual water resources per capita reach more than 30,000 m^3, while in some north and northwest regions, it is as low as 300–500 m^3/head. Table 1.1 shows the proportion of land, population, and water resources in the different regions as a proportion of China's total. Seasonal disparities mean that most of the runoff is concentrated in the rainy season, causing serious flooding while in the other periods, drought occurs. For example, in northwest China, rainfall from June to September accounts for 60–70 % of the yearly total but in May and June when crops need most water, the rainfall is only around 20 %, which can barely meet their water demand. As a result, drought is frequent, causing a reduction of crop yield and even threatening livelihoods and food security.

When China entered the new century, it was facing serious challenges with regard to water resources management:

- According to the Criteria of Safe Rural Water Supply issued by the Ministry of Water Resources and Ministry of Health in 2004 (see Box 1.1), by 2010, the water supply for 298 million rural population was not safe (State Development and Reform Commission 2012);
- Four hundred cities out of 668 are lacked enough water and of these 108 cities are suffered from serious water shortage that amounted to 6 billion cubic meter

Table 1.1 Proportion of land, population, and water resources to China total, %

River basin	Land	Water resources	Population	Cultivated land
Endorheic rivers in northwest China	35.4	4.6	2.1	5.8
Rivers in northeast China	13.1	6.9	9.8	19.8
Haihe River	3.3	1.5	9.8	10.9
Yellow River	8.3	2.6	8.2	12.7
Huaihe River, etc.	3.5	3.4	15.4	14.9
Yangtze River	18.9	34.2	34.8	24
Rivers in southeast China	2.5	9.2	7.4	3.4
Rivers in south China	6.1	16.8	11	6.8
Rivers in southwest China	8.9	20.8	1.5	1.7

Source Planning and Designing Institute for Water Resources and Hydropower 1989

annually (China Ministry of Water Resources 2003). Water shortage caused annual loss of 25 billion USD of industrial production and 25 million tons of grain;

- According to the report by the State Council to the 12th Standing Committee of the People's Representative Congress on poverty alleviation, to the end of 2012, the impoverished rural population comprises about 100 million, and water scarcity is often one of the root causes of poverty. In the poorest regions about 55 million people lacked safe drinking water.
- According to the State Statistics Bureau, in 2006, among 411 monitoring stations in the seven main river basins in China, 32 % of the river courses were categorized as having water quality of Class IV or V and 17 % worse than Class V (China Ministry of Water Resources 2003). In the 28 largest lakes, water quality in 11 lakes was Class IV or V, and ten lakes had water quality worse than Class V. The groundwater quality has also been deteriorating due to industrial and agricultural pollution and over-exploitation.

Box 1.1 Criteria for Safe and Healthy Rural Domestic Water Supply
In 2004, China Ministry of Water Resources and Ministry of Health have issued a document stipulating index system for evaluating the safe and healthy rural domestic water supply. The index system includes 4 parts, namely water quality, water quantity, convenience for fetching water, and the water supply reliability. A safe and healthy rural domestic water supply should meet the 4 criteria simultaneously. If any of the four is not met, then the water supply is deemed as unsafe or unhealthy.

- Water quality: the water supply is safe and healthy if the water meets the demand of <Standards for Drinking Water Quality (GB 5749-2006)> issued by the Ministry of Health in 2007. If the water cannot fully meet the GB5794 but can meet the <Criteria in Implementing the Standards for Drinking Water Quality (GB 5749-2006) in Rural Area>, then it is deemed as qualitatively safe in the main;
- Water quantity: the whole country is divided into 5 zones in terms of the climate, topography, water availability, and culture conditions. For each zone, the allowable lowest water supply quantity is set up, see Table 1.2.
- Convenience of fetching water: time of round trip for fetching water should be less than 10 min. If the time is between 10 and 20 min; the water supply is deemed as meeting safety criteria in the main;
- Supply reliability: the reliability of safe water supply should be 95 %, and reliability of 90 % is deemed as meeting safety criteria in the main.

Water resources have become a major restricting factor with respect to social and economic development as well as environment conservation in China. Furthermore, in the areas listed below, water shortages either on an annual or seasonal basis are even more critical and threaten livelihoods and food security, namely in

Table 1.2 Criterion of water supply quantity for safe and healthy rural domestic water supply

Zone	I	II	III	IV	V
Quota for safe water supply, L/day/person	40	45	50	55	60
Quota of water supply meeting safety criteria in the main, L/day/person	20	25	30	35	40

Note Zone I is the arid and semi-arid areas in the western China. Zone II is part of the semi-arid and sub-humid areas in the north and northeast China. Zone III is part of the semi-arid and sub-humid areas in north China. Zone IV is the humid area of western China. Zone V is the humid area of east and south China

- Semi-arid loess plateau regions of Gansu, Ningxia, Shaanxi, Qinghai provinces, and Inner Mongolia Autonomous Region of the northwest China;
- Semi-arid mountainous areas in Shanxi and Hebei provinces of north China;
- Karst regions in Guizhou, Guangxi, and Yunnan provinces of southwest China;
- The dry hilly areas in Sichuan and Henan provinces of middle China;
- The islands and rocky coastal areas, where the fresh water sources are lacking around the coast.

The first 2 areas belong to semi-arid area zone, where the annual precipitation is less than 550 mm. The last 3 areas have yearly precipitation more than 550 mm, even more than 1000 mm but owing to the unfavorable rainfall regime seasonal droughts are common.

These critical areas are characterized by the following features:

- Serious water shortage and water insecurity: most of the rural populations without access to safe water supply are concentrated in these 5 areas.
- Low economic status of local inhabitants: water shortage caused widespread poverty. About 60 % of China's most impoverished counties locate here.
- Frequent drought causes low agriculture productivity and food insecurity. Water shortages also limit crop varieties to those with low production values.
- Serious soil erosion, degraded land, and degraded environments.

1.1.3 Water Scarcity Problem in the Loess Plateau of Gansu Province

Gansu, located in the interior in Northwest China, is one of the driest and poorest provinces in the country. The annual precipitation is only 306 mm while the potential evaporation ranges from 1500 to 2500 mm. Water resources per capita are only 52 % and cultivated land per capita only 32 %, of the national average. The situation is even worse in the loess area located at the central and eastern parts of the province. It has a total area of about 145,000 km^2 with population of 13.7 million (Luo 1987). The area lies within the Yellow River Basin, and the key geomorphologic feature is the extensive loess plateau and hills which are crisscrossed by numerous gullies and ravines. Loess soil is highly permeable with depth up to 100–300 m. Most of the precipitation is absorbed by the loess soil. The runoff coefficient is around 0.1, compared with an average of 0.45, nationally. The surface and subsurface water are scarce. The annual precipitation ranges between 300 and 550 mm, and total water resources equate to less than 1000 m^3 per capita. In the middle and north part of loess area, water resources are as low as only 230 m^3 per capita. Moreover, the precipitation is unfavorably distributed: 50–70 % occurring in July to September, with only 19–24 % in May and June that is the main crop growth period. Owing to the topographic and geological conditions, water conveyance systems are very difficult to build. Agriculture is mainly rain-fed making it vulnerable and unreliable. Historical records show that 749 droughts occurred in the period 206 BC-1949 AD (2155 years), but in the 41 years from 1949 to 1990, there were 31 droughts happened, and seven of these were severe. The Yellow River Basin in Gansu was particularly badly affected with 13.8 million ha, 79 % of the provincial total affected (Editorial Board of Disaster of Flood and Drought in Gansu 1996). The droughts caused low crop yields, less than 1 t/ha for wheat in the dry years, which could not even provide sufficient for seeds for the next crop.

Water scarcity also resulted in a monotype agricultural structure as most cash crops could not be grown. Ninety percent of the land was planted with grain crops. The annual income per capita was as low as 100–150 USD in the 1980s.

Another result of water scarcity was the serious soil erosion. On the loess plateau of Gansu, erosion rates as high as 5000–10,000 t/km^2 were recorded. This is equivalent to a surface layer of soil 5–10 mm thick being stripped off each year. Owing to the low productivity of the land, people tried to reclaim as much as possible. Even steep sloping land was cultivated, causing land degradation and environmental deterioration as well as intensifying soil and water loss. There is a saying: "The poorer the local people were, the more they reclaimed; the more they reclaimed, the poorer they would be." Figure 1.2 shows the vicious circle between water scarcity, agriculture production, and environment deterioration. In such a

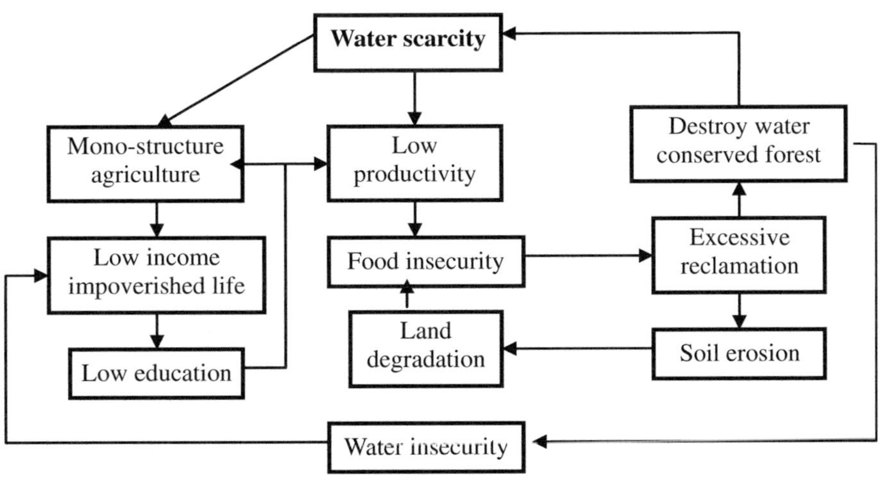

Fig. 1.2 Flow chart of vicious circle of water, land and environment

cycle, the agriculture production was kept in very low level, and people suffered from impoverishment for generations.

The most serious impact of water scarcity is that before the rainwater harvesting project, over 3 million people in the area had no access to a safe water supply. In a normal year, their water sources were in the bottom of valleys, small springs, or ponds shared by humans and animals, etc. In the dry years, the government had to send trucks carrying water from great distances, and people had to stand in a line waiting for a bucket of water that was a two or three day ration for family use. This water shortage had a serious impact of the health and welfare of the affected populations.

1.1.4 How to Address Water Scarcity Problem?—A Case Study of the Loess Plateau of Gansu

The local farmers and agriculturalists and technicians of Gansu have made great efforts over the years to improve the agriculture production and livelihoods. Most of these efforts were focused on the traditional dry-land farming practices. According to the Britannica, dry farming is referred to "The cultivation of crops without irrigation in regions of limited moisture, typically less than 20 in (50 cm) of precipitation annually. Dry farming depends upon efficient storage of the limited moisture in the soil and the selection of crops and growing methods that make the best use of this moisture" (www.encyclopedia.com).

On the loess plateau of Gansu, the unfavorable topographical and geological conditions make irrigation systems very difficult to build. Besides, because of the unique features of the loess soil, irrigated land would be at risk of subsidence, slips, or even collapse if excessive irrigation caused the soil to become too wet. The natural precipitation is the only water source readily available for cultivation in the area. But most of this valuable rainwater is lost due to the steep slopes, sparse vegetation cover and poor land management. The key principles for better utilization of the rainfall are as follows: first, to keep the rain in the soil as much as possible, second, to prevent soil moisture from evaporating, and third, to select suitable crops that can better adapted the natural rainfall conditions. Following these principles, the following adaptation measures for enhancing agriculture production have been adopted;

- Reforming the land to retain more rainfall, such as terracing, contour planting;
- Cultivation measures to keep soil moisture from evaporation such as deep plowing, harrowing and tillage, mulching, etc.;
- Laying plastic sheeting or straw mulch on the fields to reduce evaporation loss;
- Fertilizing to increase resistance of crop to water stress;

- Breeding new varieties that can have higher resistance to water stress and adaptability to the rainfall condition; and
- Construction of water conservation structures such as micro-catchments, floodwater harvesting and spate irrigation, etc.

Among these measures, terracing was the first one being widely used on the loess plateau since the 1950s. In 2010, the terraced land amounted to 2.07 million ha, 52 % of the slope cultivated land (Gansu Provincial Government 2011). According to a report on terracing (Soil and Water Conservation Bureau 1994), terraced land can retain 750 m^3/ha of runoff annually and reduce soil erosion 60–150 t/ha. Yield measurements on 3934 ha of terraced land and 5282 ha of slope land from year 1986 to 1990 in the Jinchuan County showed that the yields on the terraces were 43.7 % higher than on sloped land. Therefore, terracing has become one of the most important measures for enhancing crop yields in the loess region of Gansu. However, much of the soil moisture retained on the terraced land in the wet period from July to September is lost between harvest time and the sowing period next spring due to intense evaporation. Tests conducted by GRIWAC showed that moisture in the upper 30 cm soil layer reduced from 14.8 to 11.5 %, reducing by 22 % (GRIWAC et al. 2002). In the recurrent droughts in the 1990s, crops in the terraced land also suffered from significant water stress and the yields were still low.

Another widely adopted measure is the plastic sheeting. Results of irrigation experiments in Gansu showed that in the crop evapo-transpiration, evaporation from the soil accounts for more than 50 %, of which most can be reduced by plastic sheeting. Tests by GRIWAC showed that corn yield with plastic sheeting can be increased by 22 % (GRIWAC et al. 2002). However, this depends on whether there is rainfall during the critical growing period, if not, the soil moisture is still not enough for a normal growth.

Corn has much higher yield than wheat because of its longer growing period and its adaptation to the natural rain, see Fig. 1.3. Before the 1980s, its planting was restricted due to the risk of early frosts before harvest in late September. Since the 1980s, corn could be sowed earlier because the widespread use of plastic sheeting, which can raise the ground temperatures by 2–3 °C thus preventing corn suffering from early frost. Since then planting corn instead of wheat became

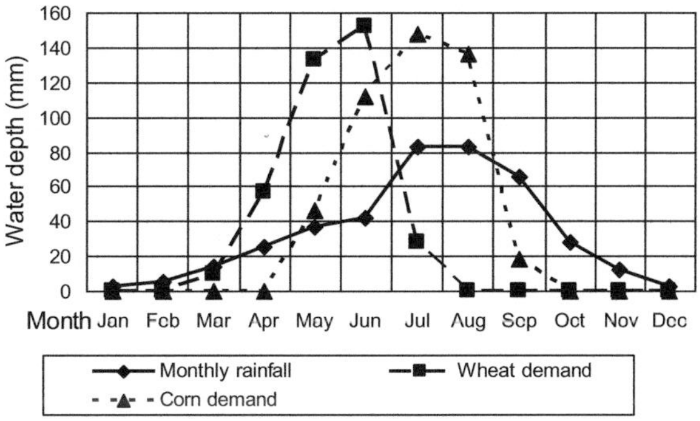

Fig. 1.3 Crop water demand versus rainfall

one of the measures for enhancing the agriculture productivity. However, with the unfavorable rainfall distribution, corn still suffered from water shortage. From Fig. 1.3, it can be seen that the biggest water shortage for corn is in June when the corn grows to the big bell-mouthed period, the most critical period affecting the yield.

To sum up, traditional dry-land farming methods have shown to have limitations especially during drought years. From Fig. 1.3, it can be seen that the crop water demand cannot be met, either on an annual or seasonal basis. When the rainy season arrives in July to September, the two major crops have already passed through their main growing periods and are approaching harvest time. The surplus rainwater in the soil can be used only in the following spring when crop needs the water most.

The principle of water management for dry-land farming is to store as much rainwater in the soil as possible and keep it available until when the crop needs the water. In the loess area of Gansu due to the adverse conditions, the retention capacity of soil moisture is insufficient even when measures such as terracing and plastic sheeting are taken. It can hardly bridge the long gap between the time that the crop needs water most and time when the most rain falls. This is also a period of intense evaporation. Water shortage is greatest in May and June which are the most critical time for crop growth: wheat is passing through stages of heading and milking, and corn is passing through stages of booting and flowering. These stages have the greatest effect on the yield.

The above analysis indicates that the inadequacy of the adaptation capability of the traditional dry-land farming systems to the climate variability is mainly due to the low retention capability of the soil to the natural rain. This is the main reason why crop yields has been low for such a long time.

Theoretical analysis and practical experience bring us to the conclusion that only through combining an artificial water supply to the crop with the traditional dry farming measures would it be possible to boost crop yields in the dry-land farming system.

The problem then is how to get water? The conventional solution is to divert water from another area—for example, through inter-river-basin water diversion. But the local conditions present many barriers to this solution:

- For high input and O&M costs, even though the government would invest the initial cost, the O&M cost is still difficult for the water users to bear;
- The long lead time to bring such a large project into play;
- The scattered nature of settlement in the remote mountainous communities makes a centralized solution unrealistic and uneconomic. Environmental problems including ground subsidence, slips, and land collapse would associate with these large-scale irrigation projects in the Gansu context.

This expensive, large-scale, high technology solution can be realized only through a long-term painstaking efforts. The most potential, readily available, and easy-to-use water resource alternative in the area is the rainwater harvesting.

1.2 Innovation—Rainwater Harvesting Development in Gansu

1.2.1 The Potential and Unfavorable Condition for Rainwater Harvesting in the Loess Area of Gansu

Although the rainfall in Gansu is low, the total amount of rainwater falling on the whole loess area of almost 140,000 km^2 annually is equivalent to 57 billion cubic meter, more than the yearly runoff of the Yellow River at Lanzhou. Of this total rainfall, it is estimated that 20.7 % forms river runoff and renewable groundwater (blue water), 18.6 % is directly used by the crops (green water) and the sparse vegetation in the area, and rainwater harvesting now uses 0.086 %. The remaining 60.6 % is consumed by evaporation. So there is a high potential for greater rainwater utilization.

The local people have traditionally used rainwater. The history of rainwater utilization can be traced back a thousand years. The *Shuijiao* and *Shuiyao* (two major types of underground tank in the area) have been widely built by the local people to store runoff, and rainwater harvesting is widely accepted. The local population has also mastered the techniques of building the traditional water cellar, water cave, and upgraded underground water storage tank.

Rainwater utilization in Gansu does face some challenges.

- The low yearly rainfall: the annual rainfall ranges between 250 and 550 mm and averages only about 400 mm;
- The uneven distribution of rainfall in a year and between years makes the rainwater difficult to collect and use;
- In the semi-arid areas of Gansu, half of the rain events are classified as small (rainfall amount and intensity in each event being less than 10 mm and 0.02 mm/min, respectively), which cannot produce runoff on a natural soil surface.

Owing to the above unfavorable conditions, the efficiency of rainwater harvesting in the past was very low, and it could not even meet the demand for domestic use, let alone for supplemental irrigation. To enhance the rainwater harvesting efficiency, the best strategy is to concentrate the rainwater both spatially and temporally.

The spatial rainwater concentration involves raising the rainwater collection efficiency (RCE) by treating the catchment surface and enlarging the rainwater catchment area. Or as the local people say: "Use several pieces of sky for the people under one piece of sky." Temporal rainwater concentration involves the storage of the rainfall in the rainy season for use in the dry season.

In the late 1980s, owing to the serious water shortage in the loess area of Gansu and its impact on the people's welfare due to the low agricultural productivity and environmental degradation, urgent action was needed. Experts and leaders from the Gansu Bureau for Water Resources and GRIWAC focused on how to address the problems by using the only potential water source, the Rain, in an efficient way, thus, started more than 20 years development of the rainwater harvesting in the loess area of Gansu.

1.2.2 A Brief Description of the Development of RWH Projects in Gansu

Rainwater harvesting projects in Gansu has passed through 3 stages, namely, the Research, Demonstration and Extension Project, the "1-2-1" Rainwater Catchment Project and the Rainwater Harvesting Irrigation Projects.

1.2.2.1 The Research, Demonstration, and Extension Project

An experimental research, demonstration, and extension project were first established by the GRIWAC and the water resources bureaus in relevant counties in 1988 under the support of the Gansu Provincial Government and the Gansu Water Resources Bureau. Before carrying out the project, a survey of the existing traditional RWH system was made. It was found at that time that each family in the area had on average 2.2 water cellars. But many of the storage tanks were only partly filled, or even empty, because the catchment for collecting rainwater were mainly the country roads and natural slopes that have a very low rainwater collection efficiency (RCE). Water stored in the tank could not meet the demand for drinking purposes. Also the quality of the stored water was very bad: silt and dirt were usually found inside the tank. The water cellar has the advantages of low cost, less construction material needed, and easy to build. But many of them were lined with straw clay mud for seepage control, which was of poor quality resulting in seepage and sometimes collapse of the tanks.

The research project, using the findings from the survey, aimed to enhance the rainwater use efficiency, raise the water quality, and study the technical and economic feasibility of RWH systems for both domestic water supply and supplemental irrigation. For this purpose, the rainfall-runoff relationship under different rainfall characteristics and on different rainwater collection fields was studied, the appropriate rainwater utilization patterns were tested, and the design procedure for the RWH system was formulated.

Firstly, 28 testing plots with 8 different materials were set up, and the rainwater collection efficiency (RCE) was tested with both natural and artificial rain (produced using sprinklers). Figure 1.4 shows the test sites for rainwater collection efficiency in the Huining County. Materials for the rainwater collection surfaces included 8 kinds: concrete, tiles (cement tile, machine, and manual made clay tile), plastic film, cement soil, lime soil, and the compacted soil. During the years of

Fig. 1.4 The experiment sites for RCE test in Huining County of Gansu Province

1988–1990, there were 421 rain-runoff tests conducted. A total of 766 sets of data including rainfall, intensity, runoff, and water content of the catchment material before rain were obtained. Based on these tests, the relationship between RCE and rainfall amount and intensity was worked out for the 8 materials. In addition, effect of the moisture of the catchment material before rain on the RCE for compacted soil and natural soil has also been studied.

With the data from the 421 rain events, the yearly RCE was calculated using the data of 9 typical gages selected from 100 gages over about 80,000 km^2 of land, representing areas with mean annual precipitation of 300, 400, and 500 mm. For the three areas, yearly RCE in the normal, dry, and extremely dry years was calculated for all 8 different materials. The detailed methodology for the calculation will be introduced in the Chap. 2.

The rainwater harvesting pattern was studied on the basis of traditional practices and upgraded with modern knowledge. In the past, people regarded rainwater harvesting as just a matter of building the storage tank and did not pay much attention to having an efficient and healthy catchment. During the study, an integrated RWH system was designed, composed of three parts, namely, the catchment or rainwater collection field, the water storage, and the water supply and irrigation system. To ensure the water quality and to enhance RCE, a tiled roof and concrete lined courtyard were adopted as the standard catchments for domestic use. The underground tank or *Shuijiao* (water cellar) located in the courtyard of the household was used for storage. It had the traditional style, but the straw clay mud for seepage control in the original design was replaced with cement materials. The most commonly used equipment for extracting the water was a bucket plus rope or hand pump. For irrigation, a highway, threshing yard, sport ground, etc., were used, and the storage tank was the same as that in domestic use except its location was closer to the field. Simple but effective methods suggested by the local people were adopted for supplemental irrigation. Drip and other modern systems were also used for the demonstration.

During the project in the early 1990s, the research team worked out guidelines for the design and construction of the RWH system. These were later developed into a local technical standard, in which the design procedure and quality control of construction as well as the requirements for operation and maintenance were formulated (Gansu Bureau of Water Resources and Gansu Bureau of Technical Supervision 1997). It has also been a basis of the National Code of Practice for the Rainwater Collection, Storage, and Utilization compiled by several relevant institutes and issued by the State Ministry of Water Resources in 2001 (Ministry of Water Resources 2001).

On the basis of the above research findings, demonstration projects and extension were carried out since 1989. The resulting household RWH system for daily supply included an 80–150 m^2 tiled roof and/or concrete paved courtyard (depending on the local annual rainfall), one or two water cellars with a storage capacity of 15–20 m^3 each and a hand pump or bucket plus rope. Pilot projects for irrigation purpose were also set up including the RWH system and the greenhouse (Fig. 1.5).

Fig. 1.5 RWH irrigation greenhouse built in 1991 in Beishan, Yuzhong County

By the end of 1994, 2.4 million m^2 of catchment and 22,280 water cellars with improved design had been built. Investigations showed that RCE of these systems reached 67–70 %. In 1991, a serious drought seen only once in 20 years occurred in Xifeng Municipality. Farmers who owned a RWH system had enough water to use till the next rainy season and even shared spare water with their neighbors, meanwhile the other farmers without the new RWH systems had to rely on the relief water from government tankers. So the new RWH systems were warmly welcomed by both the farmers and the government. Progress was also made with respect to the rainwater irrigation project. Twenty-two pilot projects were established with total irrigated area of 1.4 ha. In the mountainous area of Yuzhong County, 16 greenhouses were also built, and the first "fine vegetables" such as cucumbers, peppers, and tomatoes were produced in this mountainous area (Zhu and Wu 1995).

1.2.2.2 The "1-2-1" Rainwater Catchment Project

In 1995, a serious drought occurred in the Northwest China. From October 1994 to June 1995, little rain fell on the whole of the loess plateau in Gansu. There were about three million people and two million animals who lost their water supplies. As a result, four hundred thousand people had to fetch water from a distance of 10–40 km. In some places, the market price for one m^3 was as high as 100 yuan (12 USD). Most of the summer crops in the rain-fed areas died out.

During his trip to the areas suffered from drought, the Governor of Gansu visited the RWH pilot projects in the Yuzhong County and found them to be very effective in addressing the water shortage. After the visit, he published a paper on the Gansu Daily, in which he made high evaluation to the RWH projects and suggested to replicate this successful experience. Later the Provincial Government made a decision to implement the "1-2-1" Rainwater Catchment Project aiming at addressing the water shortage problem in the area where no surface water or groundwater exists but only rainwater was available. The project name of "1-2-1" meant that the government would support each family in the area to build ***one*** rainwater catchment with an area of 80–150 m^2, two underground water tanks each with storage capacity of 15–20 m^3, one for domestic supply and another for irrigation, and ***one*** piece of land by the side of the house to be irrigated by the stored rainwater (Zhu and Li 2000).

The Project was implemented by the united efforts of local farmers, the technical agencies, and the government. The households contributed labor and provided local building materials. Relevant institutions, including GRIWAC provided free technical advice. The government provided support to each household with cement for building the catchment and the storage tank. Based on the experience with the demonstration projects, the government decided to support each family with 1.5 tons of cement equivalent to CNY 400 (US$ 50). The population in the area for whom rainwater was their sole water source totaled to 1.2 million. To achieve the goal of "1-2-1" project—solving these rural people's water problems, one hundred and nine million Chinese Yuan (13.4 million USD) of subsidies were needed. The government was able to provide 53 million CNY. To meet the balance a broader fund-raising campaign was mobilized by the government and the media. This resulted in a total of 56.43 million Yuan (6.9 million USD) being raised from the whole society, including some from other provinces and municipalities as well as foreign friends.

From July of 1995 to the end of 1996, implementation of the 1-2-1 Project became the central task of the government at different levels in the relevant counties. A strict audit was undertaken to monitor the use of funds to ensure a fair allocation of the resources (primarily cement) to the villages and households. Technical institutions at the provincial and county level provided training/education for the technicians in the townships and to the farmers. Since the RWH system was based on traditional technology that the local farmers soon learnt and mastered, the new construction methods and almost all the RWH systems were built by the farmers themselves. The policy issued by the government was "who builds, who owns, who manages." This greatly stimulated and motivated the farmers to build their own systems. After the RWH system was finished, the county water resources bureau would make inspection to see if it met the quality standard. After its acceptance, the information pertaining to the system would be kept on file and the certificate of ownership issued to the household.

After just one and a half years, the project had met its preset goals. The beneficiaries totaled 1.31 million (264,000 households) in 2018 villages under the jurisdiction of 27 different counties. People in these households upgraded their roofs and paved the courtyard with concrete slabs for collecting rainwater. The total area of these two forms of catchment was 37.16 million square meter. There were 286,000 newly designed water cellars built. Furthermore, additional water cellars provided supplementary irrigation for 1330 ha of land for developing the courtyard (household) economy, and helping farmers generate income. Figure 1.6 shows a typical household 1-2-1 Rainwater Catchment Project.

1.2.2.3 Rainwater Harvesting Irrigation Project

Based on the successful experience of the "1-2-1" Project, the government and the farmers realized the great value of rainwater as a resource. In late 1996, the government initiated the Rainwater Harvesting Irrigation Project. While "1-2-1" project supplied domestic water for meeting the basic needs of people, the irrigation project opened up the way for the farmers to develop their household economy and livelihoods in this dry mountainous area, Fig. 1.7.

To lower the cost, people made full use of the less permeable surface of existing structures like paved highways, country roads, threshing floors, and natural slopes for collecting rainwater. In areas located far away from the highway, concrete lined or plastic film covered surfaces were also used as catchments for planting high value crops. The type of storage tank was the same as that in the system for domestic water supply but had a larger volume. Usually the capacity ranged from 30 to 70 m^3.

Fig. 1.6 A typical household 1-2-1 Rainwater Catchment Project

Fig. 1.7 Rainwater tank for supplemental irrigation

Because of the very limited amount available, rainwater for irrigation was used in the highly efficient way, which was realized as follows:

- Firstly, the deficit irrigation approach was adopted. That is to say, water was applied only in the critical periods of crop growth. Compared to the conventional irrigation practice, the frequency of irrigation with RWH systems was less. Normally, one water application during seeding with only 1 or 2 further water applications during the growing season for wheat or corn.
- The water use quota in each application was only 5–15 m^3/Mu (75–225 m^3/ha, Mu is a land unit in China, equals to 1/15 ha or 667 m^2.) while in the conventional irrigation, it is usually 40–60 m^3/Mu for each application. Simple but highly efficient irrigation methods were used, such as irrigating during seeding, manual irrigation, and irrigating through the holes on the plastic sheeting, etc. Sometimes, when the farmers had money or they can get loan from the bank, they also used modern drip and mini-spray systems.
- Irrigation was targeted only to the crops' root zone to avoid evaporation losses.

Chapter 5 includes a detailed description on how to irrigate crops using the RWH system.

Although irrigation water with the RWH irrigation has a very low capacity, its effect on increasing the crop yield is significant. Investigations and irrigation tests showed that crop yields could be raised by between 22 and 88 %, or 40 % on average. Besides, with water in the tank, farmers were able to diversify crop production with a much wider range of produce to meet the market needs. Cash crops that were impossible to grow in the past could be developed using the RWH irrigation systems. This led to increasing household incomes and poverty alleviation in the region.

RWH irrigation projects developed rapidly in Gansu. By the end of 2001, there were 2.2 million newly built storage tanks, allowing supplemental irrigation on about 80,000 ha of land.

1.3 Evaluation of the RWH System in Gansu

1.3.1 Significance of RWH Projects in Gansu

1.3.1.1 Effective Way for Domestic Water Supply

With the RWH system, the local farmers get a reliable and cheap water source. In the 1-2-1 Project, due to the funding limitations and the small area of household roofs and courtyard at that time, the water supply available was limited. In a normal year, it was 10 L/person/day, while in an extremely dry year with 20 years recurrence, the design supply quota was only 6 L/person/day. This level of water supply could only meet the basic needs for human existence (drinking, cooking and limited personal hygiene). Despite this, the farmers were very happy with the change in their domestic water supply situation both in quantity and quality when compared to the conditions experienced before the RWH project, and said, "The dirty water becomes clear, the bitter water becomes fresh and the far away water becomes nearby."

Along with the improvement in the economic conditions of the households and the increase support from the government in the New Village Construction Program, many households' living condition and their rainwater harvesting system have been gradually improved. In the Daping Village of Qingran Township, one of the poorest townships in the province, the annual rainfall is about 400 mm. Partially subsidized by the government, villagers has re-built their house and improved their domestic water supply in the year 2003. In the home of one villager, Ran Xiong, the area of his tiled roof and concrete lined courtyard amounts to 200 and 180 m^2, respectively. He has 2 water cellars with total capacity of 80 m^3. Water for drinking and cooking is now abundant. His family now has enough water to use a washing machine twice a week and has hot (solar heated) water for showering. Besides this, he has surplus water to irrigate 8 pear trees and a small vegetable garden, see Fig. 1.8. The Daping's example shows that in the semi-arid climate condition, the RWH system can enable the rural people to have leisure water condition.

On the water quality aspect, although the villagers are much satisfied at the quality of the stored water as compared to the past. However, owing to the pollution during collection and storage, especially in most situations, runoff from the roof is mixed with surface runoff. Quality of the stored rainwater is technically defined as poor. According to testing, there was no or very little chemical and toxicological pollution in the cellar water. The quality problem was mainly on the microbiological aspect and the turbidity. Several measures for improvement were

Fig. 1.8 Ran Xiong's RWH system and his showering equipment

suggested by the GRIWAC, and some have been accepted by the farmers. When building the RWH system, the roof and lined courtyard are used for collection, and the water cellar is put inside the courtyard. Animals and poultry are not allowed to go in the courtyard. Further, the farmers clean their courtyards when they expect that the rains will come. GRIWAC also suggested using a "first flush" system. This was not well accepted because rainfall in the area is so rare and farmers never know how long the rain will last and when the next rain would come, they were reluctant to even waste a single drop.

An effective and affordable measure for water treatment has been boiling of water before drinking. To avoid dependence on firewood or other biomass, a simple solar cooker was developed in the area. This can boil water in a 3-L container within 30 min during sunny conditions. The cost was only 20–25 USD. It is estimated that there are about 10 factories for producing solar cooker in the area with total yearly production capability of 300,000 sets (Zhu et al. 2012). Figure 1.9 shows a solar cooker in use. However, farmers boil water only for making tea. In most cases, they just drink water from a jar in the kitchen without treatment.

Although there have been no reports of outbreaks of water borne decreases after more than 20 years of people drinking from the water cellars, there are still concerns about if there is the long-term impact on the farmers' health. GRIWAC has therefore introduced some measures such as installing gutters on newly built roof catchments for separating the roof runoff from surface flow. Farmers have also been advised to put chemical agents in the water to better settle down and to kill pathogens with chloride. But the farmers do not like putting chemicals in their water because of the bad smell.

Fig. 1.9 A solar cooker in a rural household

Since entering the new century, Chinese authorities have paid more and more attention to the quality of rural water supplies. To meet the new demands, GRIWAC has undertaken a project to develop a purification system for the rural dwellers drinking rainwater. The equipment is composed of a water flocculation container and a ceramic filter with opening of only 0.1 μm, which can screen out all the bacteria in the water (Zhu et al. 2012). However, a longer time of operation to test the long-term performance is needed, and the price is still a little too high for the farmers to buy.

Another important benefit of RWH is that the water tank is just by the side of the house. The farmers do not need to fetch water from a long distance. It is estimated that each family with 4 members had to spend 70 labor days each year for carrying water for daily use before the RWH project, and this was mainly a burden of women and children. Implementing of the RWH enables rural laborers to find temporary jobs in towns and get additional income.

Furthermore, by tiling the roof and paving the courtyard, the surrounding of the farmers' houses have been greatly improved. A better environment and higher income from the cash crop and/or fruit trees have raised the quality of life and future outlook. The previously impoverished farmers and their families have started to thrive.

1.3.1.2 Enhancing the Rain-fed Agriculture

By integrating the traditional dry-land farming methods with RWH systems that can store and artificially supply water to the crops, the effect of drought has been greatly mitigated. Although water is applied in very low amounts and just 2 or 3 times, the resulting 40 % average yield increases are promising. The water supply efficiency (WSE = yield increase divided by irrigation water supply) can reach 1.5–6 kg/m^3 (GRIWAC et al. 2002; Zhu et al. 2012).

Practical experiences have shown that rainwater harvesting irrigation project can change the agriculture situation significantly by increasing crop yields and raising the people's income. Mr. Bao Haiji, a farmer in the Zhenggou Chuan Village of Qingran Township, started building RWH system for irrigation in 1996. He collects runoff from a paved highway passing through his field allowing him to fill his 9 tanks with total capacity of 430 m^3. He uses these for irrigating 2 greenhouses and 1 ha of land. In 2009, his maize was suffered from the effects of drought. By giving two applications amounting to 20 m^3 /Mu of irrigation, his maize yielded 11 t/ha, much higher than his neighbor. Harvesting rainwater in the past 15 years made the fortune of this poor farmer in the past, accumulated to a sum of 160,000 CNY. He updated his house and bought many production machines. He owns color TV, washing machine, and 2 mobiles. His life has been dramatically changed. In Luoma village of Baiyiin Municipality, with rainwater in the tanks, farmers have changed the cropping system by planting maize instead of summer wheat. Annual yields have increased from the previous 975 kg/ha of wheat, to 3950 kg/ha of corn in 1998. Villager Luo Zhenjun's family was one of the poorest in the community before the project, but has since built six Shuijiaos with total storage capacity of 120 m^3. Using his RWH system, he planted 0.4 ha of plastic-sheeted maize and doubled production of his small orchard. His total food production has increased from 900 to 3675 kg per year, and his annual net income increased from 190 to 700 USD (Zhu and Li 2003).

The role of RWH for agriculture is not only in raising the yield but also in creating condition for modifying agriculture structure and generating income. In the loess area, with rainwater in their tanks, farmers started to change their cropping patterns. Production of vegetables, watermelon, fruit trees, and other cash crops, which were impossible to grow before, has become a good business, and farmers have greatly increased their incomes. As a result, the use of greenhouses developed rapidly in this loess mountainous area. Since the first built 16 greenhouses in the year 1991, it is estimated that greenhouses in the area now total around 100,000. The simplified greenhouse built with bamboo and steel rod as framework and covered with a roof of plastic film and cost about US $1000, while the annual net profit can be up to 350–500 USD. At this rate, the investment can be returned within 2–3 years. Investigations have shown that when planting wheat or corn, the yield increase by applying one m^3 of supplemental irrigation is about 2–4 kg, valued at US$ 0.7–1.4. When planting vegetables or fruit, the return on applying 1 m^3 of rainwater can be as high as 8–10 US$. In Liuping Village of Qin'an County, farmers grew nectarine trees inside their greenhouses. The fruit can thus ripen 2 months earlier than those grown in the fields and command a premium price at the market 6–20 times the normal value. In this case, the return on each m^3 of water amounts to CNY 150 (US$ 25) (Zhu et al. 2012). Figure 1.10 shows the vegetables and fruit trees planted in these greenhouses.

From the above examples, it can be seen that rainwater harvesting can not only significantly raise the crop yield thus ensuring food security, but also change the agriculture from a monocultural pattern to a diverse one and thus increase its

Fig. 1.10 Greenhouse vegetable (*left*) and nectarine trees in the greenhouse (*right*)

production value. The farmers have some sayings: "Water equals to grain, water equals to money", "To be rich, build a cellar in your field." Rainwater harvesting therefore becomes a strategic measure for both poverty alleviation and subsequently building a prosperous society in these dry mountainous areas.

1.3.1.3 Promotion of the Ecological and Environmental Conservation

Before the RWH project, due to its low fertility, farmers had to utilize as much land as possible even on the steeper slopes. This caused serious soil erosion, land degradation, and further decline in fertility. To break this vicious cycle, it was necessary to raise the fertility and productivity of the land, for which, the key factor was water. As productivity increased after the start of the RWH project, farmers began to abandon low-yielding land on the steeper hill slopes and were encouraged to do so through a favorable government policy. These marginal lands were switched to planting trees and grasses to reforest the steepest slopes. In addition, as the agricultural structure changed, development of animal husbandry required more grassland. All these factors had a positive effect on re-vegetating marginal land and thus improving the overall ecological condition.

The land conversion program was a fundamental policy issued by the State Government as part of its Strategy of Great Development of West China. It encouraged farmers in West China to change the land use on the steeper slopes, subject to serious soil erosion from cereal crops to trees and grasses in order to reduce soil erosion and to restore the ecosystem. Each year the Government provided a subsidy in money and grain to farmers for planting trees or grasses. The amount of the subsidy linked to the area of converted land in a certain period (2, 5, and 8 years for converting to planting grass, economic trees and eco-trees, respectively). The RWH project has greatly enhanced the land conversion program by enabling farmers to participate.

1.3.2 Evaluation of the RWH Project

1.3.2.1 Economic Evaluation

In 2005, an interim assessment of the RWH project in Gansu to evaluate its economic viability was carried out by GRIWAC. The study included an economic evaluation of the domestic water supply and the RWH for irrigation components of the project. The evaluation covered the period from 1995 to 2005.

(1) Parameters for evaluation

The calculation period for the evaluation was taken as 20 years. According to the <Methodology and parameters of economic evaluation for the construction project Version III> compiled by the State Development and Reform Commission and

Ministry of Construction in 2006, the social discount rate is taken as 7 %. The reference year is 2003.

(2) Benefits estimation

- Benefits of the domestic water supply: two aspects were included in the benefit estimation: namely, labor saved for fetching water before the RWH project and avoidance of water tariff payments. As previously stated, before the RWH project, each family had to spend in average 70 labor days for fetching water. After the project, this labor could be used to find work in the city and get paid. In the evaluation, it is assumed that 70 % of the labor days can find job and get cash payment. However, in 2005 evaluation, 15 CNY was used to estimate cost of labor day, which was later considered to be underestimated. Another benefit is that in the case of RWH domestic supply, no water tariff needed to be paid by the farmers as compared with the pipe water supply. Farmers can therefore save payment of about 2.5 CNY/m^3 (the price of piped water in 2005).
- Benefits for irrigation relate to the agricultural production increase by irrigation using rainwater compared to purely rain-fed production. Since the benefit of production increase was also partially from other agriculture improvements, the benefit shared to RWH irrigation is taken as the total production increase multiplied by a sharing factor, which is taken as 0.8 for wheat and corn, 0.6 for vegetables, and 0.4 for greenhouse and orchard.

(3) Cost estimation

- Investment: the investment of the project included fund provided by the government and the householders' contributions mainly in the form of labor and local construction materials. According to the statistics from the water resources bureaus of relevant counties, in the past 10 years, investment amounted to 807 million CNY (119 million USD) and 2.14 billion CNY (315 million USD) for domestic supply and RWH irrigation project, respectively. Out of the total investment, households input 83 % toward the domestic supply and 62.6 % toward irrigation projects.
- Yearly cost: for domestic supply, this includes labor and energy costs for delivering water from the RWH system, repair costs and depreciation costs. For irrigation, it includes the energy and labor cost for pumping water for irrigation, the repair cost as well as the depreciation cost. The depreciation cost is calculated by taking the service life of the RWH system as 20 years.

(4) Results

The results of the evaluation are shown in Table 1.3. It can be seen that RWH projects for both the domestic supply and irrigation are economically feasible.

The indirect benefits of the domestic water supply should also include improvement of health and reduction of cost for medical treatment, which are more difficult to express in monetary terms. Actually, since access to safe water supply

Table 1.3 Results of economic evaluation for the RWH project in Gansu

Indexes	Internal return rate (IRR) (%)	Economic present value 10^6 CNY	Economic benefit cost ratio	Conclusion
Criterion for economic feasible	≥7	≥0	≥1	
Domestic supply	19.7	676.5	1.56	Economically feasible
Irrigation	20.8	1800	1.25	Economically feasible

Fig. 1.11 RWH for irrigation of young trees on the converted land (*left*) and trees planted in the pits for concentrating rainwater (*right*)

is crucial for human existence, it cannot really be evaluated in terms of money alone. In the benefits of RWH-based irrigation, the costs for poverty alleviation and emergency relief incurred by the government during severe droughts are also needed to be included. From the above figures and these additional considerations, the RWH project is most definitely a highly cost-effective approach for rural development (Fig. 1.11).

1.3.2.2 Social Impact of the RWH Project

It is apparent that the RWH project has many positive social impacts including the following.

- Over the past 20 years, since the introduction of the RWH program, the situation regarding domestic supply has been dramatically changed for 1.25 million people in the rural mountainous regions of Gansu. The water problems of past generations have been solved. This has greatly promoted the development of these impoverished villages and allowed them to keep pace with the country as a whole.
- The RWH project for domestic water supply has freed women and children from the burden of fetching water. Now children have time to go to school, and women have more time to participate in the public and household activities.

- The RWH Irrigation Project has greatly enhanced crop yield of the dry-land farming areas in Gansu. The experiences are relevant to the development of the rain-fed agriculture elsewhere in China, where it accounts for 55.6 % of total cultivated land.
- The RWH irrigation systems have created the opportunity for households to change and diversify their agricultural production patterns according to market needs thus helping them increase their income and to escape from poverty.

1.3.2.3 Environment Impact Assessment

- The RWH project, by increasing crop yields and improving food security, made it easier for farmers to engage in the State initiated "Land conversion" program aiming at restoration of ecosystem in the West China. The RWH irrigation project has enabled farmers to diversify their production patterns. This has encouraged more orchards and grass planting for raising animals. Forest and vegetation cover in the mountainous areas is recovering as less marginal land is cultivated and trees and shrubs are planted. Any concerns that RWH in the mountains might reduce runoff, and hence, river flows downstream are unfounded. Actually, the rainwater collected by the RWH system in Gansu only accounts for just 0.086 % of the total rainfall so has a minimal impact, while any kind of water resources development (building reservoir, diverting water from river, exploiting groundwater, etc.) will have impact on the hydrological cycle and RWH is no exception. In fact, the rainwater used for either domestic use or irrigation would otherwise mostly return to the air by evaporation. Although the RWH projects would reduce surface runoff and groundwater charge in some extent but if the RWH project did not exist, it would be necessary to build other kinds of water projects for supplying water to these mountain communities with which the impacts on downstream water resources would be much bigger.
- Tests showed RWH irrigation scheme can raise overall water use efficiency WUE (output/water consumption) by 15–59 % for wheat and maize compared to the pure rain-fed agriculture. According to investigations, the WUE of pure rain-fed agriculture is only 0.39 kg/m^3 for wheat and 1.7–1.9 kg/m^3 for corn. With RWH and by adopting the highly efficient irrigation methods, the WUE amounts to 0.48–0.62 kg/m^3 for wheat and 1.96–2.4 kg/m^3 for corn, an increase of 23–59 % and 14.6–34.6 %, respectively. RWH also created the opportunity to change agricultural cropping patterns allowing the value of crops produced by a unit of water to be much higher than from pure rain-fed agriculture. This fact is significant when the world is facing challenge of meeting growing demands for food and water security while at the same time to keep water exploitation within sustainable limits. The only way to meet the challenge is to enhance the crop (value) per drop or the WUE.

From the above analysis, it can be seen that RWH is an environmentally friendly solution to addressing food security and poverty alleviation.

To sum up, one can conclude that rainwater harvesting is a strategic measure for integrated rural development. It is characterized by the following features:

- RWH is decentralized;
- RWH uses indigenous resources;
- RWH uses appropriate and acceptable technology;
- RWH has low input and O&M fees;
- Household/community ownership of RWH projects enables wide participation and high motivation;
- RWH is environmentally friendly;
- RWH is not only a temporary measure to fight drought but also a strategic intervention for sustainable development.

1.4 Understanding the Rainwater Harvesting System

1.4.1 Rainwater Utilization and Rainwater Harvesting

Precipitation is the source of all renewable fresh water on earth. All water on the planet originates from precipitation. Part of it forms surface runoff and groundwater flow and another part evaporates, returns to the sky. The first part comprises the water resources as commonly understood. However, water is a natural resource. In addition to the water resource development, rainwater utilization is another way to use water. Unfortunately, even in the new century, when the world is facing a serious water crisis, rainwater utilization is not yet being paid enough attention. Table 1.4 shows the total precipitation and water resources in the world. We can see that the water resources only occupy about half of the precipitation, the total renewable water in the world.

The figures in Table 1.4 remind us that apart from the water resources (surface runoff and groundwater flow), there is another source of water that can be used by humans. For those countries/regions, short of water resources should think of how to use this part of renewable fresh water, especially for those regions with a low ratio of water resources/precipitation. This can be taken as one of the strategies of water management.

Then how should humans use the rain?

When rain falls, we can directly use the rain drops, e.g., raindrops intercepted by plants. When rain reaches the ground, part of it becomes runoff or infiltrates into the soil and we can use the runoff soon after it forms or use the moisture for crops. Therefore, rainwater utilization denotes direct use of rain or use the runoff and moisture immediately transformed from the rain. When the runoff converges into a river or the infiltrated water recharges the groundwater flow, then use of these kinds of water can no longer be regarded as rainwater utilization although they originate from rain. To make it clearer, we give some examples in the following.

Table 1.4 Global precipitation and water resources by region

Region	Area (FAOSTAT 1999) 10^6 km^2	Population 10^6 (FAOSTAT 1999)	Precipitation (1961–1990) IPCC (km^3/year)	Internal water resources (km^3/year)	Internal water resources/inhabitant (m^3/year)	Internal water resources/precipitation
Northern America	21.9	409.895	13,384	6662	16,253	0.5
Central America & Caribbean	0.749	72.430	1506	781	10,784	0.52
Southern America	17.854	345.737	28,635	12,380	35,808	0.43
Western and Central Europe	4.898	510.784	4096	2170	4249	0.53
Eastern Europe	18.095	217.051	8452	4449	20,498	0.53
Africa	30.045	793.288	20,415	3950	4980	0.19
Near East	6.348	257.114	1378	488	1897	0.35
Central Asia	4.656	78.563	1270	261	3321	0.21
Southern and Eastern Asia	21.191	3331.938	24,017	11,712	3515	0.49
Oceania and Pacific	8.059	25.389	4772	911	35,869	0.19
World	133.795	6042.188	107,924	43,764	7243	0.41

Source quoted from <Review water resources by country> FAO, 2003 except the "Internal water resources/precipitation" was calculated by the author using data in the table

- When rain falls, it is absorbed by the soil. Then the crops can grow with suitable moisture condition. To make efficient use of the natural rain, people adopt cultivation measures to raise efficiency in rainwater retention and soil moisture utilization, like the traditional dry-land farming practices: deep plowing, tillage, mulching, land leveling, etc.;
- To retain more rain in the ground, sloping land is leveled, including terracing, contour planting, fish-scale pitting, and other soil and water conservation measures;
- To store excessive rain in the stormy season, people build dams, ponds, and other retention structures in small watersheds;

- To use the storm water, spate irrigation (diversion of the flood flow into large fields surrounded by high bunds) is adopted in many places. Water stored in the soil can provide moisture to meet crop need for part of the growing season. The runoff on the slope can also be diverted to recharge groundwater;
- Micro-catchments are a method of rainwater concentration. People set aside sloping spaces between crop lines so runoff concentrates along the cropped areas. The rain received by the crops can be doubled or tripled depending on the ratio between spaces and cropped area (Zhu et al. 2012).
- Rainwater harvesting.

Rainwater harvesting (RWH) is a specific kind of rainwater utilization. As defined by the <Chinese Code of Practice for Rainwater Collection, Storage and Utilization>, RWH is a mini-scale water resources project that collects and stores rainwater by structural measures and regulates and makes use of it for domestic and/or productive use (Ministry of Water Resources 2001). This definition represents how rainwater harvesting is understood in China.

The principle of rain-fed agriculture where rain is the only source of water for plant growth can be summed up as "collect, store and use" of rain. In conventional rain-fed agriculture, collection and storage of rain are natural process. Collection of rain usually is on a natural slope. In most cases, the area of the collection field is about the area of crop field while sometimes the collection area can be enlarged, namely, diverting water from non-cropping areas on to the cropping area. Storage is essential for the rain-fed agriculture in order to mitigate the gap between time when rain falls and time when crops need water. In the conventional rain-fed agriculture, soil moisture is the only form of storage for the rainwater. This is the easiest way for rainwater storage. However, storage in the soil has its limitation because soil moisture is subjected to loss by evaporation and its capacity is also limited. Often in semi-arid areas like the loess plateau of Gansu, most of the soil moisture from the rainy season would be lost in the long, dry periods which follow and what remains is far from enough to meet the crops water demand in the dry spell. Because of these limitations, the rainwater use efficiency would be low.

In rainwater harvesting, the rainwater collection efficiency is enhanced by both treating the catchment and enlarging its area. The rainwater storage system needs to have an appropriate capacity and measures to prevent evaporation loss. The water in the storage structure can be fully controlled by human thus enabling on-time and higher efficient use. Just like the local people said, in the past (before RWH project), we were passively fighting drought because we could only be waiting for the rain to come, but now we are actively fighting drought since we can use water from the tank. The capability for mitigating drought was greatly increased after RWH was introduced.

The differences between RWH and other kinds of rainwater utilization are shown in Table 1.5.

Table 1.5 Comparison of RWH and the other kinds of RWU

Item	Conventional RWU	RWH
Rainwater collection	On the natural slope	Area can be enlarged and seepage control measures taken
Rainwater storage	In the soil voids	In man-made tank
Time of water use	Water use only at time when rain falls	Able to control time of use according to need
Utilization efficiency	Low	High

1.4.2 Components of the Rainwater Harvesting System

RWH is an integrated system, composed of three indispensable components, namely, rainwater collection subsystem, storage subsystem, and the water supply (for domestic and production) subsystem. A chart of a typical RWH system is shown in Fig. 1.12. An illustration of the RWH system is shown in Fig. 1.13.

Someone regards the RWH system simply as a storage tank only. This is a wrong understanding. Without the rainwater collection subsystem, the water source of the RWH system, the storage tank would not be filled up, and the RWH system cannot work. This was what happened before the RWH project taking place in Gansu. The households' tanks were not full or even empty because the collection efficiency of their catchment was too low. Without the water supply

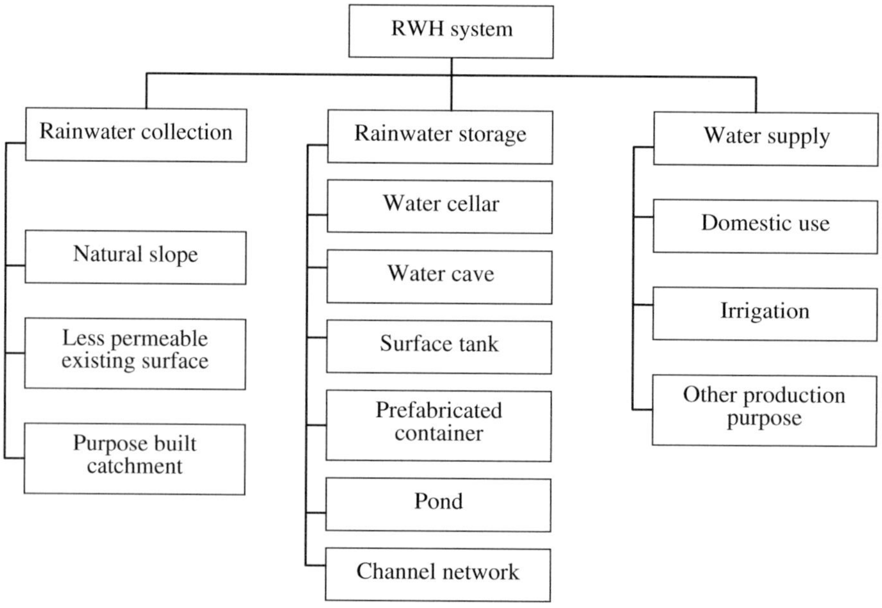

Fig. 1.12 Chart of RWH systems

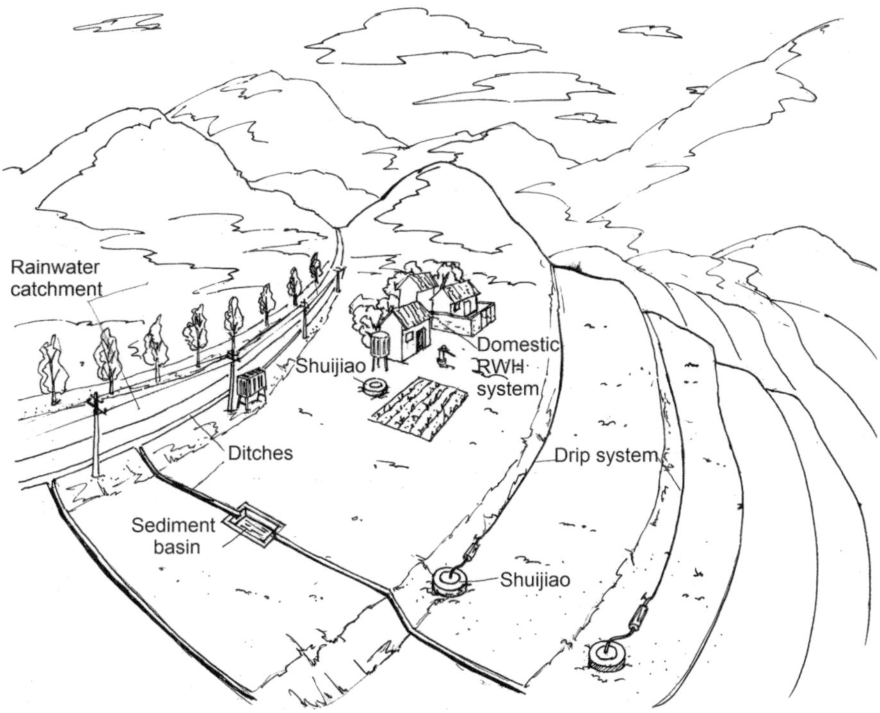

Fig. 1.13 Illustration of RWH system

system, the RWH cannot be brought into full play. This was one of the reasons that some of the RWH irrigation projects in Gansu were not in use. And lack of proper equipment for domestic use would cause water quality problem.

1.4.2.1 Rainwater Collection Subsystem

Rainwater collection subsystem is the water source of RWH. It includes the rainwater collection surface (catchment) as well as interception, collection, and conveyance ditches.

In China, the rainwater catchment can be divided into three types, namely, natural surface, less permeable surface of existing structure, and the purpose-built catchment. The catchment should preferably be impervious or have low permeability to get high rainwater collection efficiency (RCE). It should also be affordable. Besides, runoff from it should not wash contaminants into the tank. For economic reasons, it makes sense to first consider making use of the less permeable surface of existing structure such as tiled roofs, courtyards, paved highways, sports grounds, country roads, threshing floors, etc. In humid and semi-humid areas, natural slopes are often moist and thus have low infiltration capacities so are commonly used as catchments.

In dry regions, however, in order to achieve a high enough RCE for collecting sufficient water to meet demand, it is sometimes necessary to construct a purpose-built catchment. In this case the ground is paved with impermeable materials like cement concrete, cement soil, or covered with plastic film, etc.

Different kinds of collection fields are described as follows.

(1) Roof

Roof is one of the most commonly used surfaces for collecting cleaner rainwater. In China, the roofs of many rural houses are built with straw-mud. This kind of roof not only has the disadvantage of having a low RCE but also can cause pollution of runoff so it is best replaced with a tiled roof. Tiles are made of clay mud dried in the sun and baked in a kiln or from cement. In Gansu, it was found that tiles made using cement mortar have a higher RCE than clay tile. The price of cement tile is sometimes the same with even cheaper than clay tile if the coal is not available locally.

(2) Courtyard

The surfaces of courtyards normally become compact by the constant trampling by people and as a result have a higher RCE. It is easy to clean the courtyard frequently and thus to get relatively cleaner water. In Gansu since the rainfall is rare, part of the courtyard is usually paved with concrete to increase the RCE.

(3) Concrete or asphalt-paved highway

An existing paved highway is a very cost-effective form of rainwater collection surface with high RCE. Although there have been only a limited amount of tests conducted on them in rural Gansu, and these have shown that there are no significant elements harmful to human health being found in the runoff, it is still recommended not to use water from the asphalt-paved highway for domestic supply. To collect water from a highway, a drainage ditch by the side of the road can be used. The conveyance canal is linked to the ditch at certain place to divert water to the tanks.

(4) Country earth road and threshing yard and other compacted surfaces

These surfaces are normally well compacted and much less permeable than a natural soil-covered slope and can be used for rainwater collection. However, since the RCE is not that high, the number of tanks supplied by this kind of catchment is small.

(5) Natural slopes

In humid areas where rain is frequent, the soil is often moist, and the RCE even for natural soil-covered slopes high, these can be used for collecting rain. The rocky slopes are even better due to their high RCE and higher resistance to erosion. In semi-arid regions, the RCE of natural slopes is usually lower; however, owing to the expansive bare slopes in these regions, this can be compensated for by using very large of collection area. In both cases, a sedimentation basin has to be installed to prevent sediment entering the tank.

(6) Purpose-built catchment

When the climate is too dry and runoff small, it may be necessary to treat the natural catchment slope to raise RCE. The treatment could include paving the ground with concrete, cement soil, or covering it with plastic film.

An interception ditch is used to intercept the rainfall-runoff on the slope and the collection ditch collects water from the interception ditches. The conveyance ditch transports water to the tank.

1.4.2.2 Water Storage Subsystem

The purpose of a water storage subsystem is to store the rainwater in rainy season to meet water demand of users in the dry season. In China, owing to the wide range of climates, topography and geology, there are different types of storage systems for different regions. Basically, six types can be identified, namely, the water cellar, water cave, surface tank, pond, prefabricated container, and channel network.

The water cellar is an underground tank, known locally as a "*Shuijiao.*" The underground tank is commonly used for the purpose of domestic water supply. This system has several advantages including (i). In addition to roof runoff, the Shuijiao has the capacity to store surface runoff as well; (ii) Water inside of the underground tank can have a lower temperature so that the quality can be kept better; and iii) In semi-arid areas, the design helps to avoid the high evaporation losses.

Among all types of underground tank, Water cellar (*Shuijiao*) is most commonly used in the north and northwest China.

Water cave is also a kind of underground tank and has a local name of *Shuiyao*. It is like the cave dwelling (cave on a cliff), which was common in the past time in North and Northwest China. Water caves have the advantage of not occupying the land because they are constructed in cliff sides.

Surface tank is popular in the humid and sub-humid areas in mid, southwest, and south China, where the precipitation is large and the evaporation is low. The surface tanks in China are round or rectangular in shape and usually have diameters up to 10 m or more.

Ponds are water storage using natural depressions. These are very popular in the mountainous areas of southern China. In southeast China, people built weirs in the rivers to store water. The weir can also facilitate water collection or diversion for irrigation.

Factory-made water tanks are used widely for rainwater storage for domestic use in the southeast China. The tank is typically made of reinforced high-strength cement mortar (Ferro-cement tank). In the recent years, tanks made of polymers or stainless steel have also become available on the market. Plastic tanks with volumes up to 50 m^3 are also available.

In the plain area of southeast China, channels and ditches were linked together forming a network in the ancient China. Originally these were mainly for transportation. Later these were also used for flood mitigation and more recently for rainwater storage. In the past, the water quality was good and suitable for domestic use. People cleaned the rice and vegetables in the channels. However, due to the pollution from industry and the over use of chemicals in agriculture, water quality has deteriorated in recent years and can no longer used for domestic purposes, just for irrigation. During recent urbanization, many channels have been filled to reclaim land for development, and the traditional role of these canal networks has been reduced. Today, however, measures have begun to be taken to preserve this ancient waterway networks.

1.4.2.3 Water Supply Subsystem

The water supply subsystem is divided into three categories: domestic water supply, irrigation water supply, and water supply for other production like husbandry, small industry use, etc. In rural areas, water supply equipment for domestic use can be rope plus bucket, hand pump, or electric pump.

1.4.3 Typical Layout of a Rainwater Harvesting System in China

In China, the layout of a number of typical RWH system types can be categorized as follows.

1.4.3.1 Roof-Courtyard RWH System

Figure 1.14 shows the typical layout of RWH for water supply to the rural household in the semi-arid area in northwest China. The tiled roof and courtyard with concrete pavement are often used together as a rainwater catchment surface to collect cleaner water. If the rainfall is high or the roof area is sufficient to catch enough water, then courtyard will not need to be used. In Gansu, to improve the water quality, farmers are advised to put the underground water tank in the courtyard and exclude animals from entering it. This layout also helps to greatly reduce the labor of carrying water. It is also important to sweep the courtyard before rain comes and to maintain the system to protect the water quality.

1.4.3.2 Paved Highway Catchment System Supplying a Number of Tanks

Paved highways are designed to be impermeable, have a high RCE, and can make excellent catchments. A section of sloping highway is set aside to collect runoff and a small dike built along the drainage ditch at the side of the highway which

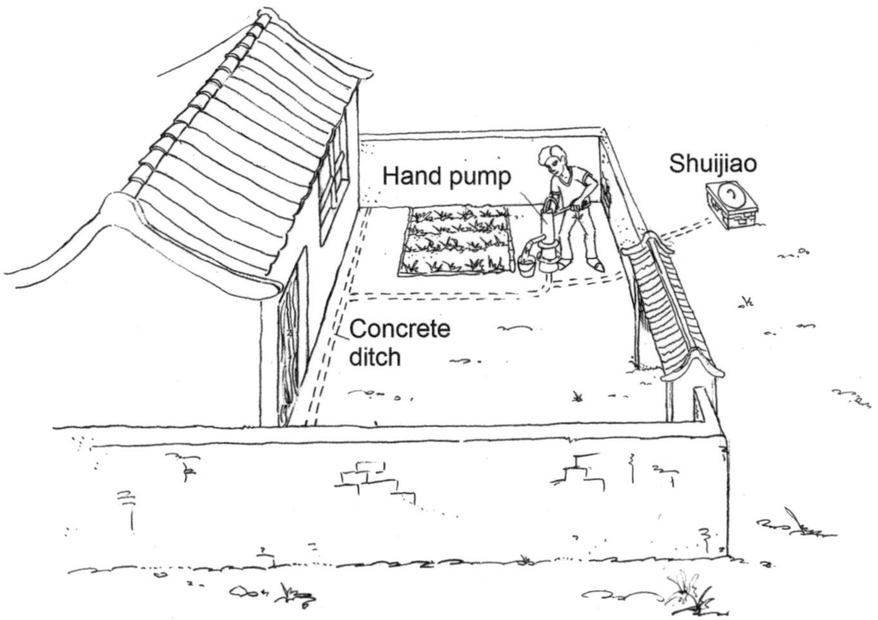

Fig. 1.14 Layout of roof-courtyard RWH system

is used to divert runoff into a number of tanks located down slope by the side of highway. This kind of RWH system is used for irrigation and stock watering purposes, and has become quite common in China, see Fig. 1.15.

1.4.3.3 Rainwater Distributed by Concrete Lined Canal to a Number of Storage Tanks

When the storage tanks that are scattered over a wide area are supplied by a highway catchment, a water conveyance canal from the highway to the tanks is needed. The canal is usually lined with concrete slabs or masonry to avoid seepage loss. This type of layout is illustrated in Fig. 1.16.

1.4.3.4 A purpose-Built Concrete-Lined Catchment Supplies Water to a Number of Tanks

In the dry areas where highway or other existing impermeable surfaces are unavailable, to irrigate high value crops/trees, a purpose-built catchment lined with concrete can be constructed. Usually these catchment are located on a hill top or a high place so the runoff from it can flow by gravity to the tanks, see Fig. 1.17.

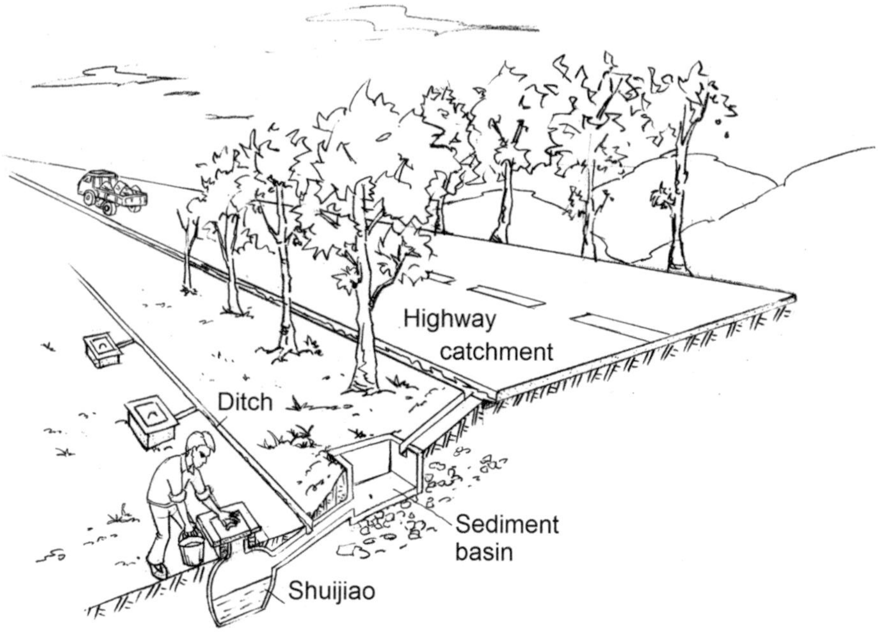

Fig. 1.15 Highway catchment supplies water for a number of tanks

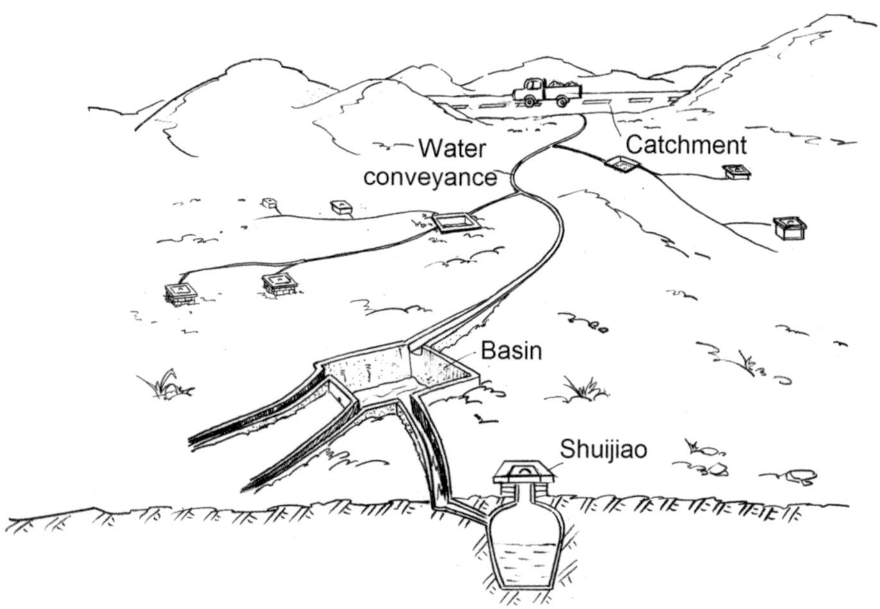

Fig. 1.16 Long concrete-lined canal supplies water to many tanks

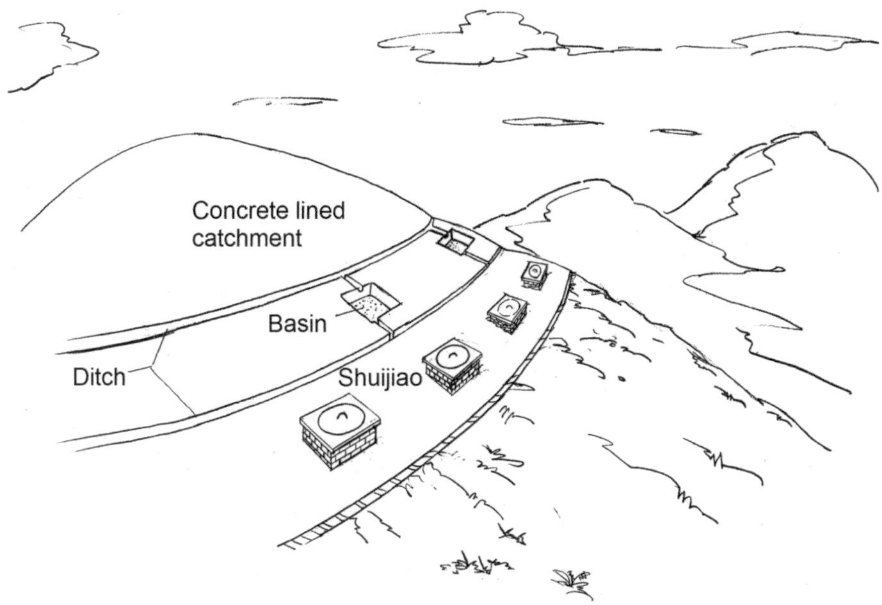

Fig. 1.17 A purpose-built concrete catchment on a hill top supplied a number of tanks

1.4.3.5 A Small Catchment Supplies Water to 1-2 Tanks for Irrigation

A country road or a threshing yard that can produce small amount of runoff supplies water for one or two tanks capable of irrigating a small piece of land. Figure 1.18 is a photo of threshing floor being used as a catchment.

1.4.3.6 Tank Rice Irrigation RWH System

In the mountainous areas of southwest China, surface tanks storing rainwater are used for paddy irrigation. A 50 m^3 tank can typically irrigate one *Mu* of rice. This has enabled farmers to switch from cultivation of one rain-fed corn crop annually to two crops a year, one of rice, followed by another harvest of irrigated corn. Figure 1.19 shows the tank group for paddy irrigation

1.4.3.7 RWH System with Greenhouse

The roof of a greenhouse provides a good catchment with a high RCE. Water collected from thc roof is often enough for producing 2–3 harvests of vegetables if water-saving irrigation methods are adopted. In the driest areas, if water from the roof is insufficient, then the land around the greenhouse can be lined with concrete

Fig. 1.18 Threshing yard used as catchment

Fig. 1.19 Tank for paddy irrigation

Fig. 1.20 RWH System with Greenhouse

and also used as a catchment to provide additional runoff. In case that there are greenhouse group, enough space should be kept between greenhouses for to avoid shading and for collecting rainwater. The storage tanks can be put inside and/or outside the greenhouse. Figure 1.20 illustrates a typical layout.

The above layouts of RWH systems show some common types found in China. In general, a wise arrangement of the RWH system can enhance the system efficiency and reduce the cost. It is important to arrange the RWH system according to the local climatic, topographic, geological, and economic conditions. The following points are suggested to ensure the adoption of a good RWH layout.

- Use suitable types of catchment and storage tank for different uses. For example, only use the roof and hardened courtyard catchments for rainwater collection for domestic use;
- Always remember to have an integrated RWH system, in which the capacity of catchment and storage are matched to each other. Either inadequate catchment area or storage volume will result in the low efficiency of the system;
- To maximize the economic advantages existing impermeable surfaces such as roads or tiled roofs should be used if possible; the storage tank should also be located close to where water is used and/or close to catchment. It is preferable to put the tank at higher location so that the water supply can be delivered by gravity flow.

References

China Ministry of Water Resources. Country report of the People's Republic of China, 2003. France: World Water Council; 2003.

Editorial Board of Disaster of Flood and Drought in Gansu. Disaster of flood and drought in Gansu. Zhengzhou: Yellow River Water Resources Press; 1996 (in Chinese).

FAO-AQUASTAT. www.fao.org/; 2003.

Gansu Bureau of water Resources and Gansu Bureau of Technical Supervision. Technical standard for rainwater collection, storage and utilization in Gansu Province. Lanzhou, China: Gansu Science and Technology Press; 1997. (in Chinese).

Gansu Provincial Government. Plan for terrace construction (2011-2015), http://wenku.baidu.com; 2011. (in Chinese).

Gansu Soil and Water Conservation Bureau, Gansu Soil and Water Research Institute, Soil and Water Research Institute of Pingliang Prefecture and Soil and Water Research Institute of Dingxi Prefecture, 1994, Experiment and Research Report on Leveled Terrace, (unpublished, in Chinese).

GRIWAC (Gansu Research Institute for Water Conservancy), Dryland Agriculture Institute of Gansu Academy of Agriculture Sciences, Gansu Agriculture University, and Dingxi County Government. Technology integration and innovation study on highly efficient concentration of natural rainfall in the semi-arid mountainous area, GRIWAC, Lanzhou; 2002. (unpublished report, in Chinese).

Hydrological Bureau, China Ministry of Water Resources and Electric Power. Assessment of China's water resources. Beijing: Water and Power Press; 1987 (in Chinese).

Luo M. Water resources development in the dry area of Gansu. Xinjiang Water Res Sci Technol. 1987; 1987(4–5). (in Chinese).

Ministry of Water Resources. Code of practice for rainwater collection, storage and utilization. Beijing: China Water Resources and Hydropower Press; 2001

Renewable internal freshwater resources per capita (cubic meters), data.worldbank.org/indicator/infrastructure.

Shiklomanov IA, Rodda JC. World water resources at the beginning of the 21st century. Cambridge: Cambridge University Press; 2003.

Shiklomanov IA, et al. World water resources at the beginning of the 21st century. 1999. webworld.unesco.org/water/ihp/db/shiklomanov/.

State Development and Reform Commission, Ministry of Water Resources, Ministry of Health and Ministry of Environment Protection. Plan of rural safe domestic water supply project in the 12th Five Year Plan. 2012. http://wenku.baidu.com. (in Chinese).

Zhu Q, Li Y. Rainwater harvesting for survival and development. Waterlines. 2000;18(3):11–4.

Zhu Q, Li Y. Rainwater harvesting in Gansu Province. Low-External-Input Sustain Agric (LEISA) J; 2003.

Zhu Q, Li Y. Rainwater harvesting: the key to sustainable rural development in Gansu, China. Waterlines. 2006;24(4):4–6.

Zhu Q, Wu F. A lifeblood transfusion: Gansu's new rainwater catchment systems. Waterlines. 1995;14(2):5–7.

Zhu Q, Li Y, Gould J. Every Last Drop: Rainwater Harvesting and Sustainable Technologies in Rural China. Rugby: Practical Action Publications; 2012 142.

Chapter 2
Dimensioning the Rainwater Harvesting System

Qiang Zhu

Keywords Water supply quota · Water supply reliability · Rainwater collection efficiency · Dimensioning rainwater harvesting system

In this chapter, the methodology for dimensioning the RWH system including the area of catchment and the storage capacity will be introduced. In the past, people regarded RWH as a temporary measure for mitigating drought. Water professionals paid little attention to the design procedure for the RWH system. The catchment area and the volume capacity of storage tank were determined in a rather haphazard way without scientific analysis. This leads to poor design of the RWH systems. These were either too small to meet the water supply demand, or too large and inefficient, wasting precious resources. In the past decade, the author made a systematic study to formulate a strict design procedure based on principles from hydrology and water resources engineering. In addition, since most of the RWH projects are implemented at on a small-scale by the grassroots communities, simplified methods that can be easily mastered by the local technicians at the township and county level and even by farmers themselves have been developed. Furthermore, calculation sheets using the excel software were provided.

Before starting the introduction of the calculation procedure, we will first discuss the water demand of the RWH system and several parameters for the calculation.

Q. Zhu (✉)
Gansu Research Institute for Water Conservancy, Lanzhou, China
e-mail: zhuq70@163.com

Q. Zhu et al. (eds.), *Rainwater Harvesting for Agriculture and Water Supply*,
DOI 10.1007/978-981-287-964-6_2

2.1 Water Demand

2.1.1 Domestic Water Demand

Domestic water demand includes drinking, food preparation and cooking, personal hygiene (body and hand washing), washing of clothes, house cleaning, washing pots, pans and other utensils, cattle use and toilet use, etc. (Gould and Nissen-Petersen 1999). Since the water storage capacity of RWH system is limited, using rainwater for toilet flushing is neither feasible nor sustainable in the Chinese context. Rather it would be preferable to adopt flush toilets reusing gray water, or use of composing 'dry toilets' based on the concept of ecological sanitation.

The quantity of domestic water use varies greatly depending on the local climatic, cultural and economic conditions. Considering the RWH system is used mostly in areas with serious water scarcity, a relatively low quota is adopted here. The following factors for determining the volume of the local water supply quota include as follows:

- Availability of water resources
- Local culture/practice of water use
- Economic affordability: Scale of RWH system relates to the water use quota. It thus has to conform to the local economic condition
- To meet the basic need for human existence in the beginning and create conditions for the further development of the RWH system.

In China, the typical quota for domestic supply from RWH systems has been gradually increasing in tandem with improving rural economic conditions. Table 2.1 shows the design quotas in different situations for domestic supply using RWH systems which have been adopted in standards since 1997 in China [Gansu Provincial Bureau of Water Resources (GBWR) and Gansu Provincial Bureau of Technical Supervision (GBTS) 1997; Ministry of Water Resources (MWR) 2001; Ministry of Housing and Urban–Rural Development (MHURD) and General Administration of Quality Supervision, Inspection and Quarantine (AQSIQ) 2010].

Table 2.1 Design quota for domestic water supply of RWH system (L/person day)

Name of standard	Quota for domestic water supply		Year of issue
	Semi-arid area	Humid and sub-humid area	
Technical standard of rainwater catchment and utilization project in Gansu province	10		1997
Technical code of practice for rainwater collection, storage and utilization	10–30	30–50	2001
Technical code for rainwater collection, storage and utilization	20–40	40–60	2011

Table 2.2 Quota of water supply for domestic use in RWH system

Region with mean annual precipitation (mm) of	Quota (L/person day)
250–500	20–40
>500	40–60

The latest design standard was based on the <Evaluation indexes of safe and health rural drinking water> issued by the Ministry of Health and Ministry of Water Resources in 2004 (see Box 1.1 of Chap. 1). Table 2.2 shows the quota for domestic water use for a RWH system recommended in the "Technical code for rainwater collection, storage and utilization" (MOHURD and AQSIQ 2010).

These quotas for domestic water use are based on the China's current situation and are not necessarily relevant for determining the domestic water use in other countries and regions. Different countries/regions need to establish their own quotas based on local conditions.

The annual domestic water use for any RWH system can be obtained simply by multiplying number of family members by the quota volume and by 365 days.

2.1.2 Supplemental Irrigation Water Use

Since the volume of rainwater in a RWH system is limited, irrigation using this water has to be carried out in a highly efficient and water-saving way. In China, this is done through three ways.

First, through adopting deficit irrigation for the RWH-based irrigation systems. This approach to irrigation involves applying irrigation only at critical periods of crop growth. The role of this limited irrigation water is to help the crops tide over periods of severe water stress so that crops could avoid severe damage at these critical stages. Once the rainy season arrives the crop can then use natural rain effectively and efficiently.

Second, by using small amounts of water for each application to the crops. Efficient but simple and affordable irrigation methods need to be adopted, including the locally innovated methods and the modern micro-irrigation techniques.

Third, irrigation water needs to just moisten a small area surrounding the crop root zone. Most of ground surface should remain dry to avoid as much evaporation loss as possible. As the farmers say, "We are irrigating the crop but not the field."

In Chap. 5, detailed irrigation techniques used in the RWH irrigation will be introduced.

Estimates of irrigation requirements for different crops when designing a RWH system, on the following table based on investigations in China, has been summarized in the <Code of Practice for Rainwater Collection, Storage and Utilization> (MWR 2001).

Figures in Table 2.3 can be used as a rough reference when calculating the capacity of a RWH system. During the actual operation, the irrigation water use

Table 2.3 Application frequency and quota for RWH irrigation in China

Crop	Irrigation method	Application frequency under annual precipitation (mm) of		Quota of each application, m^3/ha
		250–500	≥500	
Corn, Wheat	Irrigation during seeding	1	1	45–75
	Manual irrigation	2–3	2–3	75–90
	Irrigation through holes on plastic film	1–2	1–2	45–90
	Injection in root zone	2–3	1–2	30–60
	Furrow irrigation under plastic film	1–2	2–3	150–225
Vegetable (one harvest)	Drip irrigation	5–8	6–10	120–180
	Micro-sprinkler	5–8	6–10	150–180
	Manual irrigation	5–8	8–12	75–90
Fruit tree	Drip irrigation	2–5	3–6	120–150
	Bubble irrigation	2–5	3–6	150–225
	Micro-sprinkler	2–5	3–6	150–180
	Manual irrigation	2–5	3–6	150–180
Paddy	Dry cultivation		6–9	300–400

Source (MWR 2001), <Technical code of practice for rainwater collection, storage and utilization>

should be determined according to the irrigation experiments and/or consulting the experts and experienced farmers based on the real local climatic and environmental conditions.

The total irrigation water use can be calculated by multiplying the number of application in the growing season with the quota for each application and the land area (in hectares).

The irrigation quota in Table 2.3 is very low when compared to conventional irrigation. However, practical experiences showed that even with this low amount of irrigation, crop yields can be enhanced significantly, see Chap. 5. The irrigation water use quotas and application frequency shown above are based on the climate, soil and crop type as well as economic conditions in the area. These figures cannot be directly applied to other countries, especially where the climate and soil conditions are different.

Example for estimating the irrigation water amount

Suppose a household has an area of 12 Mu, or 0.8 ha to be irrigated, divided between 0.75 ha of corn and 0.05 ha of vegetables.

The irrigation frequency and water amounts shown in Table 2.3 can be used for determining the total water use. For corn, one water application during the seeding with application amount of 50 m^3/ha and two applications with 90 m^3/ha each by method of "irrigation though holes on plastic film" is adopted. For vegetables, it is

Table 2.4 Daily water use for each animal/bird, liter/head

Name of animal/poultry	Large animals	Pig	Sheep/goat	Poultry
Consumption	40–50	10–20	5–10	0.5–1

planned to have 14 applications with 120 m^3/ha each for two harvests using drip irrigation.

Then we can estimate that the water amount for corn irrigation is

$$0.75 \times (50 + 2 \times 90) = 172.5\ m^3.$$

Irrigation for 2 harvests of vegetable is

$$0.05 \times 14 \times 120 = 84\ m^3.$$

Total amount of irrigation is

$$172.5 + 84 = 256.5m^3.$$

2.1.3 Animal Husbandry Water Use

The water use for animal husbandry can be calculated by taking the number of animals/poultry and multiplying by the water use quota per animal/bird. The water supply quotas for farm animals and poultry supplied by the RWH systems in China are shown in Table 2.4 for a small family-sized farm. For large-scale farms, the water requirements for washing barn floors should also be taken into consideration.

2.1.4 Other Water Use

Water use in the rural areas also includes the small industrial and farm processing industries. Table 2.5 shows the water demand of several types of small industry in China.

Table 2.5 Water consumption for small industries

Industry	Oil press	Bean process	Sugar	Cannery	Brewery	Brick making	Slaughter	Skin process
Unit	m^3/t	m^3/t	m^3/t	m^3/t	m^3/t	$m^3/10^4$ pieces	m^3/piece	m^3/piece
Number	6–30	5–15	15–30	10–40	20–50	7–12	0.3–1.5	0.3–1.5

2.2 Parameters for Dimensioning RWH System

2.2.1 Water Supply Reliability and Design Rainfall

(1) **What is Water Supply Reliability**

Water demand for domestic consumption or productive use should be met not only in wet and normal years but also in the dry years. We call the percentage of years in which the water supply can be met for a long period the water supply reliability. It can be expressed with the following equation:

Water supply reliability = Years in which water demand is met ÷ Total years in long period.

(2) **Water Supply Reliability in China**

High reliability of supply is essential for stable industrial and agricultural production. A reliable domestic water supply is also crucial to ensure a good household living standard. On the other hand, high reliability requires greater inputs in terms of water resources infrastructure, operation, and maintenance cost. From the perspective of the overall benefit of the society, there should be a balance between the water supply reliability and the input and O&M cost.

In China the reliability of water supply for different purposes is different. For a safe rural domestic water supply, the <Evaluation indexes of safe and healthy rural drinking water> requires a reliability of 90–95 % [MWR and Ministry of Health (MOH) 2004].

For the irrigation purposes, the reliability depends on the crop type, irrigation method, and climatic conditions as well as overall water availability. Table 2.6 shows the reliability for irrigation purposes (General Administration of Technical Supervision and Ministry of Construction 1999).

The reliability of water supply for other purpose like husbandry, small industries in the rural area is usually taken as 90 %.

Based on the above data, the <Technical code for rainwater collection, storage and utilization> stipulated the water supply reliability for different water uses of the RWH project in China, as shown in Table 2.7.

Table 2.6 Water supply reliability for irrigation

Irrigation method	Climate and water availability	Crop type	Reliability (%)
Surface irrigation	Arid area or area with water scarcity	Dry cultivation	50–75
		Paddy field	70–80
	Semi-arid and sub-humid area with unstable water availability	Dry cultivation	70–80
		Paddy field	75–85
	Humid area with rich water resources	Dry cultivation	75–85
		Paddy field	80–95
Micro-irrigation			85–95

Source <Technical Code for Design of Irrigation and Drainage Engineering (GB 50288—99)>

Table 2.7 Water supply reliability for different water uses of the RWH project

Water use	Domestic	Irrigation	Husbandry	Small industry
Reliability %	90	50–75	75	75–90

(3) **Design Rainfall**

The yearly rainfall for dimensioning the RWH system is called the design rainfall. The yearly water inflow of the RWH system is the runoff on the catchment during rainfall over the whole year. The yearly runoff should meet the demand of water supply over the year with a given reliability. It is apparent that a higher yearly runoff/rainfall requires a smaller catchment area when the water supply demand is fixed and vice versa.

To determine the design rainfall, we will first discuss the stochastic property of a hydrological variable and the methodology to deal with this kind of variable. As we know, the rainfall and also the runoff vary from event to event and also with time and there are a number of random factors affecting the value of the rainfall/runoff. We cannot exactly know how much rainfall we will have in a certain rainfall event or in a certain year using a deterministic model. However, based on enough rainfall data series, we can evaluate the variability of the rainfall and estimate the value in the future with certain probability by using the mathematical stochastic methodology. This will be done by setting up a frequency distribution curve of the stochastic variable, for instance the yearly rainfall.

Table 2.8 shows a continuous rainfall data series over 30 years. We will rank the yearly rainfall from biggest to smallest. It can be easily done by the Excel sheet.

In the table, the empirical frequency of the yearly rainfall can be calculated with Eq. 2.1.

$$P = \frac{m}{n + 1}, \tag{2.1}$$

where P is the empirical frequency, m is the magnitude ranked order of the yearly precipitation from large to small, n is the number of years of the series. The ranked data are shown in a probability paper, see Fig. 2.1. In the table we can find, for example, the empirical frequency of 74.2 % (column VI) relating to yearly rainfall of 443.1 mm (column IV) in the year 1980 (column III). The frequency of 74.2 % means that in the 30 years, there are 23 years with yearly rainfall no less than 443.1 mm. Or we can say that we have 74.2 % of chances to have a yearly rainfall equal or greater than 443.1 mm. If we take design rainfall equal to 443.1 then the size of the RWH system can ensure a water supply reliability of 74.2 %. This also is seen from Fig. 2.1. All the rainfall before frequency of 74.2 % is larger than 443.1 mm, so if we take 443.1 mm as the design rainfall then we can have runoff greater than or equal to the water demand with reliability of 74.2 %.

From the above example we can say that the design rainfall is the rainfall with hydrological frequency equal to water supply reliability. The following is the procedure to determine design rainfall using an empirical frequency distribution curve.

First the yearly rainfall data of preferably a 30-year series is to be prepared. Then rank the rainfall as shown in Table 2.8. The empirical frequency curve of yearly rainfall can then be calculated using Eq. 2.1. The yearly rainfall with

Table 2.8 Rainfall data and the calculation of the frequency distribution curve

Year	Rainfall (mm)	Sort of rainfall (mm)		Rank	Empirical frequency (%)
1961	784.7	1967	910.6	1	3.2
1962	469.6	1983	821.4	2	6.5
1963	711.4	1964	792.4	3	9.7
1964	792.4	1961	784.7	4	12.9
1965	398.1	1982	769.1	5	16.1
1966	453.7	1986	729.3	6	19.4
1967	910.6	1968	718.2	7	22.6
1968	718.2	1987	715.8	8	25.8
1969	596.5	1963	711.4	9	29.0
1970	634.6	1975	691.6	10	32.3
1971	420.6	1970	634.6	11	35.5
1972	473.9	1981	624	12	38.7
1973	313.1	1977	618.8	13	41.9
1974	600.7	1988	612.6	14	45.2
1975	691.6	1974	600.7	15	48.4
1976	583.8	1990	597.7	16	51.6
1977	618.8	1969	596.5	17	54.8
1978	431.3	1989	586.6	18	58.1
1979	365.5	1976	583.8	19	61.3
1980	443.1	1985	509.3	20	64.5
1981	624	1972	473.9	21	67.7
1982	769.1	1962	469.6	22	71.0
1983	821.4	1966	453.7	23	74.2
1984	357.3	1980	443.1	24	77.4
1985	509.3	1978	431.3	25	80.6
1986	729.3	1971	420.6	26	83.9
1987	715.8	1965	398.1	27	87.1
1988	612.6	1979	365.5	28	90.3
1989	586.6	1984	357.3	29	93.5
1990	597.7	1973	313.1	30	96.8
I	II	III	IV	V	VI

hydrological frequency equal to the water supply reliability for different purposes of water use (Table 2.7) is the design rainfall.

The empirical frequency curve sometimes cannot give the rainfall with frequency exactly equal to the design reliability. For example, if the reliability is 75 % but in Table 2.8 only frequency of 74.2 and 77.4 % corresponding to the year 1966 and 1980 and rainfall of 453.7 and 443.1 mm can be found. To get the rainfall at 75 % percentile, an interpolation calculation is necessary as shown below:

$$R_{75} = R_{77.4} + (R_{74.2} - R_{77.4}) \times (77.4 - 75) \div (77.4 - 74.2) = 451\text{mm}$$

where R_{75}, $R_{77.4}$, and $R_{74.2}$ is the rainfall at frequency of 75, 77.4, and 74.2 %, respectively.

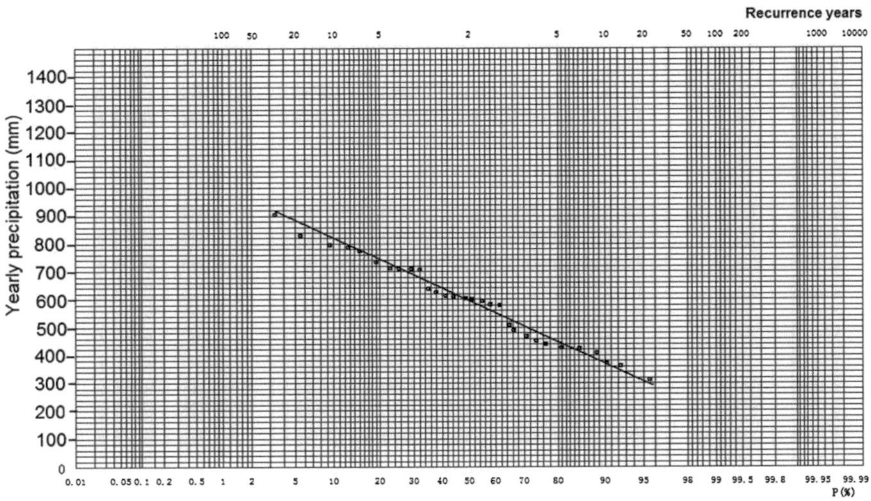

Fig. 2.1 Frequency distribution curve of yearly rainfall

Using the same method the rainfall at frequency of 90 % is 386.6 mm.

The frequency of the yearly rainfall can also be calculated by a theoretical method. This is explained as follows:

There are many kinds of frequency distribution model. In China, two models are usually used in the hydrological stochastic analysis, namely, the Normal Distribution and the Pearson Distribution Type III. The latter is a non-symmetric curve with one limited end and another infinitive end. If we have a yearly rainfall data series of n years, $x_1, x_2, x_3 \ldots x_n$, we can get the mean of the data series x_a. For a certain variable x_p, the probability of the expression $P(x \geq x_p)$ can be found in special table that is available in many hydrology textbooks, see Eq. 2.2.

$$x_p = (\Phi_p C_v + 1)x_a \tag{2.2}$$

where Φ_p is a variable related to the probability P, C_v is the variability coefficient of the data series; x_a is the mean of the series. To get the value of Φ_p, the skewness coefficient C_s should also be determined. The parameters C_v and C_s can be calculated with Eqs. 2.3 and 2.4.

$$C_v = \frac{1}{x_a}\sqrt{\frac{\sum_{i=1}^{n}(x_i - x_a)^2}{n-1}} \tag{2.3}$$

$$C_s = \frac{\sum_{i=1}^{n}(x_i - x_a)^3}{x_a^3(n-1)C_v^3}, \tag{2.4}$$

with C_v and C_s, frequency P for ($x \geq x_p$) can be found from the table of the Pearson III Distribution Model in any hydrological manual. Then the frequency distribution curve for variable x can be set up.

Practically, the frequency distribution curve is plotted using the "curve fitting method." First, plot all the real yearly precipitation versus the empirical frequency on a probability paper and you can get an empirical curve of frequency distribution. Then plot several theoretical curves with the calculated C_v and values of C_s equal to 1.5, 2.0, 2.5, and 3.0 times of C_v. From the different values of C_s, select a curve that can best fit the empirical curve by visual judgment. Then with C_v and C_s we can determine the theoretical frequency of rainfall by using the table of the Pearson Distribution Type III. To find out the theoretical curve that can best fit the empirical one some computer models have been developed.

The following is the summary of the steps for determining design rainfall using the theoretical frequency distribution curve.

Step 1 Prepare the yearly rainfall data with length of preferably 30 years.
Step 2 Rank the yearly rainfall data from the largest to smallest and calculate the empirical frequency curve using Eq. 2.1. Plot the empirical frequency distribution curve on a piece of probability paper.
Step 3 Calculate variability coefficient C_v using Eq. 2.3.
Step 4 Find several Pearson Distribution curves from hydrological manual with calculated C_v and assumed C_s equal to 1.5, 2.0, 2.5, and 3.0 C_v and plot them on the probability paper.
Step 5 Select one C_s value and the related Pearson Distribution curve that best fits the empirical curve.

Determining design rainfall with theoretical frequency analysis is quite complicated. Since RWH is a small-scale project, design rainfall using this empirical method (Eq. 2.1) to determine the hydrological frequency for yearly rainfall can give a satisfactory result.

2.2.2 Water Collection Efficiency

(1) Factors Affecting Rainwater Collection Efficiency (RCE) in a Rainfall Event

When a rainfall event happens part of the rain will infiltrates into the soil, part of it evaporates and the remainder forms the runoff, see Fig. 2.2.

The rainwater collection efficiency (RCE) is defined as the ratio between runoff and rainfall in a rain event from the following equation:

$$\mathrm{RCE} = \frac{W_{\mathrm{Runoff}}}{W_{\mathrm{rain}}} \cdot 100\,\%, \tag{2.5}$$

where W_{runoff} and W_{rain} are the runoff and rainfall in a rain event, respectively.

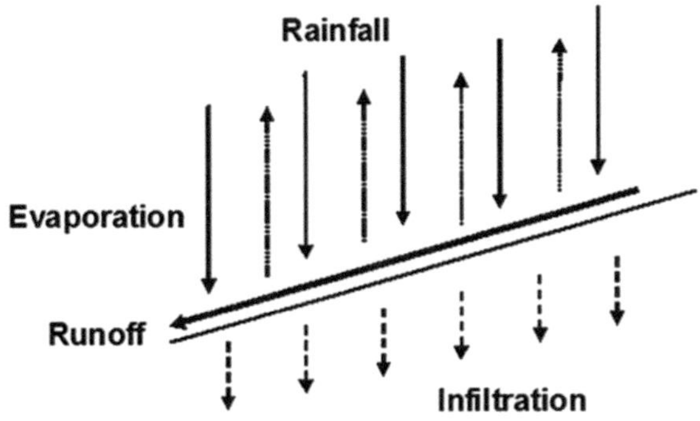

Fig. 2.2 Illustration of runoff on a catchment

RCE is affected by the following factors.

- The rain characteristics including the amount and intensity: a larger rainfall amount and intensity results in a higher RCE (Rain intensity is the rainfall amount divided by the rainfall duration, usually in mm/min).
- The permeability of the catchment surface: RCE will be higher the more impermeable the catchment
- Slope of the surface: The RCE is higher the greater the slope.
- Water content of the surface layer before rain: a wetter surface produces a higher RCE.
- Length of the catchment in the direction of the slope: RCE of a catchment with shorter slope is higher.

(2) **Determining RCE of a Rain Event**

Since there are so many factors affecting RCE for each rain event, RCE is usually determined by experiment. The GRIWAC has conducted a systematic testing on the RCE of different surface materials. During these tests, rainfall, intensity, and runoff were measured during each rain event and for some catchments with higher permeability the water content of the surface material of the catchment before rain was also measured. RCE is calculated by using Eq. 2.5.

This research project lasted for 4 years, from 1988 to 1991. To test the RCE on different catchments, 28 test plots with eight different catchments were established. These included ground surface lined with concrete, cement soil, and lime soil, tiled roofs (including cement tiles, baked clay tiles made in factory and clay tiles from a village workshop), plastic film buried with sand/mud soil, compacted loess soil and a natural soil catchment. To accelerate testing, in addition to the natural rain, the artificial rain (using a sprinkler system) was also used for testing. In total tests with 421 rain events (natural and artificial rain) were conducted. Figure 2.3 shows the test plots in the Huining County.

Based on the experimental data, correlation equations for the RCE versus rainfall and rainfall intensity were worked out using the least square method (see Box 2.1). For the catchment with compacted soil, moisture content before rain was included in the correlation variables.

Fig. 2.3 RCE test in Huining County

Box 2.1 Brief introduction of "Least square method"

Least square method is a mathematical tool used to work out the correlation or regression equation between random variables.

For two variables Y and X: If we have a number of measurements: (X_1, Y_1, X_2, Y_2...X_n, Y_n) then a linear correlation equation can be set up between X and Y as follows:

$$Y = a + bX \tag{B1.1}$$

Using Eq. B1.1, we can get an approximation of Y_i: $y_i = a + bX_i$.

The deviation between the result from the regression equation and the practical value is ($y_i - Y_i$). The sum of deviation square expressed in Eq. B1.2 denotes the overall error between the calculated results from regression equation and the practical values of Y_i.

$$\mathrm{SD} = \sum_{i=1}^{n} (y_i - Y_i)^2. \tag{B1.2}$$

The coefficients a and b in Eq. B1.1 can be obtained by minimizing the sum of deviation square SD, that is to let the derivation of SD to a and b equal to zero.

Let $\partial \mathrm{SD}/\partial a = 0$ and $\partial \mathrm{SD}/\partial b = 0$, we can get the following equations for a and b.

$$b = \frac{\sum X_i Y_i - n\bar{X}\bar{Y}}{\sum X_i^2 - \bar{X}^2} \tag{B1.3}$$

$$a = \bar{Y} - b\bar{X}, \tag{B1.4}$$

where $\bar{X}$ and $\bar{Y}$ are the mean of series X_i and Y_i, respectively.

The regression coefficient R is used to test the significance of correlation between these two variables.

$$R = \frac{\sum X_i Y_i - n\bar{X}\bar{Y}}{\sqrt{\left(\sum X_i^2 - n\bar{X}^2\right)\left(\sum Y_i^2 - n\bar{Y}^2\right)}}. \tag{B1.5}$$

The necessary condition for correlation between X_i and Y_i being significant at the level of 99 % is that the regression coefficient R calculated by Eq. B1.5 should not be less than the values in Table 2.9.

For derivation of other forms of regression equation such as logarithm, exponential and power function, the linearization of the variable can be used by taking the following transfer equation:

For logarithm equation, let $Z_i, = \mathrm{Ln}X_i$, then derive the linear regression equation between Y_i and Z_i using the above method. The regression equation is

$$Y = a + bZ.$$

Substitute Z with X, we get $Y = a + b\mathrm{Ln}\,X$.

Also to get exponent equation, just let $Z_i = e^{X_i}$.

For the regression equation with two variables (X and Y), the Excel sheet can provide regression equation in the form of linear, logarithm, exponent, and power function.

For variables more than two then correlation between Y and X_1, X_2...X_m can be established with the following equations:

$$Y = a + b_1X_1 + b_2X_2 + b_3X_3 + \cdots + b_mX_m.$$

There are $m + 1$ factor a, b_1, b_2...b_m to be determined. To minimize the sum of square deviation (Eq. B1.2), we will have a set of $m + 1$ multiple-element algebra equations. By solving these equations, we can get value of a, b_1, b_2...b_m.

Table 2.9 Lowest value of R for correlation significance of 99 %

Number of sample N	5	6	7	8	9	10	12	15	18	20
R ≥	0.94	0.88	0.84	0.8	0.76	0.73	0.68	0.62	0.57	0.56

Following are some examples of the resulted regression equation of RCE versus rain characteristics and moisture content of catchment surface material before rain.

For concrete slab in Tongwei testing station

$$\mathrm{RCE} = 1 - 0.0711R^{-0.0739}I^{0.467}. \tag{2.6}$$

For concrete in Xifeng station

$$\mathrm{RCE} = 1 - 0.314R^{-0.12}e^{2.82I}. \tag{2.7}$$

For cement tile

$$\mathrm{RCE} = 1 - 0.208R^{0.424}I^{-0.261}. \tag{2.8}$$

For machine made clay tile

$$\mathrm{RCE} = 1 - 0.551e^{-0.0485R}I^{-0.127}. \tag{2.9}$$

For plastic film covered with sand

$$\mathrm{RCE} = 1 - 1.65R^{0.508}e^{-0.668I}. \tag{2.10}$$

For lime soil

$$\mathrm{RCE} = 1 - 1.68R^{-0.235}e^{-3I}. \tag{2.11}$$

For compacted soil with slope of 2 %

$$\mathrm{RCE} = 1 - 3.95R^{-0.408}I^{-0.32}e^{9.55W}. \tag{2.12}$$

For compacted soil with slope of 5 %

$$\mathrm{RCE} = 1 - 0.0281R^{-0.244}e^{-1.79I}W^{-2.2}. \tag{2.13}$$

In the above equations, R is the rainfall, in mm, I is the intensity, in mm/min, and W is the water content in weight of surface material of the catchment, %. Two examples of these equations are shown in Figs. 2.4 and 2.5.

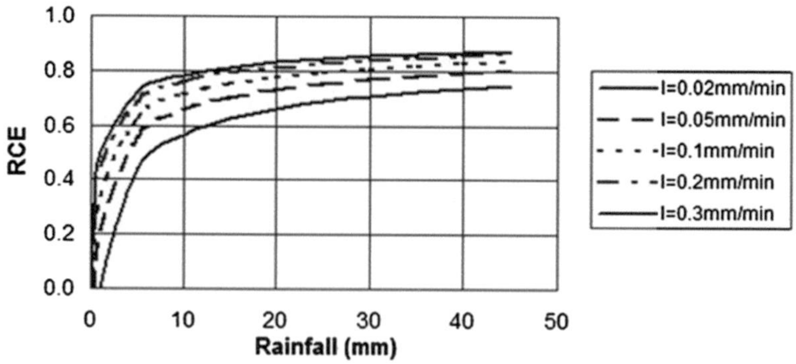

Fig. 2.4 RCE versus rainfall and rain intensity, concrete in Yuzhong station

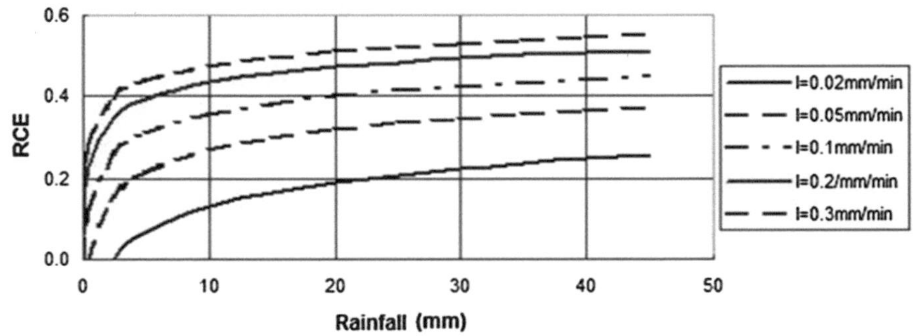

Fig. 2.5 RCE versus rainfall and rain intensity, compacted soil, Tongwei station

(3) Yearly Rainwater Collection Efficiency in the Loess Area of Gansu Province

To calculate the area of a catchment, we need to determine the yearly RCE. The yearly RCE is the mean of RCE in a year, which can be calculated with the following equation.

$$RCE_y = \frac{\sum_{i=1}^{n} RCE_i \bullet R_i}{\sum_{i=1}^{n} R_i}, \tag{2.14}$$

where RCE_y and RCE_i are the yearly RCE and RCE of the *i*th rain event, respectively, n is the number of rain event in the year, R_i is the rainfall in *i*th event.

Figure 2.6 shows the flow chart of the computation of yearly RCE.

Computation of the yearly RCE was carried out event by event over the whole year. For this, the data for rainfall and intensity for the year's rainfall events should be available. During the research projects on RWH undertaken by GRIWAC and its partners, data from 9 rain gauges were used. These gauges represented three regions with mean annual rainfall of 300, 400, and 500 mm. The data included the rainfall and intensity of each rain event in the year and the data series of no shorter than 30 years. In each representative gauge, rain data of three years with hydrological frequency of 50, 75, and 95 % (representing normal, dry and extremely dry years) were selected for computation. The results are listed in Table 2.10.

From Table 2.10, we can find out the main factors affecting the yearly RCE.

- The yearly RCE is higher in areas with a greater mean annual precipitation. This is because in the wetter areas, larger rainfall events occur with greater frequency than in the drier areas, and these produce a higher RCE. Analysis of rainfall data in the loess region of Gansu province, has shown that small rainfall events (less than 10 mm or intensity lower than 0.02 mm/min) account for 40–70 % of total rainfall in the normal year (50 % exceedance at the frequency

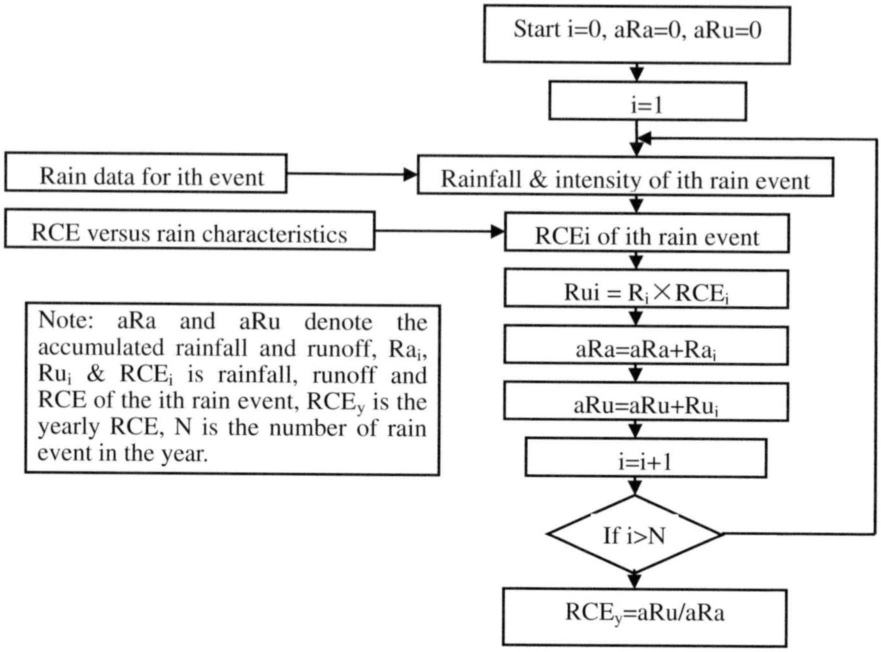

Fig. 2.6 Flow chart of computation of yearly RCE

Table 2.10 The yearly RCE (%) of various types of material in different frequency years of regions with different annual precipitations in loess area of Gansu Province

Annual precipitation (mm)	Frequency of yearly rainfall (%)	Concrete	Cement soil	Buried plastic film	Compacted loess	Cement tiles	Clay tiles machine made	Clay tiles hand made	Natural soil
400–500	50	80	53	46	25	75	50	41	8
	75	79	51	45	23	74	48	38	7
	95	76	42	36	19	69	39	31	6
300–400	50	80	52	46	26	75	49	40	8
	75	78	46	41	21	72	42	34	7
	95	75	40	34	17	67	37	29	5
200–300	50	78	47	41	20	71	41	34	6
	75	75	40	34	17	66	34	28	5
	95	73	33	28	13	62	30	24	4

distribution curve) and 58–83 % in the driest years (75–90 % exceedance). So the lower runoff collected in the dry area is not only owing to the lower annual rainfall but also to the higher frequency of smaller rainfall events that result in a lower RCE. This indicates that the above results of yearly RCE cannot be simply copied to other areas with different climatic conditions.

- The variability of the RCE between years on less permeable catchments is less than that on the areas with higher infiltration rates. For example, RCE on the concrete lined catchment varies from 78 % on average in areas with an annual precipitation of 400–500 mm to 75 % in the areas with annual precipitation of 200–300 mm, reducing only by 4 %. While RCE on compacted loess soil varies from 22 to 17 %, reducing by 23 %. The reduction of RCE on natural soil amounts to 29 %. It indicates that concrete-lined surface and cement tile roof have more stable RCE between different years.

- The type of catchment has the largest effect on the RCE. The maximum RCE occurs on concrete slab and cement tile catchments. The annual RCE usually ranges from 60 to 80 % and even up to 90 %. While compacted soil produces a RCE of only 15–20 % and sometimes even less than 15 %. Natural soil catchments under the dry conditions of Gansu province have RCE as low as 4–8 %, just 1/14 of that of concrete surface.

(4) Yearly Rainwater Collection Efficiency Recommended in the Chinese Code of Practice

The above results of annual RCE were based on experiments undertaken in the loess area of Gansu Province, a semi-arid area. In 2001, the Ministry of Water Resources decided to formulate a technical code for RWH projects in the whole of China. Unfortunately, so far there were no data on RCE testing during rain event in other part of China as the GRIWAC did. Based on the investigation of the RWH project all over China and taking the experiment results from Gansu as an important reference, RCE on different catchments were worked out for sub-humid and humid areas. These were adopted in the <Technical Code of Practice for Rainwater Collection, Storage and Utilization> (MWR 2001) as shown in Table 2.11.

Table 2.11 RCE for different catchments recommended by China code of practice

Material of catchment surface	Yearly RCE (%) with mean annual precipitation of		
	250–500 mm	500–1000 mm	1000–1500 mm
Concrete	75–85	75–90	80–90
Cement tile	65–80	70–85	80–90
Clay tile (machine made)	40–55	45–60	50–65
Clay tile (hand made)	30–40	35–45	45–60
Masonry in good condition	70–80	70–85	75–85
Asphalt paved road in good condition	70–80	70–85	75–85
Earth road, courtyard, threshing yard	15–30	25–40	35–55
Cement soil	40–55	45–60	50–65
Bare plastic film	85–92	85–92	85–92
Plastic film covered with sand/soil	30–50	35–55	40–60
Natural slope (sparse vegetation)	8–15	15–30	30–50
Natural slope (dense vegetation)	6–15	15–25	25–45

2.3 Determining the Area of Catchment

Before talking about how to calculate the catchment area, we will first discuss what are the appropriate dimensions for a RWH system to meet the water demand?

To meet the demand the RWH system should: first has a catchment with an area large enough to produce runoff under the design rainfall no less than the annual demand. Second, since the runoff is produced at different times the storage tank should have enough capacity to regulate the inflow to meet the demand at any time.

2.3.1 Determine Catchment Area by Equation

The area of catchment of a RWH system can be calculated using the following equation.

$$A = \frac{1000 \bullet W_y}{R_\mathrm{d} \bullet \mathrm{RCE}_y}, \tag{2.15}$$

where A is the area (m^2), W_y is the water demand in one year (m^3), R_d is the design rainfall (mm), RCE_y is the yearly rainwater collection efficiency (decimal). Equation (2.15) is for the situation that the RWH project has only one single purpose of water supply and single type of catchment. In case catchments of two or more types are used, the equation is as follows.

$$A_1\mathrm{RCE}_{y1} + A_2\mathrm{RCE}_{y2} + A_3\mathrm{RCE}_{y3} = \frac{1000 \bullet W_y}{R_\mathrm{d}}, \tag{2.16}$$

where A_1, A_2, and A_3 are the catchment area of type 1, 2, and 3, respectively; and RCE_{y1}, RCE_{y2}, and RCE_{y3} are the yearly RCE of the 3 types of catchment. If there are two different catchments used in the RWH project, there should be one with known area. For instance, in the RWH project for domestic use, there can be two catchments in use, namely the roof and the concrete lined courtyard. Usually the roof is to be fully used and its area is known. So another catchment area can be got by solving the equation.

When there are two or more kinds of water use, for which the supply reliability is different the design rainfall for different water uses is different. Also more than one type of catchment can be used for one kind of water use. In this case we have first to calculate the areas of the various types of catchment for various kinds of water use by using Eq. 2.16 in which the design rainfall R_d is different for different water uses. Then the total area of each catchment can be obtained by adding up the areas for the water supplies with different purposes.

$$A_i = \sum_{j=1}^{m} A_{ji}, \tag{2.17}$$

where A_i is the area of the catchment of types i, A_{ji} is the area of catchment type i for the water use of kind j, m is the number of water use.

Now we will sum up the procedure of determining the catchment area.

Step 1 Determine the water supply quota for different kinds of water supply.
Step 2 Calculate the amount of water demand for different kinds of water supply.
Step 3 Determine the reliability of different kinds of water supply.
Step 4 Calculate the design yearly rainfall for different reliabilities either by empirical or theoretical method.
Step 5 Determine the yearly RCE for different kinds of catchment either based on local experience and/or test results (refer to Table 2.11).
Step 6 Calculate the areas of the different types of catchment for each kind of water supply, then the total area for the same type of catchment by adding the areas for different water supplies.

Example 1 Assume that domestic water use is 18 m^3 per year with a reliability of 90 % and the design rainfall with frequency of 90 % is 252 mm. A tiled roof of 80 m^2 and a courtyard lined with concrete are used as the catchment. The RCE for tiled roof and concrete-lined surface are 0.45 and 0.75, respectively. The area of courtyard is to be determined.

First, rainwater can be collected from the roof is $80 \times 252 \times 0.45/1000 = 9.07$ m^3. The remaining water demand will be met by courtyard. So the area of courtyard is $(18 - 9.07) \div 0.75 \div 0.252 = 47.2$ m^2.

Example 2 In the above example, in addition to the domestic water use, irrigation water of 25 m^3 is needed with a required supply reliability of 75 %. The rainfall with frequency of 75 % is 323 mm. The courtyard is lined with concrete specifically to collect water for irrigation purposes.

The area of the concrete-lined courtyard for collecting irrigation water is $25 \div 0.323 \div 0.75 = 103.2$ m^2. So the total area of concrete lined courtyard is $47.2 + 103.2 = 150.4$ m^2.

2.3.2 Table for Determining Catchment Area

To simplify the procedure for determining the catchment area, specially designed tables have been prepared using the above methods. In the tables, the area of different types of catchment for 1 m^3 of water supply at reliability of 50, 75, and 90 % is listed. This can be used to determine the area of any particular type of catchment needed to supply a certain amount of water by multiplying the figure in the table with the water supply amount in the whole year as Eq. 2.18.

$$A_i = W_{yi} \bullet a_i, \tag{2.18}$$

where A_i is the catchment area needed for water supply W_{yi}, a_i is the area needed for collecting 1 m^3 of water at certain reliability. It can be found in the tables.

If there are two or more types of catchment, first determine area of one type of catchment A_1 according to the real situation then calculate area of the second type of catchment. For instance, suppose there is both a roof and courtyard used for collecting rainwater and the roof is to be fully utilized. The amount of water collected on the roof should first be calculated. If the water collected on the roof cannot meet the water demand then the courtyard can be used for additional catchment. The calculation is shown below. Since area of roof is known, then area of courtyard which has to be hardened with concrete to extend the catchment will be

$$A_c = \left(W_y - \frac{A_r}{a_r}\right) \bullet a_c, \tag{2.19}$$

where A_r and A_c are the area of roof and courtyard, respectively, a_r and a_c are the unit area of roof and concrete, respectively, as listed in Tables 2.12, 2.13 and 2.14. Note C_v = Variation Coefficient.

Example 3 The water demand for domestic and irrigation use is the same with Example 1 and 2. Here we assume the annual rainfall of 388 mm and $C_v = 0.26$. From Table 2.14 ($P = 90$ %), we can find the figure in the table for tiled roof (machine made tile) and concrete only at annual rainfall of 300 and 400 mm and C_v of 0.25 and 0.3. To get the figure for annual rainfall of 388 mm and C_v of 0.26, we have to use the interpolation method to calculate the area for collecting 1 m^3 of rainwater as shown in Tables 2.15 and 2.16.

From Table 2.15, we get the figure of 9.01 and 5.13 m^2 corresponding to tiled roof and concrete area, respectively for collecting 1 m^3 of rainwater for domestic use under annual rainfall of 380 mm and C_v of 0.26. The tiled roof of 80 m^2 can collect water of 80/9.01 = 8.9 m^3; the remaining water amount of (18 − 8.9) = 9.1 m^3 is to be collected by concrete lined courtyard. The area needed is $9.1 \times 5.13 = 46.7$ m^2.

From Table 2.16, we get concrete area of 4.42 m^2 for collecting 1 m^3 of rainwater for irrigation under rainfall of 380 mm and C_v of 0.26. For collecting irrigation water of 25 m^3, the area of concrete lined surface is $25 \times 4.42 = 111$ m^2. We can see the result is close to the result in Example 1 and 2 with error of 1 and 7.4 %, respectively.

In Tables 2.15 and 2.16, we can see that the result of taking C_v of 0.25 is close to the taking C_v of 0.26 in both tables. The errors are less than 1.8 %. So in these two cases, we only need to find the figure in Tables 2.13 and 2.14 for rainfall of 380 mm by interpolation and C_v of 0.25.

Table 2.12 Catchment area (m^2) for each m^3 of water supply at reliability of 50 %

C_v	Annual rainfall	Concrete	Cement tile	Clay tile M	Clay tile H	Compacted soil	Asphalt road	Bare plastic	Natural soil-1	Natural soil-2
0.2	250	5.6	6.2	10.2	13.5	27.1	6.2	4.8	50.8	67.7
	300	4.6	5.1	8.2	10.8	20.5	5.1	4.0	34.4	44.0
	400	3.4	3.8	5.9	7.5	13.0	3.8	2.9	18.8	22.9
	500	2.6	2.9	4.5	5.6	9.0	2.9	2.3	11.9	14.0
	600	2.2	2.4	3.6	4.3	6.6	2.4	1.9	8.1	9.5
	800	1.6	1.7	2.4	2.9	4.0	1.7	1.4	4.5	5.1
	1000	1.2	1.3	1.8	2.0	2.7	1.3	1.1	2.9	3.2
	1200	1.0	1.0	1.4	1.5	1.9	1.0	0.9	2.0	2.2
	1400	0.8	0.9	1.1	1.2	1.5	0.9	0.8	1.5	1.6
	1600	0.7	0.7	1.0	1.1	1.3	0.7	0.7	1.3	1.4
	1800	0.6	0.7	0.9	0.9	1.1	0.7	0.6	1.1	1.3
0.25	250	5.6	6.3	10.3	13.7	27.4	6.3	4.8	51.3	68.4
	300	4.6	5.2	8.3	10.9	20.7	5.2	4.0	34.8	44.4
	400	3.4	3.8	5.9	7.6	13.1	3.8	3.0	19.0	23.1
	500	2.7	3.0	4.5	5.6	9.1	3.0	2.4	12.0	14.2
	600	2.2	2.4	3.6	4.4	6.7	2.4	1.9	8.2	9.6
	800	1.6	1.7	2.5	2.9	4.0	1.7	1.4	4.6	5.2
	1000	1.2	1.3	1.8	2.1	2.7	1.3	1.1	2.9	3.3
	1200	1.0	1.0	1.4	1.6	1.9	1.0	0.9	2.0	2.2
	1400	0.8	0.9	1.1	1.2	1.5	0.9	0.8	1.5	1.6
	1600	0.7	0.8	1.0	1.1	1.3	0.8	0.7	1.3	1.4
	1800	0.6	0.7	0.9	0.9	1.1	0.7	0.6	1.1	1.3

(continued)

Table 2.12 (continued)

C_v	Annual rainfall	Concrete	Cement tile	Clay tile M	Clay tile H	Compacted soil	Asphalt road	Bare plastic	Natural soil-1	Natural soil-2
0.3	250	5.7	6.4	10.4	13.8	27.6	6.4	4.9	51.8	69.1
	300	4.7	5.2	8.4	11.0	20.9	5.2	4.0	35.2	44.9
	400	3.4	3.8	6.0	7.6	13.2	3.8	3.0	19.2	23.4
	500	2.7	3.0	4.6	5.7	9.2	3.0	2.4	12.1	14.3
	600	2.2	2.4	3.6	4.4	6.7	2.4	2.0	8.3	9.7
	800	1.6	1.7	2.5	2.9	4.1	1.7	1.4	4.6	5.3
	1000	1.2	1.3	1.8	2.1	2.7	1.3	1.1	2.9	3.3
	1200	1.0	1.1	1.4	1.6	2.0	1.1	0.9	2.0	2.3
	1400	0.8	0.9	1.1	1.2	1.5	0.9	0.8	1.5	1.6
	1600	0.7	0.8	1.0	1.1	1.3	0.8	0.7	1.3	1.4
	1800	0.6	0.7	0.9	1.0	1.2	0.7	0.6	1.2	1.3
0.35	250	5.8	6.5	10.5	14.0	28.1	6.5	5.0	52.6	70.2
	300	4.8	5.3	8.5	11.2	21.2	5.3	4.1	35.7	45.6
	400	3.5	3.9	6.1	7.8	13.5	3.9	3.0	19.5	23.7
	500	2.7	3.0	4.6	5.8	9.3	3.0	2.4	12.3	14.5
	600	2.2	2.5	3.7	4.5	6.8	2.5	2.0	8.4	9.8
	800	1.6	1.8	2.5	3.0	4.1	1.8	1.5	4.7	5.3
	1000	1.3	1.3	1.9	2.1	2.8	1.3	1.2	3.0	3.3
	1200	1.0	1.1	1.4	1.6	2.0	1.1	0.9	2.1	2.3
	1400	0.8	0.9	1.2	1.3	1.5	0.9	0.8	1.5	1.7
	1600	0.7	0.8	1.0	1.1	1.3	0.8	0.7	1.3	1.5
	1800	0.6	0.7	0.9	1.0	1.2	0.7	0.6	1.2	1.3

(continued)

Table 2.12 (continued)

C_v	Annual rainfall	Concrete	Cement tile	Clay tile M	Clay tile H	Compacted soil	Asphalt road	Bare plastic	Natural soil-1	Natural soil-2
0.4	250	5.9	6.6	10.7	14.3	28.5	6.6	5.0	53.5	71.3
	300	4.8	5.4	8.7	11.4	21.6	5.4	4.2	36.3	46.3
	400	3.6	4.0	6.2	7.9	13.7	4.0	3.1	19.8	24.1
	500	2.8	3.1	4.7	5.9	9.5	3.1	2.5	12.5	14.8
	600	2.3	2.5	3.7	4.6	6.9	2.5	2.0	8.6	10.0
	800	1.6	1.8	2.6	3.0	4.2	1.8	1.5	4.8	5.4
	1000	1.3	1.4	1.9	2.2	2.8	1.4	1.2	3.0	3.4
	1200	1.0	1.1	1.5	1.6	2.0	1.1	1.0	2.1	2.3
	1400	0.8	0.9	1.2	1.3	1.5	0.9	0.8	1.5	1.7
	1600	0.7	0.8	1.0	1.1	1.3	0.8	0.7	1.3	1.5
	1800	0.7	0.7	0.9	1.0	1.2	0.7	0.6	1.2	1.3

Table 2.13 Catchment area (m^2) for each m^3 of water supply at reliability of 75 %

C_v	Annual precipitation	Concrete	Cement tile	Machine made tile	Hand made tile	Compacted soil	Asphalt road	Bare plastic	Natural soil 1	Natural soil 2
0.2	250	6.4	7.2	11.6	15.5	31.0	7.2	5.5	58.1	77.5
	300	5.3	5.9	9.4	12.4	23.5	5.9	4.5	39.4	50.4
	400	3.9	4.3	6.7	8.6	14.9	4.3	3.4	21.6	26.2
	500	3.0	3.4	5.1	6.4	10.3	3.4	2.7	13.6	16.1
	600	2.5	2.7	4.1	5.0	7.6	2.7	2.2	9.3	10.8
	800	1.8	1.9	2.8	3.3	4.6	1.9	1.6	5.2	5.9
	1000	1.4	1.5	2.1	2.3	3.1	1.5	1.3	3.3	3.7
	1200	1.1	1.2	1.6	1.8	2.2	1.2	1.0	2.3	2.5
	1400	0.9	1.0	1.3	1.4	1.7	1.0	0.9	1.7	1.8
	1600	0.8	0.9	1.1	1.2	1.5	0.9	0.8	1.5	1.6
	1800	0.7	0.8	1.0	1.1	1.3	0.8	0.7	1.3	1.4
0.25	250	6.7	7.5	12.2	16.3	32.5	7.5	5.7	61.0	81.3
	300	5.5	6.2	9.9	13.0	24.6	6.2	4.8	41.4	52.8
	400	4.1	4.5	7.0	9.0	15.6	4.5	3.5	22.6	27.5
	500	3.2	3.5	5.4	6.7	10.8	3.5	2.8	14.2	16.8
	600	2.6	2.9	4.3	5.2	7.9	2.9	2.3	9.8	11.4
	800	1.9	2.0	2.9	3.4	4.8	2.0	1.7	5.4	6.2
	1000	1.5	1.6	2.2	2.5	3.2	1.6	1.3	3.4	3.9
	1200	1.2	1.2	1.7	1.9	2.3	1.2	1.1	2.4	2.7
	1400	1.0	1.0	1.3	1.5	1.7	1.0	0.9	1.7	1.9
	1600	0.8	0.9	1.2	1.3	1.5	0.9	0.8	1.5	1.7
	1800	0.8	0.8	1.0	1.1	1.4	0.8	0.7	1.4	1.5

(continued)

Table 2.13 (continued)

C_v	Annual precipitation	Concrete	Cement tile	Machine made tile	Hand made tile	Compacted soil	Asphalt road	Bare plastic	Natural soil 1	Natural soil 2
0.3	250	7.0	7.9	12.8	17.1	34.2	7.9	6.0	64.1	85.5
	300	5.8	6.5	10.4	13.7	25.9	6.5	5.0	43.5	55.5
	400	4.3	4.7	7.4	9.5	16.4	4.7	3.7	23.8	28.9
	500	3.3	3.7	5.6	7.0	11.3	3.7	2.9	15.0	17.7
	600	2.7	3.0	4.5	5.5	8.3	3.0	2.4	10.3	12.0
	800	2.0	2.1	3.1	3.6	5.0	2.1	1.8	5.7	6.5
	1000	1.5	1.6	2.3	2.6	3.4	1.6	1.4	3.6	4.1
	1200	1.2	1.3	1.8	2.0	2.4	1.3	1.1	2.5	2.8
	1400	1.0	1.1	1.4	1.5	1.8	1.1	1.0	1.8	2.0
	1600	0.9	0.9	1.2	1.3	1.6	0.9	0.8	1.6	1.8
	1800	0.8	0.8	1.1	1.2	1.4	0.8	0.7	1.4	1.6
0.35	250	7.4	8.3	13.4	17.9	35.8	8.3	6.3	67.1	89.5
	300	6.1	6.8	10.9	14.3	27.1	6.8	5.2	45.5	58.1
	400	4.5	5.0	7.8	9.9	17.2	5.0	3.9	24.9	30.3
	500	3.5	3.9	5.9	7.4	11.9	3.9	3.1	15.7	18.5
	600	2.9	3.1	4.7	5.7	8.7	3.1	2.5	10.8	12.5
	800	2.1	2.3	3.2	3.8	5.3	2.3	1.9	6.0	6.8
	1000	1.6	1.7	2.4	2.7	3.5	1.7	1.5	3.8	4.3
	1200	1.3	1.4	1.8	2.0	2.5	1.4	1.2	2.6	2.9
	1400	1.1	1.1	1.5	1.6	1.9	1.1	1.0	1.9	2.1
	1600	0.9	1.0	1.3	1.4	1.7	1.0	0.9	1.7	1.9
	1800	0.8	0.9	1.1	1.2	1.5	0.9	0.8	1.5	1.7

(continued)

Table 2.13 (continued)

C_v	Annual precipitation	Concrete	Cement tile	Machine made tile	Hand made tile	Compacted soil	Asphalt road	Bare plastic	Natural soil 1	Natural soil 2
0.4	250	7.8	8.7	14.2	18.9	37.8	8.7	6.7	70.9	94.6
	300	6.4	7.2	11.5	15.1	28.6	7.2	5.5	48.1	61.4
	400	4.7	5.2	8.2	10.5	18.1	5.2	4.1	26.3	32.0
	500	3.7	4.1	6.2	7.8	12.5	4.1	3.3	16.6	19.6
	600	3.0	3.3	5.0	6.0	9.2	3.3	2.7	11.4	13.2
	800	2.2	2.4	3.4	4.0	5.6	2.4	2.0	6.3	7.2
	1000	1.7	1.8	2.5	2.9	3.7	1.8	1.5	4.0	4.5
	1200	1.4	1.4	1.9	2.2	2.7	1.4	1.3	2.8	3.1
	1400	1.1	1.2	1.6	1.7	2.0	1.2	1.1	2.0	2.3
	1600	1.0	1.0	1.4	1.5	1.8	1.0	0.9	1.8	2.0
	1800	0.9	0.9	1.2	1.3	1.6	0.9	0.8	1.6	1.8

Table 2.14 Catchment area (m^2) for each m^3 of water supply at reliability of 90 %

C_v	Annual Precipitation	Concrete	Cement tile	Machine made tile	Hand made tile	Compacted soil	Asphalt road	Bare plastic	Natural soil 1	Natural soil 2
0.2	250	7.3	8.2	13.2	17.7	35.3	8.2	6.2	66.2	88.3
	300	6.0	6.7	10.7	14.1	26.7	6.7	5.2	44.9	57.4
	400	4.4	4.9	7.7	9.8	16.9	4.9	3.8	24.6	29.9
	500	3.5	3.8	5.8	7.3	11.7	3.8	3.0	15.5	18.3
	600	2.8	3.1	4.6	5.6	8.6	3.1	2.5	10.6	12.4
	800	2.0	2.2	3.2	3.7	5.2	2.2	1.8	5.9	6.7
	1000	1.6	1.7	2.4	2.7	3.5	1.7	1.4	3.7	4.2
	1200	1.3	1.4	1.8	2.0	2.5	1.4	1.2	2.6	2.9
	1400	1.1	1.1	1.5	1.6	1.9	1.1	1.0	1.9	2.1
	1600	0.9	1.0	1.3	1.4	1.7	1.0	0.9	1.7	1.8
	1800	0.8	0.9	1.1	1.2	1.5	0.9	0.8	1.5	1.6
0.25	250	7.8	8.7	14.2	18.9	37.8	8.7	6.7	70.9	94.6
	300	6.4	7.2	11.5	15.1	28.6	7.2	5.5	48.1	61.4
	400	4.7	5.2	8.2	10.5	18.1	5.2	4.1	26.3	32.0
	500	3.7	4.1	6.2	7.8	12.5	4.1	3.3	16.6	19.6
	600	3.0	3.3	5.0	6.0	9.2	3.3	2.7	11.4	13.2
	800	2.2	2.4	3.4	4.0	5.6	2.4	2.0	6.3	7.2
	1000	1.7	1.8	2.5	2.9	3.7	1.8	1.5	4.0	4.5
	1200	1.4	1.4	1.9	2.2	2.7	1.4	1.3	2.8	3.1
	1400	1.1	1.2	1.6	1.7	2.0	1.2	1.1	2.0	2.3
	1600	1.0	1.0	1.4	1.5	1.8	1.0	0.9	1.8	2.0
	1800	0.9	0.9	1.2	1.3	1.6	0.9	0.8	1.6	1.8

(continued)

Table 2.14 (continued)

C_v	Annual Precipitation	Concrete	Cement tile	Machine made tile	Hand made tile	Compacted soil	Asphalt road	Bare plastic	Natural soil 1	Natural soil 2
0.3	250	8.4	9.5	15.4	20.5	41.0	9.5	7.2	76.9	102.6
	300	7.0	7.8	12.5	16.4	31.0	7.8	6.0	52.2	66.6
	400	5.1	5.7	8.9	11.3	19.7	5.7	4.5	28.5	34.7
	500	4.0	4.4	6.8	8.4	13.6	4.4	3.5	18.0	21.3
	600	3.3	3.6	5.4	6.6	10.0	3.6	2.9	12.3	14.3
	800	2.4	2.6	3.7	4.3	6.1	2.6	2.1	6.8	7.8
	1000	1.8	2.0	2.7	3.1	4.1	2.0	1.7	4.3	4.9
	1200	1.5	1.6	2.1	2.3	2.9	1.6	1.4	3.0	3.4
	1400	1.2	1.3	1.7	1.8	2.2	1.3	1.2	2.2	2.4
	1600	1.1	1.1	1.5	1.6	1.9	1.1	1.0	1.9	2.1
	1800	0.9	1.0	1.3	1.4	1.7	1.0	0.9	1.7	1.9
0.35	250	9.1	10.3	16.7	22.2	44.4	10.3	7.8	83.3	111.1
	300	7.5	8.4	13.5	17.7	33.6	8.4	6.5	56.5	72.2
	400	5.5	6.2	9.6	12.3	21.3	6.2	4.8	30.9	37.6
	500	4.3	4.8	7.3	9.1	14.7	4.8	3.8	19.5	23.0
	600	3.6	3.9	5.8	7.1	10.8	3.9	3.2	13.4	15.5
	800	2.6	2.8	4.0	4.7	6.6	2.8	2.3	7.4	8.5
	1000	2.0	2.1	3.0	3.4	4.4	2.1	1.8	4.7	5.3
	1200	1.6	1.7	2.3	2.5	3.2	1.7	1.5	3.3	3.6
	1400	1.3	1.4	1.8	2.0	2.4	1.4	1.3	2.4	2.6
	1600	1.2	1.2	1.6	1.7	2.1	1.2	1.1	2.1	2.3
	1800	1.0	1.1	1.4	1.5	1.9	1.1	1.0	1.9	2.1

(continued)

Table 2.14 (continued)

C_v	Annual Precipitation	Concrete	Cement tile	Machine made tile	Hand made tile	Compacted soil	Asphalt road	Bare plastic	Natural soil 1	Natural soil 2
0.4	250	10.0	11.2	18.2	24.2	48.5	11.2	8.6	90.9	121.2
	300	8.2	9.2	14.8	19.4	36.7	9.2	7.1	61.7	78.8
	400	6.0	6.7	10.5	13.4	23.2	6.7	5.3	33.7	41.0
	500	4.7	5.2	8.0	10.0	16.1	5.2	4.2	21.2	25.1
	600	3.9	4.3	6.4	7.7	11.8	4.3	3.4	14.6	17.0
	800	2.8	3.0	4.4	5.1	7.2	3.0	2.5	8.1	9.2
	1000	2.2	2.3	3.2	3.7	4.8	2.3	2.0	5.1	5.8
	1200	1.7	1.9	2.5	2.8	3.5	1.9	1.6	3.5	4.0
	1400	1.4	1.5	2.0	2.2	2.6	1.5	1.4	2.6	2.9
	1600	1.3	1.3	1.7	1.9	2.3	1.3	1.2	2.3	2.5
	1800	1.1	1.2	1.6	1.7	2.0	1.2	1.1	2.0	2.2

Note: Natural slope 1 denotes soil with poor vegetation, natural slope 2 denotes soil with dense vegetation

Table 2.15 Calculation of area for collecting 1 m^3 of rainwater with annual rainfall of 380 mm and C_v of 0.26 using Table 2.14 ($P = 90$ %)

Material	C_v	a_0 when annual rainfall is		a_0 when annual rainfall is 380 mm
		300 mm	400 mm	
Tiled roof	0.25	11.5	8.2	(11.5 − 8.2) × (400 − 380)/(400 − 300) + 8.2 = 8.86
	0.3	12.5	8.9	(12.5 − 8.9) × (400 − 380)/(400 − 300) + 8.9 = 9.62
	0.26			(9.62 − 8.86) × (0.26 − 0.25)/(0.3 − 0.25) + 8.86 = 9.01
Concrete	0.25	6.4	4.7	(6.4 − 4.7) × (400 − 380)/(400 − 300) + 4.7 = 5.04
	0.3	7.0	5.1	(7.0 − 5.1) × (400 − 380)/(400 − 300) + 5.1 = 5.48
	0.26			(5.48 − 5.04) × (0.26 − 0.25)/(0.3 − 0.25) + 5.04 = 5.13

Table 2.16 Calculation of area for collecting 1 m^3 of rainwater with annual rainfall of 380 mm and C_v of 0.26 using Table 2.13 ($P = 75$ %)

Material	C_v	a_0 when annual rainfall is		a_0 when annual rainfall is 380 mm
		300 mm	400 mm	
Concrete	0.25	5.5	4.1	(5.5 − 4.1) × (400 − 380)/(400 − 300) + 4.1 = 4.38
	0.3	5.8	4.3	(5.8 − 4.3) × (400 − 380)/(400 − 300) + 4.3 = 4.6
	0.26			(4.6 − 4.38) × (0.26 − 0.25)/(0.3 − 0.25) + 4.38 = 4.42

Table 2.17 Comparison of results between time step of 10 days and one month

Interval	Volume (m^3)/Date of maximum storage		
	Domestic use	Wheat irrigation	Corn irrigation
1 month	13.26/31 August	12.79/30 November	9.68/31 May
10 days	13.56/31 August	17.37/10 May	14.86/10 August
Deviation %	2.2 %	26.4 %	34.9 %

2.4 Determining the Storage Capacity

In principle, the storage capacity of the RWH system can be determined by using a simulation model, which simulates the process of the inflow (runoff from the catchment), outflow (water supply), water loss (evaporation and seepage from the storage tank), and the volume of the storage. During the process, the maximum storage capacity is the design storage capacity. In this chapter, we will discuss the methodology used in a simulation model for a RWH system. Since the RWH project has been implemented over a vast area of China by the households themselves, the determination of storage capacity using a simulation model would be too difficult for farmers and local technicians to undertake. At the end of this chapter, a simplified method for determining the storage capacity, which can be easily calculated by those having only primary school mathematics knowledge, will be provided.

2.4.1 Simulation Model

(1) **Basic Equation**

A simulation model is operated over a fixed time period, a year or a longer time series. The calculated period has to be divided into several time steps. The basic equation to calculate the storage capacity is as follows.

$$V_{i+1} = V_i + \mathrm{Fl}_{\mathrm{in}} - \mathrm{Ws} - L_{i,i+1}, \tag{2.20}$$

where V_i and V_{i+1} are the volume capacity at time t_i and t_{i+1}, respectively, $\mathrm{Fl}_{\mathrm{in}}$ is the inflow into the system, Ws is the water supply to water users, and $L_{i,I+1}$ is the water loss (evaporation and seepage) during the period from time t_i to t_{i+1}, respectively.

In Eq. 2.20, Ws is determined with the water demand introduced in Sect. 2.1, which should be met with the inflow of the system and water released (in case inflow is not enough to meet the demand) from the storage. If the inflow and/or the initial V_i is too small or close to zero, water shortage would occur. On the contrary if the inflow and/or the initial V_i is too large and the storage capacity is limited then water waste (water overflow) would happen. These two situations are described by the Eqs. 2.21 and 2.22.

Water shortage happens when

$$\mathrm{Ws} < \mathrm{Fl}_{\mathrm{in}} - (V_{i+1} - V_i) - L_{i,I+1}. \tag{2.21}$$

Water waste happens when

$$V_{\max} < V_i + \mathrm{Fl}_{\mathrm{in}} - \mathrm{Ws} - L_{i,I+1}. \tag{2.22}$$

Figure 2.7 illustrates these three situations.

The simulation model can be operated in two ways, namely the typical year method and the long series method.

(2) **Time Step in the Simulation Model**

The simulation model is operated time step by time step. The length of the time step will affect the working load and also the accuracy of the results. Usually one month or ten days is used as the time step. Comparison of the results of the calculated storage using a time step of 10 days and one month is shown in Table 2.17.

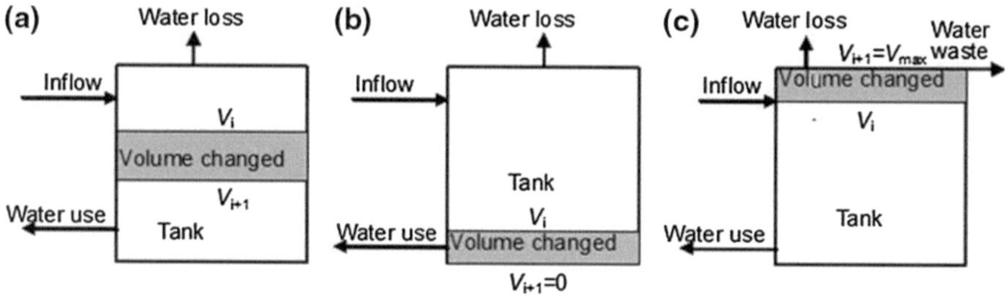

Fig. 2.7 **a** Illustration of water balance. **b** Water shortage. **c** Water waste

From Table 2.17, it can be seen that if the RWH system is for domestic water supply, the time interval can be one month without a significant error in determining the storage capacity. However, for the irrigation project, taking the time interval as one month may cause error up to 26–35 %. Therefore, it is recommended taking one month for calculating the storage of RWH project for domestic supply and 10 days for irrigation purpose to avoid the risk of a significant error.

(3) **Prepare Inflow and Outflow of RWH System**

The inflow into a RWH system is the runoff on the catchment during the rainfall event. As described in the above sections, the runoff coefficient of each rainfall event is different depending on the rainfall characteristics. So the runoff is calculated on an event basis. The inflow of the RWH system is calculated by adding the runoff of rainfall event to each time interval: 10 days or one month. When calculating the runoff the equation of RCE versus rainfall characteristics for the catchment type and the catchment area are calculated with method described in Sect. 2.3.1.

Example of calculating the inflow on a concrete lined surface in the third ten days of April 1965 with rainfall data of Huining Gauge, Gansu is shown in Table 2.18. In this calculation, the equation for RCE versus rain characteristics in a rainfall event is as follows.

$$\mathrm{RCE} = 1 - 0.314R^{-0.12}e^{-2.82I}. \tag{2.23}$$

In principle, the calculation of the inflow of the RWH system using the rainfall data and RCE on the event basis would seem scientific. However, it requires information on the relationship between RCE versus rainfall characteristics, which in most cases is not available. Measurement of the rainfall data on the event basis also needs an automatic rainfall recorder that is also usually unavailable for most rain gauges. Besides, the calculation procedure is very time consuming.

To find a simpler way, the feasibility of setting up a regressive relationship between RCE and rainfall characteristics on 10-day or monthly basis was explored. Rainfall data from Huining Gauge in Gansu with a data series of 34 years from year 1957 to 1990 were used. With rainfall and rainfall intensity data as well as RCE versus rain characteristics for catchments of concrete, clay

Table 2.18 Example of calculating the runoff on 100 m^2 concrete-lined surface, Huining gauge, (In the third ten-day period of April 1965)

Year	Date and time			Rainfall mm	Rainfall in the event mm	Intensity in the event mm/min	Runoff of concrete surface with area of 100 m^2, m^3
	Month	Date	Time hour:min				
1965	April	21	5:20				
1965		21	7:45	2.6	2.6	0.0179	1.91
1965		21	10:35				
1965		21	13:12	0.1	0.1	0.0006	0.06
1965		26	7:15				
1965		26	8:00	0.2	0.2	0.0044	0.12
1965		26	18:50				
1965		26	20:00	6.9	6.9	0.0986	5.60
1965		26	20:00				
1965		27	2:00	24.8	24.8	0.1033	20.84
1965		27	2:00				
1965		27	5:15	2.6	2.6	0.0144	1.90
1965		28	17:15				
1965		28	18:45	3.2	3.2	0.0356	2.41
1965		29	8:00	0.2	0.2	0.0012	0.12
1965		30	18:21				
1965		30	20:00	0.5	0.5	0.0050	0.33
1965		30	20:00				
1965		30	23:49	3.3	3.3	0.0183	2.45
Sum for ten days =							35.75

Table 2.19 Regression equation of RCE versus Rainfall on monthly basis

Catchment	Regression equation	Regression coefficient	Significance level
Concrete equation 1	$\mathrm{RCE}_m = 0.0224\mathrm{Ln}(R_m) + 0.6803$	0.8	0.01
Concrete equation 2	$\mathrm{RCE}_m = 0.6796R_m^{0.0312}$	0.81	0.01
Tile	$RCE_m = 0.0728\mathrm{Ln}(R_m) + 0.0572$	0.71	0.01
Compacted soil	$RCE_m = 0.0534\mathrm{Ln}(R_m) + 0.0159$	0.71	0.01

tile and compacted soil available, the runoff and rainfall of each rain events in the 34 years were calculated, and summed for each 10-day and 1-month period. The regression analysis was then conducted to find out the relationship between RCE and rainfall on a monthly and 10-day basis. The results are shown in Tables 2.19 and 2.20. The number of regressive variables is 289 and 754, for the 2 time periods, respectively. F-tests showed that all the regressive relationships are significant at a level of >0.99 ($\alpha = 0.01$).

The regressive curve of RCE versus rainfall on the monthly basis is shown in Figs. 2.8, 2.9 and 2.10. Figures 2.11, 2.12 and 2.13 show the regressive curve of RCE versus rainfall on the 10-day basis.

It should be pointed out that these equations are based on experimental data of the relationship between RCE and rainfall characteristics tested in the loess area of Gansu using locally sourced rainfall data. These equations cannot simply

Table 2.20 Regression equation of RCE versus Rainfall on 10-day basis

Catchment	Regression equation	Regression coefficient	Significance level
Concrete equation 1	$RCE_d = 0.0289Ln(R_m) + 0.679$	0.84	0.01
Clay tile	$RCE_m = 0.0864Ln(R_m) + 0.0662$	0.74	0.01
Compacted soil	$RCE_m = 0.0632Ln(R_m) + 0.0233$	0.7	0.01

Note Significance level α denotes the correlation between two variables being at a confidential probability of (1 – α). It was worked out using the F-test

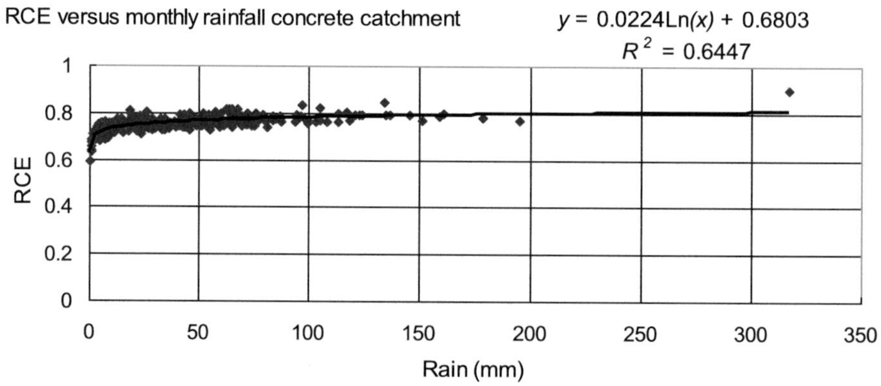

Fig. 2.8 RCE versus monthly rainfall on concrete catchment

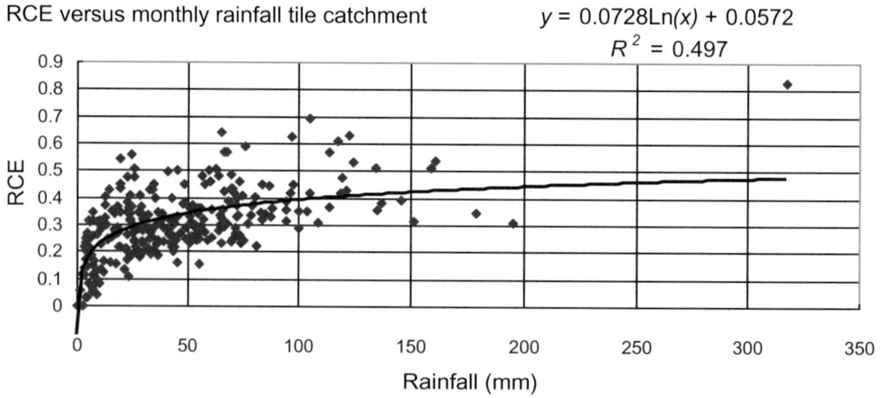

Fig. 2.9 RCE versus monthly rainfall on clay tile catchment

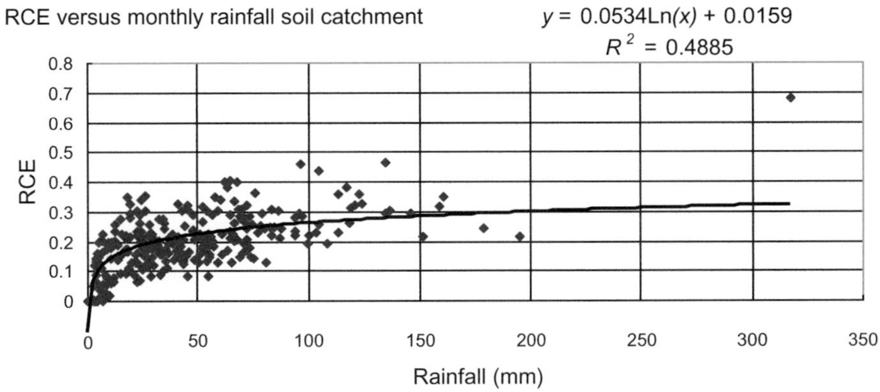

Fig. 2.10 RCE versus monthly rainfall on compacted soil catchment

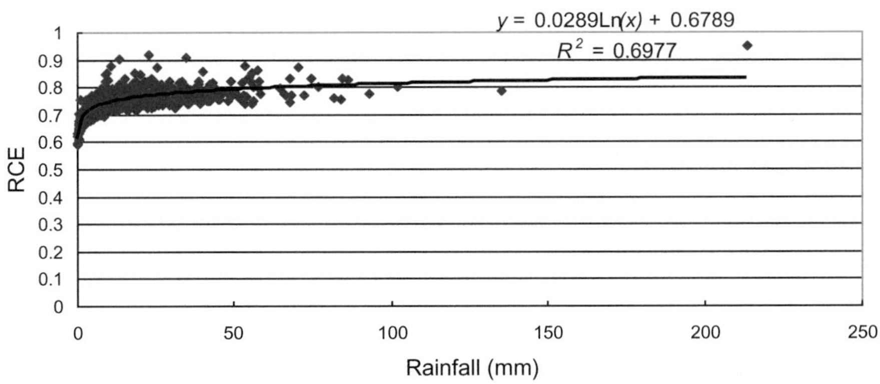

Fig. 2.11 RCE versus 10-day rainfall on concrete catchment

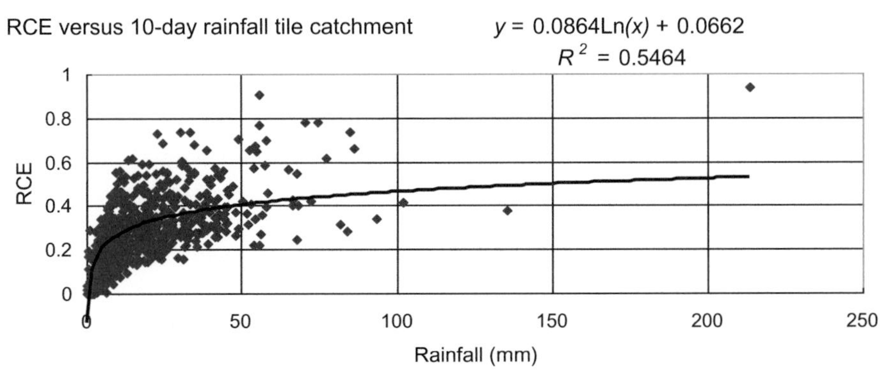

Fig. 2.12 RCE versus 10-day rainfall on tile roof catchment

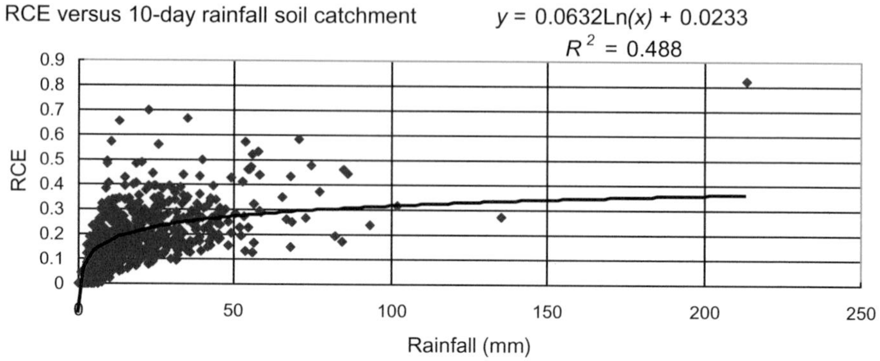

Fig. 2.13 RCE versus 10-day rainfall on soil catchment

be copied and used elsewhere. But from the above regression equation, it can be found that most of the equations are in a logarithmic form. It is suggested that the RCE versus rainfall on either a monthly or 10-day basis may take the same form as the following Eqs. 2.24 and 2.25, except that the coefficient in the equation should be derived based on the specific local conditions.

$$\mathrm{RCE}_m = a_m \mathrm{Ln}(R_m) + b_m \tag{2.24}$$

$$\mathrm{RCE}_{10\mathrm{day}} = a_{10\mathrm{day}} \mathrm{LN}\left(R_{10\mathrm{day}}\right) + b_{10\mathrm{day}}. \tag{2.25}$$

RCE_m and R_m are the mean of RCE and rainfall in the month, respectively. $\mathrm{RCE}_{10\mathrm{day}}$ and $R_{10\mathrm{day}}$ are the mean of RCE and rainfall in the 10 days, respectively and a_m, b_m, $a_{10\mathrm{day}}$, and $b_{10\mathrm{day}}$ are the coefficients that should be calibrated according to the local conditions. To make the calibration, first we may preset a yearly RCE for a certain catchment according to the local practical experience or testing. The monthly or 10-day rainfall data of four representative years with frequency of 25, 50, 75, and 90 % are prepared. Then assume a set of coefficients a_m and b_m. The monthly RCE over four years is calculated and the yearly RCE is obtained by averaging the RCE in each of the years. If the yearly RCE is not consistent with the preset one then re-assume the coefficient a_m and b_m and conduct the above calculation again until the resulting yearly RCE consistent with the preset one.

Later we will see that the error in the resulting storage capacity caused by this simplification is in an acceptable range.

When the regression equation between RCE and rainfall on a 10-day or monthly basis is obtained the calculation for yearly RCE and the inflow of the RWH system is much simpler than using data on a rainfall event basis. We only need the monthly or 10-day rainfall data, which are usually available for most rain gauges.

The outflow of the system is composed of two parts: the water supply for domestic, husbandry, irrigation uses, etc. and the water loss in the storage tank. The water supply and water loss should also be allocated to time steps: monthly or 10-day.

The water supply for domestic use and husbandry can be taken as evenly distributed over the whole year, or if data are available a slight variation of seasonal water demand may be considered. Allocation of irrigation water supply with time is determined using the irrigation schedule that will be described in detail in Chap. 5.

Water loss is composed of evaporation loss from the water surface in the tank and seepage loss through the tank structure. Obtaining an accurate estimation of these two losses is quite complicated since they depend on many factors. Evaporation is related to the air temperature, wind, solar radiation, humidity, and the area of the exposed surface of the tank, etc. Seepage loss depends on the quality of the construction, permeability of the soil, etc. When estimating the water loss, it will be roughly taken as 10 % of the water supply.

(4) **Typical Year Method**

(i) Selection of typical year

In the typical year method, the storage calculation is conducted for a typical year. This should represent a year with hydrological frequency close to the supply reliability. Therefore the typical year is selected following two points:

- First, the yearly rainfall of the typical year should be equal to the design rainfall;
- Second, the rainfall distribution should be representative. Since the rainfall distribution is one of the most important factors that is affecting the storage capacity, it should be selected carefully. Usually rainfall distributions for two or three years with hydrological frequency close to the water supply reliability on the empirical frequency distribution curve are used to form the typical year.

For example, if the supply reliability is 75 %, the rainfall distribution of two or three candidate years is used. In Table 2.8 the years 1962, 1966 and 1980 have yearly rainfall frequency of 71, 74.2 and 77.4 %, respectively, which are closest to the supply reliability of 75 % in the data series. The rainfall distribution of these 3 years will therefore be used for typical rainfall distribution as shown in Table 2.20. First the yearly rainfall is set to be design rainfall, which is 451 mm. The monthly distribution of the typical year is formed by multiplying each monthly rainfall of the 3 years with an adjustment factor equal to the ratio between the design rainfall and the actual yearly rainfall of these 3 years. In Table 2.21, the factor equals to 0.960, 0.994 and 1.018 for year 1962, 1966 and 1980, respectively.

The three candidate years will be used to calculate the storage capacity of the RWH system. The maximum calculated result of storage capacity from the three candidate years is taken as the design capacity of the storage and the rainfall distribution of the corresponding year is taken as that for the typical year.

The typical year can also be selected according to the rainfall in the critical period of water shortage. The critical period means the period when the gap between water supply and demand becomes largest. It depends on the specific conditions in different areas. In Gansu, May to June is often the critical period when

Table 2.21 Monthly rainfall distribution of three candidates of typical year

Year	Year sum	Adjust factor	January	February	March	April	May	June	July	August	September	October	November	December
1962	469.6		0	0	38.8	29.6	29.5	4.8	103.2	129.5	57.5	41.5	35	0.2
Typical	451	0.96	0	0	37.3	28.4	28.3	4.6	99.1	124.4	55.2	39.9	33.6	0.2
1966	453.7		0	115.7	7.5	97.7	8.4	36.6	47.2	81.5	55.8	3.3	0	0
Typical	451	0.994	0	115	7	97	8	36	47	81	55	3	0	0
1980	443.1		79.2	0	16.1	61.6	21.1	23.8	49.6	92.5	51	48.2	0	0
Typical	451	1.018	80.6	0.0	16.4	62.7	21.5	24.2	50.5	94.1	51.9	49.1	0.0	0.0

the crops needs water most while the rainfall is still scarce. To do this we have to first prepare the empirical frequency curve of rainfall from May to June. Also years with rainfall in the critical period having frequency close to the supply reliability are selected to form the rainfall distribution for the typical years.

(ii) Calculation of storage volume for single water use

The calculation is carried out for the whole year month by month or 10 days by 10 days. The maximum volume found in the calculation is the volume that we want to find out. The calculation is started from the time when the storage drops to zero. The trial and error method is used to find out the zero-storage time. The calculation example is shown in Table 2.21. The rainfall data in Table 2.21 are the same with Table 2.8. Runoff is obtained either with rainfall data and RCE on the event basis or with the monthly/10-day RCE. The water supply in this example is for domestic use. It is taken as evenly distributed in the whole year. Water loss is taken as 10 % of the water supply. Water balance equals to runoff minus water supply and water loss. It can be seen that from the beginning of January to the end of March there is continuous negative balance. It means in this period water has to be released from the storage tank to meet the water supply. It is reasonable to assume that the storage capacity will drop to zero in the end of March. V_{end} is the volume of storage at the end of the time step. It can be calculated with the following equation.

$$V_{\text{end}i+1} = V_{\text{end}i} + \text{Balance}_{i+1}. \tag{2.26}$$

In Eq. 2.26, $V_{\text{end}i}$ and $V_{\text{end}i+1}$ are the volumes of the storage tank at end of time step i and time step $i + 1$, respectively, Balance_{i+1} is the water balance in the time step $i + 1$. For January, December is taken as the preceding time step. Volume at the end of January equals to the volume at the end of December plus water balance in January. The second to last column "Water Shortage" equals to (V_i + Balance_{i+1} − V_{i+1}) and should be zero theoretically. However, owing to some calculation error it could be greater or less than zero. If the absolute value of "Water Shortage" is small compared to the water supply in the whole year, then it can be neglected. In this example it is only about 1.2 % of the water supply in the year.

Determination of time of zero-storage is a process of "trial and error." During calculation if negative value occurs in some time steps, it means the assumed time of zero-storage is not appropriate and should be changed. This causes some additional work load. Another way to find out the storage capacity is shown in the last column of Table 2.21 in which the accumulation of water balance (Column 7) is calculated. The storage capacity can be calculated with the following equation.

$$V = \text{Max (Accumulated water balance)} - \text{Min (Accumulated water balance)}. \tag{2.27}$$

In column "Accumulated water balance" of Table 2.22, the maximum and minimum Accumulated water balance is 16.00 and (−23.37), so the Storage Volume equals to 16.0 − (−23.37) = 39.37.

Table 2.22 Calculation of storage capacity, "typical year" method (Domestic use, reliability = 90 %, design rainfall = 368.6 mm, year 1965, catchment: tiled roof)

Month	Real rainfall	Design rainfall	Inflow	Demand	Loss	Water balance	V_{end}	Water shortage	Accumulated water balance
1	2	3	4	5	6	7	8	9	10
January	0	0	0.0	7.30	0.730	−8.03	14.27	0.00	−8.03
February	0	0	0.0	7.30	0.730	−8.03	6.24	0.00	−16.06
March	3.9	3.61	0.7	7.30	0.730	−7.31	0.00	−1.07	−23.37
April	186.7	172.84	47.4	7.30	0.730	39.37	**39.37**	0.00	16.00
May	0	0	0.0	7.30	0.730	−8.03	31.34	0.00	7.97
June	6.8	6.30	1.3	7.30	0.730	−6.71	24.63	0.00	1.26
July	30.7	28.42	6.8	7.30	0.730	−1.24	23.39	0.00	0.01
August	83.1	76.93	19.9	7.30	0.730	11.85	35.24	0.00	11.86
September	38.6	35.74	8.7	7.30	0.730	0.67	35.90	0.00	12.53
October	29.4	27.22	6.5	7.30	0.730	−1.55	34.35	0.00	10.98
November	18.9	17.50	4.0	7.30	0.730	−4.02	30.33	0.00	6.96
December	0	0	0.0	7.30	0.730	−8.03	22.30	0.00	−1.07
Sum	398.1	368.56	95.29	87.60	8.760	−1.07			39.37

Using the rainfall data of three candidate years (Table 2.21), we can get three storage capacities. The maximum one is adopted as the design storage capacity.

(iii) Calculation of storage volume for multiple water use

When the RWH system has more than one water use, for example, the domestic water supply and irrigation, then there will be more than one water supply reliability. In most cases, we can divide the multiple water use into two groups: one group with reliability of 90 % (like domestic use and small industry use) and another group with reliability of 75 % (irrigation and husbandry). Correspondingly, two typical years are formulated for calculation: one with frequency of 75 % and another with frequency of 90 %. For each typical year, there may be two or three candidate years for selection.

In the typical year with frequency of 75 %, the domestic water supply and the irrigation/husbandry water use should all be fully met. While in the typical year with frequency of 90 %, the domestic water use should be fully met but the irrigation water is not necessary to be fully met. Some water shortage for irrigation use can be happened to some extent. What is considered an acceptable level of water shortage can only be determined based on a comprehensive consideration of the local social, economic, and environmental factors. Here we make a simplified treatment. For example, in a year with frequency of 90 %, it is allowable that the irrigation and husbandry water use may have a maximum 20 % water shortage.

We will first set two groups of years with frequency close to 75 and 90 %. From Table 2.8, the design rainfall for reliability of 75 and 90 % are 368.6 and 451.1 mm, respectively. For a frequency of 75 % year 1966 and 1980 can be selected for calculation year and for 90 % year 1965 and 1979 be selected.

The storage is calculated in the years with yearly rainfall having frequency of both 75 % and 90 %.

Tables 2.23 and 2.24 are examples of storage calculation with two groups of water use.

Calculation in Tables 2.23 and 2.24 is similar with that in Table 2.21 except:

- The water supply is divided into two groups with reliability of 75 and 90 %
- In the year with 75 % frequency, the domestic water, husbandry use, and irrigation use should be fully met. The water balance equation is

$$V_{i+1} = V_i + \mathrm{Fl}_{\mathrm{in}} - (W_\mathrm{d} + W_\mathrm{h} + W_\mathrm{i})_{i,i+1} - \mathrm{WL}_{i,i+1} \quad (2.28)$$

  where W_d, W_h, and W_i are the water use of domestic, husbandry, and irrigation, respectively
- In the year with 90 % frequency, the domestic water should be fully met, the husbandry, and irrigation water use are to be partially met with certain water shortage. Generally, 20 % shortage in the extremely dry year is allowed. Water balance equation is

$$V_{i+1} = V_i + \mathrm{Fl}_{\mathrm{in}} - (W_\mathrm{d} + 0.8W_\mathrm{h} + 0.8W_\mathrm{i})_{i,i+1} - \mathrm{WL}_{i,i+1} \quad (2.29)$$

- In Table 2.23, the frequency of the year is 75 % so both the two groups of water use should all be hundred percent fully met (Eq. 2.28) while in Table 2.24, frequency of the year is 90 %, water use with reliability of 90 % should be fully met but that with reliability of 75 % is to be met only by 80 % (Eq. 2.29).

Calculation results of the candidate years will be compared and a maximum storage capacity will be adopted as the design volume of the tank.

From the above description, it can be seen that selection of the typical year has a great effect on the result of the calculation for the storage capacity using the "Typical year method." The resulting storage capacity using rainfall data from different years can be very different. For example, with the rainfall data of year 1965 and 1971, the resulting capacity was 39.37 and 31.22 m^3, respectively. The deviation between these two results was 21 %. The advantage of the "Typical year method" is the smaller work load. Only two or three years of data are used for the calculation. But the error can be significant. The "Long series method" avoids this shortcoming.

(5) **Long Series Method**

(i) Long series method with single water use

The rainfall data used for the "Long-term series method" should preferably cover a period of more than 30 years. Determination of the catchment area uses the same method as described in the Sect. 2.3. Preparation of the inflow and water demand over time for the RWH system is the same as in the typical year method. The data and RCE versus rain characteristic can be done on a rainfall event basis, on a monthly or 10-day basis. But all of them should be available for the long series.

The first step is to assume the volume of storage tank at the start of the calculation. During the water storage process, the storage capacity will vary between zero and V_m. Here V_m is the assumed storage capacity. Table 2.25 shows an example of the calculation over 3 years. The calculation is carried out in the same way as

Table 2.23 Calculation of storage capacity for RWH system, typical year method (Water use reliability of 75 and 90 %, frequency = 75 %)

Month	Real rainfall	Design rainfall	Inflow	Water supply		Water balance	V_{end}	Waste/ shortage	Accumulated water balance
				$R = 90$ %	$R = 75$ %				
1	2	3	4	5	6	7	8	9	10
January	0	0	0.00	4.02	13.64	−17.66	0.00	11.17	−17.66
February	115.7	115.02	51.57	4.02	7.87	39.69	39.69	0	22.03
March	7.5	7.46	3.13	4.02	2.09	−2.97	36.72	0	19.06
April	97.7	97.13	43.38	4.02	13.64	25.72	62.44	0	44.79
May	8.4	8.35	3.52	4.02	13.64	−14.14	48.31	0	30.65
June	36.6	36.39	15.88	4.02	13.64	−1.77	46.53	0	28.88
July	47.2	46.92	20.61	4.02	13.64	2.95	49.49	0	31.83
August	81.5	81.02	36.04	4.02	13.64	18.38	67.87	0	50.21
September	55.8	55.47	24.46	4.02	13.64	6.80	**74.67**	0	57.01
October	3.3	3.28	1.35	4.02	13.64	−16.30	58.36	0	40.71
November	0	0	0.00	4.02	7.87	−11.88	46.48	0	28.82
December	0	0	0.00	4.02	13.64	−17.66	28.82	0	11.17

Table 2.24 Calculation of storage capacity for RWH system, typical year method (Water use reliability of 75 and 90 %, frequency = 90 %)

Month	Real rainfall	Design rainfall	Inflow	Water supply		Reduced	Water balance	V_{end}	Waste/ shortage	Accumulated water balance
				$R = 90\ \%$	$R = 75\ \%$	$R_1 = 75\ \%$				
1	2	3	4	5	6	7	8	9	10	11
January	0	0	0.00	4.02	13.64	10.91	−14.93	17.61	0.00	−14.93
February	0	0	0.00	4.02	7.87	6.29	−10.31	7.30	0.00	−25.24
March	3.9	3.6	1.49	4.02	2.09	1.67	−4.20	0.00	3.10	−29.43
April	186.7	172.9	78.22	4.02	13.64	10.91	63.29	**63.29**	0.00	33.86
May	0	0.0	0.00	4.02	13.64	10.91	−14.93	48.36	0.00	18.93
June	6.8	6.3	2.63	4.02	13.64	10.91	−12.29	36.07	0.00	6.64
July	30.7	28.4	12.34	4.02	13.64	10.91	−2.59	33.48	0.00	4.04
August	83.1	76.9	34.18	4.02	13.64	10.91	19.25	52.73	0.00	23.29
September	38.6	35.7	15.59	4.02	13.64	10.91	0.67	53.39	0.00	23.96
October	29.4	27.2	11.80	4.02	13.64	10.91	−3.13	50.26	0.00	20.83
November	18.9	17.5	7.51	4.02	7.87	6.29	−2.80	47.46	0.00	18.03
December	0	0	0.00	4.02	13.64	10.91	−14.93	32.54	0.00	3.10

Table 2.25 Storage capacity calculation with "long series method", assumed $V_m = 42.2$ m^3 (Domestic use, reliability = 90 %, design rainfall = 368.6 mm, year 1965, catchment: tiled roof)

Year	Month	Rainfall	Runoff	Water use	Loss	Water balance	V_{end}	Overflow/ shortage
1	2	3	4	5	6	7	8	9
							$0.5\ V_m = 21.1$	
1961	January	0	0.0	7.3	0.73	−8.0	13.1	0.0
	February	0	0.0	7.3	0.73	−8.0	5.1	0.0
	March	62.6	15.9	7.3	0.73	7.9	13	0.0
	April	63.4	16.1	7.3	0.73	8.1	21.1	0.0
	May	38.6	9.5	7.3	0.73	1.4	22.5	0.0
	June	0	0.0	7.3	0.73	−8.0	14.5	0.0
	July	154.5	42.0	7.3	0.73	34.0	42.2	6.3
	August	210.1	58.4	7.3	0.73	50.4	42.2	56.7
	September	123.8	33.1	7.3	0.73	25.1	42.2	25.1
	October	21	4.9	7.3	0.73	−3.1	39.1	0.0
	November	110.7	29.4	7.3	0.73	21.4	42.2	18.2
	December	0	0.0	7.3	0.73	−8.0	34.2	0.0
1962	January	0	0.0	7.3	0.73	−8.0	26.1	0.0
	February	0	0.0	7.3	0.73	−8.0	18.1	0.0
	March	38.8	9.5	7.3	0.73	1.5	19.6	0.0
	April	29.6	7.1	7.3	0.73	−0.9	18.6	0.0
	May	29.5	7.1	7.3	0.73	−1.0	17.7	0.0
	June	4.8	1.0	7.3	0.73	−7.1	10.6	0.0
	July	103.2	27.3	7.3	0.73	19.2	29.9	0.0
	August	129.5	34.8	7.3	0.73	26.8	42.2	14.4
	September	57.5	14.5	7.3	0.73	6.5	42.2	6.5
	October	41.5	10.2	7.3	0.73	2.2	42.2	2.2
	November	35	8.5	7.3	0.73	0.5	42.2	0.5
	December	0.2	0.0	7.3	0.73	−8.0	34.2	0.0
…..	…..	…..	…..	…..	…..	…..	…..	…..
1966	January	0	0.0	7.3	0.73	−8.0	10.7	0.0
	February	0	0.0	7.3	0.73	−8.0	2.7	0.0
	March	3.9	0.8	7.3	0.73	−7.3	0.0	−4.6
	April	186.7	51.5	7.3	0.73	43.5	42.2	1.3
	May	0	0.0	7.3	0.73	−8.0	34.2	0.0
	June	6.8	1.4	7.3	0.73	−6.6	27.6	0.0
	July	30.7	7.4	7.3	0.73	−0.7	26.9	0.0
	August	83.1	21.6	7.3	0.73	13.6	40.5	0.0
	September	38.6	9.5	7.3	0.73	1.4	41.9	0.0
	October	29.4	7.0	7.3	0.73	−1.0	40.9	0.0
	November	18.9	4.4	7.3	0.73	−3.7	37.3	0.0
	December	0	0.0	7.3	0.73	−8.0	29.2	0.0

for the typical year method except that the volume of the storage varies within the range of (0, V_m) and the long series of rainfall data are used. Besides, we assume that at the beginning of the calculation, the tank is half filled. The figures in column 9 show the overflow (positive value) or water shortage (negative value) as calculated using Eq. 2.29. To avoid occupation of too much space, here only calculations for three years, (1961, 1962 and 1966) are listed.

$$Q = V_{\text{end}i} + \text{Water balance}_{i+1} - V_{\text{end}i+1}. \tag{2.30}$$

In Eq. 2.30, Q is the value of (overflow/shortage) in column 9 of Table 2.25, $V_{\text{end}i}$ and $V_{\text{end}i+1}$ are the tank volume at end of time step i and $i + 1$, respectively, Water balance$_{i+1}$ is the water balance in time step $i + 1$. If $Q > 0$ then overflow occurs, if $Q = 0$ then no overflow or shortage happens, if $Q < 0$ then water shortage results. For example in the year 1962, value of Q is all positive, which means no water shortage occurred. The volume at the end of July amounted to 29.9 and the inflow in August is larger than the outflow by 26.8. The maximum volume is only 42.2 so water overflow of (29.9 + 26.8 − 42.2 = 14.4 m^3) took place. In 1966, the inflow in March was less than outflow by 7.3 but the volume in the end of February was only 2.7 m^3, the maximum water amount that could be released from the tank, so a water shortage of 4.6 m^3 resulted in March and the volume dropped to zero in the end of that month.

If water shortage results during in one or more time steps in a year then this year is deemed as year with water shortage. The number of years with water shortage will then be counted in Table 2.25. The allowed number of water shortage years for certain water supply reliability can be calculated with Eq. 2.31.

$$M_0 = \text{ABS}\ (N - N \times \text{Water supply reliability}/100). \tag{2.31}$$

In the above equation, N is the length of the series, M_0 is the allowable number of years with water shortage, Water supply reliability is in percentage, ABS is a function that means to take the absolute value in parenthesis.

If the number of years with water shortage is counted from Table 2.25, $M = [M_0]$, then water supply reliability is just met. The criterion could be met with a wide range of preset storage capacity V_m. The design volume of the storage tank is taken as the smallest V_m in this range.

If $M > [M_0]$, then water supply reliability is less than what that selected, a larger V_m should be re-assumed and the calculation is to be carried out again.

If $M < [M_0]$, the water supply reliability is greater than what we need, the assumed V_m can be reduced and calculation is resumed.

Calculation using long series method can be carried out using Excel spreadsheets. With the computer aid, result can be obtained within minutes.

(ii) Long series method for multiple water uses

In this section, the method for calculating the storage capacity of a RWH system for multiple water uses will be introduced. Suppose that we have domestic water supply, irrigation water use, and water supply for animal husbandry. The water use is divided into two groups with supply reliability of 75 %, covering irrigation and animal husbandry, and 90 % for domestic use.

Then following the same method as above, we will first assume the volume of the storage tank. The calculation is carried out with the same procedure for a single water use situation. Again we will check if the number of water shortage years is within the allowable range. It should be remembered now we have two types of water use with different reliabilities. Therefore, the allowable M_0 in Eq. 2.31 will have two values: M_{01} for reliability of 75 % and M_{02} for reliability of 90 %. For reliability of 75 %, water use for domestic, irrigation as well as animal husbandry should all be fully met. But for reliability of 90 %, the water use for domestic use must be fully met while only part (for example 80 %) of the irrigation and husbandry use will be met.

Again, here only calculation of year 1961, 1962 and 1967 is listed and the tank is assumed half filled initially. The volume of the tank was 42.15 m^3 at the beginning of January 1961. Two criteria for identifying if water shortage happens are set up, which are expressed by Eqs. 2.32 and 2.33.

$$Q_1 = V_{i+1} + \text{Water balance}_{i+1} - V_i \tag{2.32}$$

$$Q_2 = V_{i+1} + \text{Water balance}_{i+1} + 0.2 \times (\text{Water use})_{75\ \%} - V_i. \tag{2.33}$$

In the table, we will check each year if water shortages occur or not using the above two criteria.

If $Q_1 < 0$, water use including domestic, irrigation and husbandry cannot be fully met. Because the domestic water supply is the priority, water shortage will happen in either irrigation or husbandry water use. Since water supply reliability for irrigation and husbandry is 75 %, for an N year series, the number of years with water shortage that can be counted in Table 2.26 with the criterion of $Q_1 < 0$ should not be greater than $M_{01} = \text{ABS}(0.25 \times N)$.

If $Q_2 < 0$ means 100 % of domestic use and 80 % of irrigation and husbandry water use cannot be met. Because the water shortage for irrigation and husbandry is 20 %, this means water shortage also happens in domestic water supply. While the water supply reliability for domestic use is 90 % so allowable number of years with water shortage in domestic use counted with criterion of $Q_2 < 0$ should not be greater than $M_{02} = \text{ABS}(0.10 \times N)$.

If the assumed storage volume V_m results in either the number of years with $Q_1 < 0$ greater than M_{01} or the number of years with $Q_2 < 0$ greater than M_{02} then it means the V_m is too small and should be increased and the assumed volume V_m should be decreased vice versa. Calculation should be done again. As described before, the above criterion can be met with a range of value for V_m. The smallest one in the range will be taken as the design storage capacity. The result of the calculation can be easily obtained with a wisely designed excel sheet.

(iii) Optimum solution using long-term method

The procedure for determining the catchment area introduced in Sect. 2.3 gives the catchment area on which the runoff amount is just equal to the water use for the whole year. This catchment area may be regarded as a minimum area (A_0) to meet the demand for the water supply. In some practical situation, the existing

Table 2.26 Storage capacity calculation with "long series method," assumed $V_m = 84.3$ m^3 (Domestic use, reliability = 90 %, design rainfall = 368.6 mm, irrigation and husbandry use reliability = 75 %, design rainfall = 451.1 mm, catchment: tiled roof, concrete, and greenhouse roof)

Year	Month	Rain	Inflow	Water use and loss		Water balance	Vend	Waste shortage/ overflow 1	Waste shortage/ overflow 2
				$R = 90$ %	$R = 75$ %				
1	2	3	4	5	6	7	9	10	11
							$0.5\ V_m = 42.15$		
1961	January	0	0	4.015	13.64	−17.66	24.49	0	2.73
	February	0	0	4.015	7.87	−11.88	12.61	0	1.57
	March	62.6	27.68	4.015	2.09	21.57	34.18	0.00	0.42
	April	63.4	28.04	4.015	13.64	10.38	44.56	0.00	2.73
	May	38.6	16.87	4.015	13.64	−0.78	43.78	0.00	2.73
	June	0	0	4.015	13.64	−17.66	26.12	0.00	2.73
	July	154.5	69.74	4.015	13.64	52.08	78.2	0.00	2.73
	August	210.1	95.50	4.015	13.64	77.84	84.30	71.74	74.47
	September	123.8	55.60	4.015	13.64	37.94	84.30	37.94	40.67
	October	21	9.05	4.015	13.64	−8.61	75.69	0.00	2.73
	November	110.7	49.59	4.015	7.87	37.71	84.30	29.10	30.67
	December	0	0	4.015	13.64	−17.66	66.64	0.00	2.73
1962	January	0	0	4.015	13.64	−17.66	48.99	0.00	2.73
	February	0	0	4.015	7.87	−11.88	37.11	0.00	1.57
	March	38.8	16.96	4.015	2.09	10.86	47.96	0.00	0.42
	April	29.6	12.86	4.015	13.64	−4.80	43.17	0.00	2.73
	May	29.5	12.81	4.015	13.64	−4.84	38.33	0.00	2.73
	June	4.8	2.00	4.015	13.64	−15.66	22.67	0.00	2.73
	July	103.2	46.16	4.015	13.64	28.50	51.16	0.00	2.73
	August	129.5	58.22	4.015	13.64	40.56	84.30	7.43	10.16

(continued)

Table 2.26 (continued)

Year	Month	Rain	Inflow	Water use and loss		Water balance	Vend	Waste shortage/ overflow 1	Waste shortage/ overflow 2
				$R = 90$ %	$R = 75$ %				
	September	57.5	25.37	4.015	13.64	7.72	84.30	7.72	10.44
	October	41.5	18.17	4.015	13.64	0.52	84.30	0.52	3.24
	November	35	15.27	4.015	7.87	3.38	84.30	3.38	4.96
	December	0.2	0.08	4.015	13.64	−17.58	66.72	0.00	2.73
…..	…..	…..	…..	…..	…..	…..	…..	…..	…..
1967	January	0	0	4.015	13.64	−17.66	9.23	0.00	2.73
	February	0	0	4.015	7.87	−11.88	0.00	-2.65	-1.07
	March	112	50.19	4.015	2.09	44.08	44.08	0.00	0.42
	April	59.9	26.46	4.015	13.64	8.80	52.88	0.00	2.73
	May	71.8	31.84	4.015	13.64	14.19	67.07	0.00	2.73
	June	13.9	5.93	4.015	13.64	−11.73	55.34	0.00	2.73
	July	175.6	79.49	4.015	13.64	61.84	84.30	32.88	35.61
	August	229.6	104.57	4.015	13.64	86.92	84.30	86.92	89.64
	September	105.2	47.07	4.015	13.64	29.41	84.30	29.41	32.14
	October	20.8	8.96	4.015	13.64	−8.70	75.60	0.00	2.73
	November	121.8	54.68	4.015	7.87	42.80	84.30	34.11	35.68
	December	0	0	4.015	13.64	−17.66	66.64	0.00	2.73

catchment area may be larger than the calculated area, for example in case of a paved highway or a natural rocky slope, etc. If the catchment area is larger than the calculated area the question arises: what will happen to the storage capacity? In other words, what is the effect of catchment area on the storage capacity? To study this situation, after we get the design storage capacity V_d, we will assume several catchment areas: $A_1 = 1.5A_0$, $A_2 = 2.0A_0$, $A_3 = 2.5A_0$… and calculate the corresponding storage capacity V_1, V_2, V_3… using the Long series method.

With data in the example shown in Table 2.25, we can get the following results shown in Table 2.27 and Fig. 2.14.

From Table 2.27 and Fig. 2.14, it can be seen that if keeping the water supply and the reliability unchanged, expanding the catchment area results in a decrease of the storage capacity of the RWH tank. Usually to build the catchment using the existing structure costs less than building a larger storage tank. We can make use of this conclusion to minimize the cost of the RWH system by enlarging the catchment area while at the same time reducing the storage capacity.

When designing a RWH project for a semi-arid area in a Middle Eastern country six sets of catchment area and storage capacity were considered as shown in Table 2.28. All these sets can meet the demand of water supply and the reliability of 75 %. Economic comparison was conducted to find out the combination with the lowest cost. The solution with lowest price is shown in italics.

(6) **Comparison of Different Calculation Methods for Storage Capacity**

In the above sections, different methods for calculating the storage capacity of RWH system have been discussed. These include the typical year method and the long series method. For determining the RCE, it includes using rainfall data and

Table 2.27 Storage capacity versus catchment area

Catchment area, m^2	Storage capacity, m^3	A/A_0	V/V_d
654	42.2	1	1
981	37.7	1.5	0.893
1308	36.8	2	0.872
1635	36	2.5	0.853

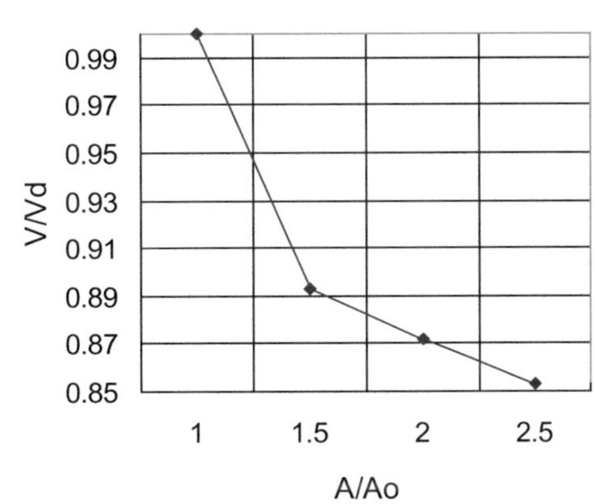

Fig. 2.14 Storage capacity versus size of catchment

Table 2.28 Economic comparison for 6 sets of catchment storage capacity

No	1	2	3	4	5	6
Catchment m^2	130,000	*110,000*	90,000	80,000	70,000	60,000
Volume m^3	5000	*5100*	5800	6200	6900	7700

the RCE on an event basis and on a monthly and 10-day basis, as well as taking RCE as constant.

The rainfall data of the Huining Gauge of Gansu Province will be used. This is because the rainfall data on the event basis are available from the Gauge. It is also because the experimental dataset for RCE using the event basis is only available in the semi-arid area in Gansu. The mean annual rainfall of Huining Gauge is 373 mm and the rainfall with frequency of 75 and 90 % are 306.3 and 236.1 mm, respectively.

In principle, calculation using the RCE data on the event basis and with the long series method is the most scientific method among those we introduced before. Therefore, the calculation result for storage capacity using RCE on an event basis and the long series method is taken as the reference for judging the accuracy of the other methods. Table 2.29 shows the comparison with different methods and data.

From Table 2.29, we can draw some useful conclusions.

- The difference between calculation methods of "typical year" and "Long series" is significant. Calculation errors can be 10 % (concrete catchment for domestic use) to 20 % (tile catchment for domestic use) and up to 50 % (soil catchment for irrigation use). The long series method is therefore recommended for use here whenever possible.
- When using the long series method with RCE data in different contexts, errors may occur but to different extents for different catchments. For concrete catchments, error produced by using RCE on the monthly basis or taking as a constant are less than 1 %. However, for the tile catchment calculation using RCE on a monthly basis and taking the RCE as constant, the storage capacity is underestimated by 2.4 and 19.9 %, respectively. While for soil catchments, the underestimation is 10 and 22 %, respectively. It means that for catchments with less permeability, the calculation with the long series method either using RCE on a monthly basis or taking it as constant produces satisfactory result. But for catchments with high permeability, using RCE on a monthly basis, results in errors of 10 %.
- For the impermeable catchments like concrete, RCE can be taken as constant and the error caused is acceptable. But for permeable catchments including soil surfaces and clay tiles taking RCE as a constant may cause large errors and therefore cannot be adopted.
- The above results are from the rainfall data in a semi-arid area. For humid areas, the RCE of catchment including the soil catchment would be higher than that in the semi-arid area. So using RCE on the monthly basis may produce better results than here.

Table 2.29 Calculation results of storage capacity with different methods and data of RCE

Type of water use and reliability	Calculation method	Data of RCE	Result of storage capacity m^3	Error (%)
Domestic supply water use amount 62 m^3/year, reliability = 90 %, concrete catchment	Typical year	Monthly basis	30.9	−9.9
		Event basis	30.6	−10.8
		Constant	30.5	−11.1
	Long series	Monthly basis	34.5	0.6
		Event basis	34.3	0
		Constant	33.9	−1.2
Domestic supply yearly water use 62 m^3/year, reliability = 90 %, tile catchment	Typical year	Monthly basis	33.8	−20.1
		Event basis	34.2	−19.1
		Constant	30.5	−27.9
	Long series	Monthly basis	41.3	−2.4
		Event basis	42.3	0
		Constant	33.9	−19.9
Irrigation, yearly water use = 200 m^3, reliability = 75 %, concrete catchment	Typical year	Monthly basis	101.2	−29.1
		Event basis	99.4	−30.3
		Constant	102.2	−28.4
	Long series	Monthly basis	141.5	−0.8
		Event basis	142.7	0
		Constant	140.9	−1.3
Irrigation, yearly water use = 200 m^3, reliability = 75 %, compacted soil catchment	Typical year	Monthly basis	94.2	−50.8
		Event basis	84.2	−56
		Constant	120.3	−37.2
	Long series	Monthly basis	172.3	−10
		Event basis	191.5	0
		Constant	148.9	−22.3

2.4.2 Simplified Model

The RWH projects in China are implemented mainly at rural households. Most of the water users have only primary education backgrounds. Besides, there have been over hundred thousand of RWH storage tanks built each year in China. Obviously it is not feasible to design storage systems for all these small RWH projects using the complicated simulation models introduced above. To enable participation in the planning and design process by the technicians and farmers at the township and county level, it is necessary to develop a simplified method. This needs to be both simple but accurate enough to produce results of practical use. For this, we have suggested a "Volume coefficient method" to determine the storage capacity of any RWH system. It is shown in the following equation.

$$V = kW_y \tag{2.34}$$

In Eq. 2.34, V is volume, W_y is the water supplied for the whole year, and k is called the volume coefficient. The value of W_y is calculated using the method described in the above and is easy to determine. The problem is how to get an appropriate value for the volume coefficient k for determining the storage capacity.

The factors that influence the value of k are listed below:

- Pattern of water inflow: when the rainfall distribution and thus the inflow pattern conform closely to that of the water use, then k is smaller and vice versa.
- Wetness of the area: in humid areas where rain is frequent, the storage tank can be filled and drained more times than in the semi-arid areas, where the storage may be full only once in the rainy season. This means in humid areas the same storage capacity can supply more water than in the semi-arid areas, so the value k is smaller.
- Pattern of water demand: since the domestic water demand requirements are more evenly distributed over the year than those of irrigation, the value k for irrigation supply is larger than that for domestic water use.

From Eq. 2.34, the value k can be derived using the following equation:

$$k = \frac{V}{W_y}. \tag{2.35}$$

In 2008, a study on the volume coefficient k was carried out. Rainfall data from 4 gauges located in both semi-arid and humid areas were used. All of the data had a time series of more than 30 years. The water supplies, for domestic use, irrigation use, and greenhouse irrigation use were studied. The domestic water use pattern was taken to be evenly distributed throughout the year. In contrast, irrigation water usage occurred only 2–4 times a year representing a concentrated pattern of water use. The irrigation in greenhouses while carried out most of the year, except for one or two months for maintenance and repair work, represented a partially concentrated water use pattern.

The storage capacity is calculated using the long series method with monthly rainfall and RCE data. In this calculation, the roof tile catchment is used for domestic water supply and the soil and concrete surface catchment is used for irrigation. In semi-arid areas, irrigation is used primarily for wheat while in the humid areas, irrigation is mainly for rice. The water supply amount for different purposes is assumed. Using the rainfall data from the four gauges, the storage capacity was obtained using the long series method introduced in the previous sections. Then the volume coefficient for different water supplies in the semi-arid and humid areas can be derived using Eq. 2.35. Table 2.30 shows the results.

Table 2.30 Volume coefficient *k* for different water uses in areas with different annual rainfalls

Annual rainfall mm	Volume coefficient *k* for different water uses and different catchments				
	Domestic use	Field irrigation		Greenhouse irrigation	
	Tile catchment	Soil catchment	Concrete catchment	Soil catchment	Concrete catchment
373	0.543	0.863	0.84	0.578	0.533
1246.7	0.506	0.742	0.763	0.318	0.324
1379.9	0.393	0.71	0.834	0.311	0.342
1409.1	0.395	0.591	0.561	0.417	0.461
Average	0.459	0.681	0.719	0.406	0.415

From the above table, several points can be seen.

- Volume coefficient *k* is greater when the water supply distribution is more concentrated. For instance, factor *k* is greater when water supply is for irrigation use than for domestic use.
- Volume coefficient is smaller in humid area than in the semi-arid area for same water use.
- Different types of catchment do not affect the value of volume coefficient *k*.

Based on the analysis and practical experiences from the whole of China, the <Technical code for rainwater collection, storage and utilization> (GB/T 50596-2010) recommended the following table for determining the volume coefficient *k*.

Data source: <Technical code for rainwater collection, storage and utilization> [Ministry of Housing and Urban–Rural Development (UHURD) and General Administration of Quality Supervision, Inspection and Quarantine (AQSIQ) 2010].

The calculation procedure for determining the storage capacity of RWH system using the volume coefficient method can be summed up as follows.

Step1 Calculate the water supply requirement for the whole year using the method introduced in Sect. 2.1 of this chapter;
Step 2 Find the volume coefficient *k* in Table 2.31 using the local annual rainfall data and the purpose of the water supply;
Step 3 Calculate the storage capacity using Eq. 2.34.

It should be pointed out that the value of the volume coefficient *k* depends on many factors that vary with the local conditions. The value of *k* in Table 2.31 is based on the rainfall and water use conditions in China and cannot be used directly in the other areas. It is thus suggested that it is necessary for the technical people working on RWH project to work out the volume coefficient for the conditions in their own country. This means using rainfall data that preferably has a series of 30 years or longer. Data on the water usage patterns throughout the year for different uses also need to be formulated. Then calculation of the storage capacity using long series method needs to be carried out and the volume coefficient can be obtained with Eq. 2.35.

Table 2.31 Volume coefficient *k*

Purpose of water supply	Area with mean annual precipitation (mm)		
	250–500	500–800	>800
Domestic use	0.55–0.6	0.5–0.55	0.45–0.55
Dry land irrigation	0.83–0.86	0.75–0.85	0.75–0.8
Paddy irrigation		0.7–0.8	0.65–0.75
Greenhouse irrigation	0.55–0.6	0.4–0.5	0.35–0.45

The methodology for calculating the capacity of RWH system introduced in this chapter is suitable for that when RWH is the sole source of water supply. In cases other than water supply schemes, for example where a piped system already exists but the supply capacity is inadequate and RWH is used as a supplemental source of supply or RWH is used for reducing the cost of the existing water supply scheme, the method for dimensioning the RWH system is different. In these cases, the shortfall in the capacity of the existing water supply scheme has first to be estimated and the capacity of RWH system needed to bridge the gap between the demand and the existing supply capacity, determined. In this second situation, the capacity of RWH system is determined based on economic considerations: the inputs for the RWH system should be paid back by a reduction of water cost over a certain period.

2.5 Annex Calculation Sheets for Dimensioning RWH System

About the calculation sheets for dimensioning RWH system

Before you use the calculation sheets, please first read this chapter carefully. This set of sheets is to show calculation examples for dimensioning the RWH system for two situations of water use. The sheets can be found at http://extras.springer.com. One is for domestic supply with supply reliability of 90 % and the catchment is concrete-lined surface. Another one is for domestic supply, greenhouse irrigation, and husbandry water use. Catchment for domestic supply includes tiled roof and concrete surface. For greenhouse irrigation, the greenhouse plastic roof is used for catchment. For livestock water use, concrete-lined surface is used for catchment.

In the calculation, two ways of determining the rainwater collection efficiency (RCE) are used. In one way, the monthly RCE_m is determined with the following equation:

$$RCE_m = A \times Rain_m + B.$$

In the equation, RCE_m and $Rain_m$ are the monthly RCE and Rainfall, respectively. *A* and *B* are two parameters to be calibrated.

In the second way, RCE is taken as constant.

Inputs for these sheets include as follows:

- Monthly rainfall data for a continuous period of 30 years. In case your data series is shorter than 30, you still can carry out the calculation. However, the result may be less accurate:
- Number of family member and livestock and their daily water supply quota,
- Area to be irrigated and the irrigation quota in the year,
- Monthly distribution of water use in the year,
- Supply reliability of water use, and
- Catchment type and the yearly RCE.

All these data should be input into the cells with blue color then the results in the cells with yellow color will be calculated automatically. The results include as follows:

- Design rainfall related to water supply reliability,
- Catchment area, and
- Storage capacity.

Calculation of the storage capacity is done with two methods: the typical year method and the long-term method. In the typical year method, 3 monthly rainfall distributions will be used to calculate the storage capacity, including monthly rainfall distribution of two years with frequency close to the supply reliability and the average monthly rainfall distribution. When there are two kinds of water supply with different reliabilities, then two sets of monthly rainfall (each set contains two years) are to be input for the two reliabilities. The average monthly rainfall distribution is automatically input.

In the typical year method, the storage capacity is found using two methods. One way is to first find the month when the zero storage happens. Usually it happens at the end of a period with continuous negative water balance. Sometimes it will be done with a "trial and error" method. Second way is to calculate the accumulated water balance from January to December. The storage capacity V can be calculated with the following equation.

$$V = \text{Max (AWB January to December)} - \text{Min (AWB January to December)}.$$

Here, AWB is the accumulated water balance.

When there are two water supply reliabilities, for example 90 % for domestic supply and 75 % for irrigation and husbandry. Then in the year with frequency equal to or less than 75 % all the water demand should be met and no water shortage is allowed. In the year with frequency equal to 90 % then the domestic water supply should be fully met but the irrigation and husbandry water use cannot be fully met. Certain amount of deficit in water supply is allowed. In these calculation sheets, we allow a 20 % discount in irrigation and husbandry water supply.

In the long-term method, a long series of rainfall data are used. First, a storage capacity V_a is assumed, which means during the long time series, the storage capacity will be changed in the range of zero and V_a in the whole series. With this storage capacity, we can find out in how many years the water shortage will happen. Here if water shortage happens in one month then this year is deemed as year with water shortage. The total number of years Ns with water shortage in the series is counted and is compared with the allowable number of years with water shortage Na. If Ns > Na, then the assumed storage capacity V_a is too small and should be added. If Ns < Na, then the assumed storage capacity V_a is too large and should be reduced. In both cases, the calculation will be carried on until a smallest capacity is found for Ns = Na.

In these sheets, the calculation examples are provided. For users to practice the calculation with their own data, the blank sheets are provided. To repeatedly use these blank sheets, you may first save the blank sheet with other file name and then carry out the calculation. The original blank sheets can be kept for the next use.

References

Gansu Bureau of Water Resources and Gansu Bureau of Technical Supervision. Technical standard of rainwater catchment and utilization (DB62/T495—1997). Lanzhou, China (in Chinese): Gansu Science and Technology Press; 1997.

General Administration of Technical Supervision and Ministry of Construction. Code for design of irrigation and drainage engineering. China Planning Press; 1999 (in Chinese).

Gould J, Nissen-Petersen E. Rainwater catchment systems for domestic supply—design, construction and implementation. London, UK: Intermediate Technology Publications; 1999.

Ministry of Housing and Urban-Rural Development (MOHURD) and General Administration of Quality Supervision, Inspection and Quarantine (AQSIQ). Technical code for rainwater collection, storage and utilization (GB/T 50596—2010). Beijing, China: China Planning Press; 2010 (in Chinese).

Ministry of Water Resources. Technical code of practice for rainwater collection, storage and utilization. Beijing, China: China Water and Power Publications; 2001 (in Chinese).

Chapter 3
Structural Design of the Rainwater Harvesting System

Qiang Zhu

Keywords Rainwater catchment · Rainwater storage · Structural design

3.1 Rainwater Collection Subsystem

As introduced in Chap. 1, the RWH system should be regarded as an integrated system, composed of three indispensable parts, namely, the rainwater collection, storage, and water supply subsystems.

3.1.1 Composition of Rainwater Collection Subsystem

Rainwater collection subsystem acts as the water source of the RWH system. Its role is to supply enough water of appropriate quality to meet the water demand. It includes the rainwater collection surface (catchment) as well as the interception, collection, and conveyance ditches.

The collection surface (catchment) should meet the following criteria:

- The quantity of water collected from the catchment should be sufficient to meet the demand. So the catchment area should be adequate and the rainwater collection efficiency (RCE) is as high as possible.

The original version of this chapter was revised: The artwork of the Figure 3.39 was corrected. The erratum to this chapter is available at DOI: 10.1007/978-981-287-964-6_12

Q. Zhu (✉)
Gansu Research Institute for Water Conservancy, Lanzhou, China
e-mail: zhuq70@163.com

Q. Zhu et al. (eds.), *Rainwater Harvesting for Agriculture and Water Supply*,
DOI 10.1007/978-981-287-964-6_3

- To ensure high water quality, the catchment should not seriously contaminate the water.
- The cost of the catchment should be low and affordable to rural households.

In China, the rainwater catchment can be divided into three types, namely, natural surfaces, less permeable surfaces of existing structures (e.g., roads), and purpose-built catchments.

Natural surfaces include earthen and rocky slopes. Bare, unjointed rock slopes can form an ideal catchment because of their high RCE. However, rocky slopes are not found everywhere and many have fissures and cracks, causing low RCE. In the humid and sub-humid areas, earthen slopes are also commonly used as catchments because the moisture content of the soil is often high, and thus, the infiltration rate is low leading to plentiful runoff. In the dry areas, the moisture content of the soil is usually low so the infiltration rate of the earthen surface is high, leading to a low RCE. However, this shortcoming sometimes can be compensated by setting up larger catchment area when the land resources are available.

For economic reasons, we should always consider making use of the less permeable surfaces of existing structures such as roofs, courtyards, highways, sport grounds, country roads, threshing yards, greenhouses, etc. This kind of catchment is commonly used in China's semi-arid areas. Relatively little work is needed to convert the surfaces of existing structures to catchments, for example, by installing gutters and downpipes for roof catchments, or building small dikes for diverting runoff from highways.

In dry regions, to get a high RCE for collecting enough water to meet the demand of domestic supply or irrigation use for high-value crops, it is sometimes necessary to set up purpose-built catchments. In this case, the ground is paved with an impermeable material like cement, concrete, cement soil, or covered with plastic film.

An interception ditch is used when earthen slopes are used as a catchment to reduce infiltration loss. The interception ditch is built along the contour line and the runoff from the slope is intercepted into the ditch and flows to the collection ditch. The conveyance channel diverts runoff from the catchment to the storage tanks which are arranged in a scattered way. The layout of the ditch and channel will be discussed later.

Figure 3.1 shows the composition of the rainwater catchment subsystem of the RWH system.

3.1.2 Natural Slope as Rainwater Catchment

3.1.2.1 Semi-arid Areas

The natural slope is sometimes used as a catchment in semi-arid areas. To increase RCE of natural slopes, it is better to compact the surface soil to a depth of 30 cm. However, in semi-arid areas, using natural earthen slopes as catchments has some

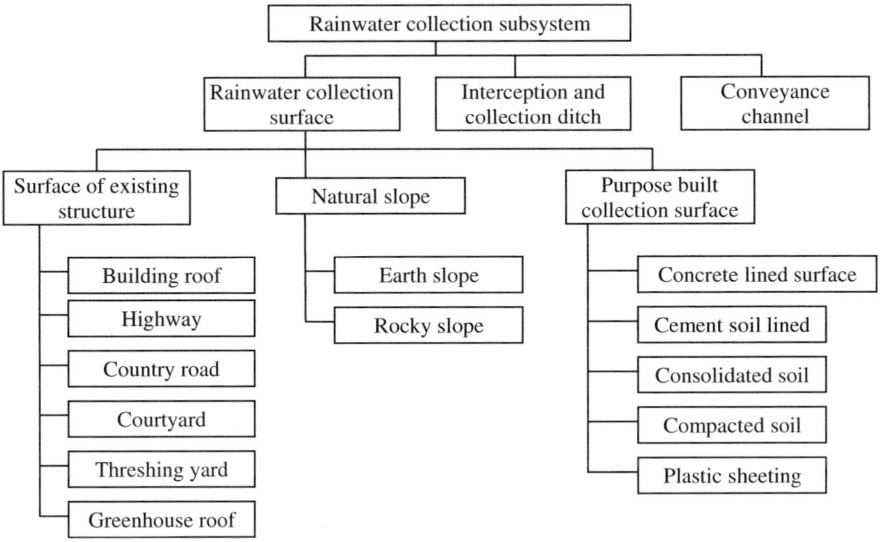

Fig. 3.1 Composition of the rainwater collection subsystem

problems. In addition to the low RCE, the vegetation on the slope is usually sparse so earthen slopes are easily eroded by runoff during heavy rain. Furthermore, when the slope is long, the high infiltration rate of the soil results in significant water loss and further lowers the RCE. To avoid these problems, it is necessary to build interception channels along the contour line at an interval of 20–30 m. The interception channels are linked to the collection channels arranged in vertical to the contour lines, see Fig. 3.2. When the catchment is large and can feed a number of storage tanks, a conveyance channel is built to divert the runoff to the tanks.

The collection channel usually has steep slope and it is necessary to line it with concrete or masonry. To reduce the seepage loss, sometimes the interception and conveyance channels are also lined even when their slope is flat.

The sections of the channel are illustrated in Fig. 3.3. In most cases, the size of the channel is small. The bottom width and the depth of the channel are of the order of 10–15 cm. For the U-shaped section, the radius is about 10–15 cm.

For the rocky slope, the RCE is high and comparable with concrete if the rock surface is integrated and less fissure and crack exists. However, the size of the channel should be large enough to avoid damage by the flood during a storm. Some spillway for drainage of the flood has to set up along the channel.

3.1.2.2 Humid and Sub-humid Areas

The natural slope land in the humid areas usually has a dense vegetation cover, and erosion of the soil is generally not a problem. Also the infiltration rate of the

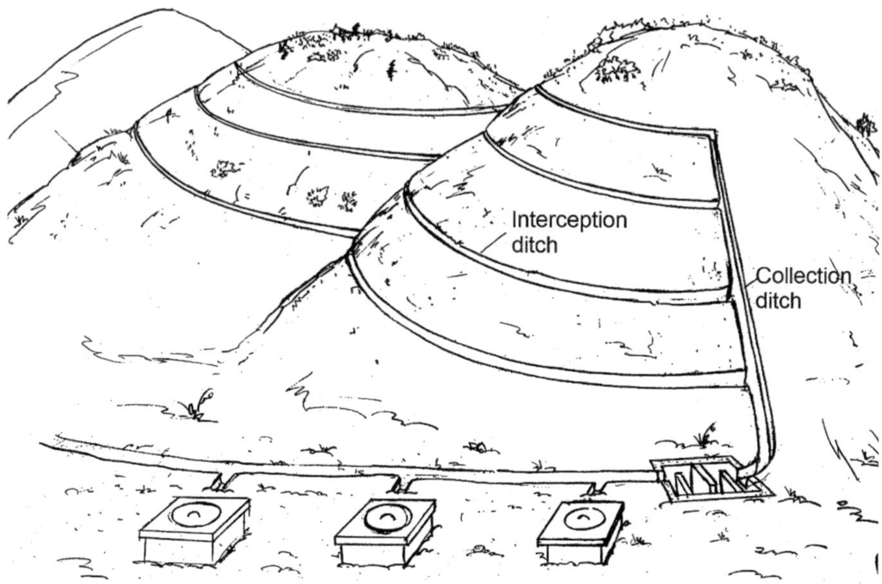

Fig. 3.2 Layout of catchment using natural slope and the interception, collection, and conveyance channel

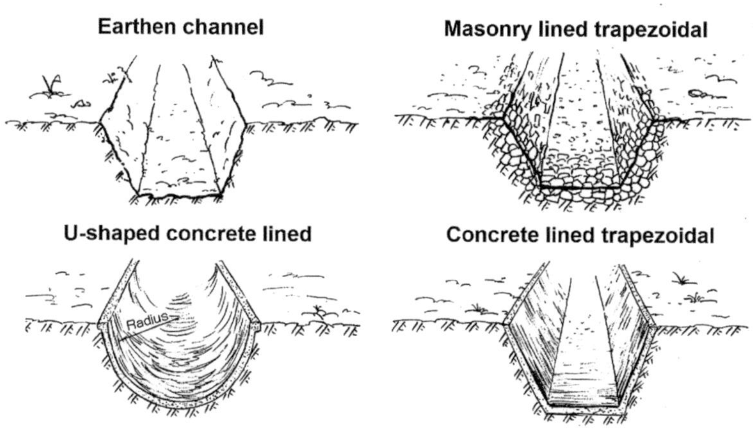

Fig. 3.3 Section of different types of channel

soil is low. The interception and collection channel on the slope are not necessary. But conveyance channels are needed when the tanks are far away from the catchment.

3.1.3 Catchments Using Existing Structures

3.1.3.1 Roofs

The roof is one of the most commonly used surfaces for collecting rainwater. In Gansu, roofs of many rural houses were formerly built with straw mud. In the southern China, people also used thatched roofs. These two kinds of roof not only have the disadvantage of low RCE but also caused pollution to the runoff. In the RWH project, they were replaced with tiled roofs. Most of the tiles are made of loam or clay soil and baked in a kiln. Tiles can also be made of cement mortar. In the semi-arid areas of Gansu Province, it is found that RCE of the cement tile can be 0.7–0.8, while the clay tiles have an RCE of around 0.35–0.4. For clay tiles made in the small village workshop, the RCE is even lower. In some places, where the coal that is necessary for baking the tile is not available, the price of cement tiles is almost the same as for clay tiles. However, the cement tile is more easily broken during handling so is not as popular. Figure 3.4 shows a photo of cement tile.

There are two ways to collect rainwater from the roof. One is to use gutters installed under eaves of the roof and linked with down pipe leading the rainwater from the roof to the tank. The ground runoff flows separately to another tank. Another way is when the courtyard is lined with concrete as part of the catchment, the runoff from the roof eave drops down to the ground, mixes with surface runoff, and then flows through a shallow ditch to the underground tank by gravity. These two kinds of arrangement are shown in Fig. 3.5.

Fig. 3.4 Cement tile

Fig. 3.5 Two kinds of collecting domestic water to the storage tank: with gutter and downpipe (*left*); without gutter and downpipe (*right*)

Fig. 3.6 A new type of roof structure for rainwater catchment Courtesy of M. Cheng of Inner Mongolia Research Institute for Water Resources

In the first way, the roof runoff is stored separately with the ground flow so the water quality can be better and sometimes the tank can be put at a higher position to make a simple tap system. However, the cost of the gutter and downpipe comprises a considerable fraction of the total cost. In the initial stages of RWH project implementation in Gansu, China, funds were insufficient to cover the cost of installing gutters and downpipes. However, in many of the newly built RWH projects, gutters and downpipes are installed and the runoff from the roof and ground stored in separate tanks.

Recently, in Inner Mongolia Autonomous Region, people have invented a new roof structure for rainwater catchment. First, the roof is leveled carefully then a PVC film with thickness of 0.08 mm is put on it. A layer of straw mud plaster with thickness of 2–3 cm is then added. After drying, a layer of cement mortar with thickness of 1.5 cm is plastered on top, see Fig. 3.6. The straw mud and cement mortar prevent the plastic sheet from aging due to exposure to the sun's radiation. This kind of roof has RCE of 0.7 and the cost is comparable with the traditional tiled roof (Cheng et al. 2009).

3.1.3.2 Paved Highway

Paved highway usually uses asphalt or concrete and therefore has a low permeability. The RCE of asphalt-paved highways is comparable with concrete. To receive runoff from the highway, the storage tanks should be put along the downslope sections of the highway. On most highways, there are drainage ditches by the sides of the highway, so a small dike can be put in the drainage ditch and the runoff diverted to storage tanks through a conveyance channel. If a drainage ditch does not exist, then it is necessary to build a collection ditch along the highway. Figure 3.7 shows a highway catchment built for irrigation purposes. The tanks are located higher than the land so water can easily flow by gravity through a siphon tube.

The runoff collected from the highway may be polluted from two possible sources: the exhaust gases from the traffic and the asphalt surface. Although some limited tests have not revealed any significant chemical- and petroleum-related pollutants, for safety reason it is recommended not to use water collected from highways for drinking and cooking purposes.

3.1.3.3 Greenhouse Roofs

Recently, greenhouses have developed rapidly in the mountainous areas of Gansu as a result of implementation of the RWH project. The greenhouse roofs are built with plastic sheeting that is impervious and has a high RCE. Tests have shown that

Fig. 3.7 Highway catchment for collecting rainwater for irrigation

Fig. 3.8 Greenhouse roof as catchment

greenhouse roofs can provide about 40 % of the irrigation water needed by the vegetables grown in the greenhouse (Zhu et al. 2012).

When using the greenhouse roofs as a catchment, a ditch in front of the greenhouse is built to divert the runoff into the tank either inside or outside the greenhouse, see Fig. 3.8.

3.1.3.4 Country Road, Threshing Yard, and Other Earthen Surfaces

These surfaces are often compacted and thus are less permeable than natural slopes and can usefully be used for collecting rainwater. However, since the RCE is round 0.2–0.25, this kind of catchment can supply only a couple of tanks.

3.1.4 Design of Purpose-Built Catchment

For irrigating high-value crops when there is no nearby highway or other kinds of less permeable surface for collecting rainwater, it is sometimes necessary to construct purpose-built catchments by paving the natural surface with impermeable materials. In China, commonly used materials for doing this include concrete and cement soil. The surface can also be covered with plastic sheet or the surface soil consolidated with a chemical agent.

3.1.4.1 Concrete-Paved Surface

Concrete is the most commonly used material for paving the surface of a purpose-built catchment. The concrete-lined collection surface is mainly for reducing the infiltration loss and it is not be subjected to a high load, so the thickness of the concrete pavement only needs to be 3–4 cm. Test have shown the RCE of concrete-paved surfaces amount to 0.75–0.85.

Concrete is a mixture of cement, aggregates (gravel and sand), and water as well as sometimes some agents for improving its properties. Requirements for these properties are different for different uses. For water resources projects, the most important properties are the strength, duration, and anti-permeability performance. Since RWH projects are small-scale projects, the key property for the concrete used is only the compressive strength. Concrete of Grade C15 or C20 is commonly used. It means that the concrete has compression strength of 15 or 20 MPa, respectively. Proportion between the components of concrete is essential to the performance of the concrete and is determined by testing. Because the RWH project is in big number, the proportion of concrete is impossible to determine by test for each project; it is determined by the local practical experiences. The concrete used in the RWH project is low grade with 32.5 cement low-grade cement most commonly used. According to the technical code for concrete structures, the upper limit of the gravel diameter should not exceed 1/4 of the slab thickness; 10 mm is the maximum diameter of the gravel. Table 3.1 shows the proportion of concrete components that can be taken as a reference.

The concrete for catchments is usually cast in situ. To avoid cracking after hardening of the fresh concrete, the concrete is cast in 1.5 m × 1.5 m or 2.0 m × 2.0 m blocks arranged in a separate form, see Fig. 3.9. The number in the figure shows the order of casting the concrete blocks. The joint between the blocks should have a width of 1–1.5 cm and should be filled with asphalt mortar, clay, or timber to prevent seepage. The sealed material can be prefabricated and installed in place when casting the concrete block or poured into the joint after concrete is cast.

Slump denotes the workability of the fresh concrete. To test the slump, fresh concrete is put in an inverted hopper-like container without bottom and compacted a little. Then the container is lifted up and let the concrete subside. The value of the subsidence is the slump expressed in cm.

Table 3.1 Proportion of concrete components

Grade concrete	Grade cement	Weight of component in 1 m^3 of fresh concrete mixture (kg)			
		Cement	Sand	Gravel	Water
C15	32.5	339	685	1166	193
	42.5	292	757	1135	193
C20	32.5	411	608	1181	193
	42.5	345	683	1163	193

Note (1) aggregate includes gravel and coarse sand; (2) maximum diameter of gravel is 10 mm; (3) slump: 30–50 mm

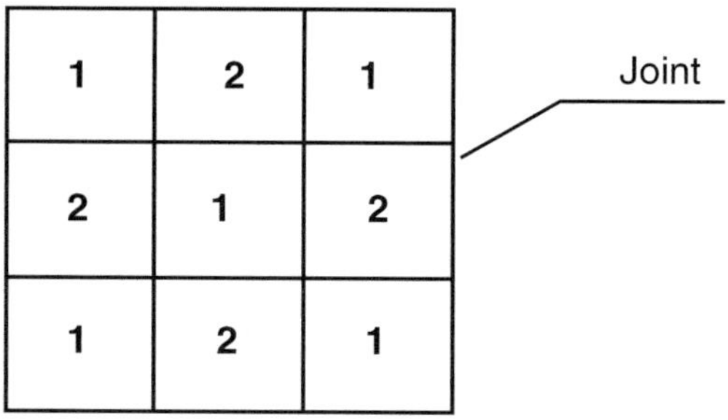

Fig. 3.9 Order of casting concrete in situ

3.1.4.2 Cement Soil

Cement soil is the mixture of cement, soil, and water. When there is no aggregate (sand and gravel), available locally using cement soil can avoid the need to transport aggregate from a distant place. Soil can be of any kind except heavy clay (with clay particle content of more than 30 %). Proportion of cement is 8–12 % of the mixture by weight. The water content is about 14–16 % of the mixture. For 1 m^3 of cement soil, cement used is about 125–190 kg.

The cement soil used in catchment of RWH project is compacted in situ either manually or using a small machine. The expected density of cement soil is 1550–1600 kg/m^3. According to tests by GRIWAC, the RCE of cement soil is 0.4–0.5.

3.1.4.3 Plastic Sheeting

In general, plastic film is impervious material. However, seepage may happen when there is a rip or tear on the sheet due to mechanical damage or prolonged exposure to solar radiation. Seepage can also happen through joints between two sheets. There are two ways of using plastic film for water catchment: the exposed and the buried plastic sheeting.

Owing to damage by solar radiation, exposed plastic sheeting can only be used as a temporary measure for seepage control. In the arid and semi-arid areas, exposed plastic sheeting is commonly used in the field to prevent soil moisture loss or to divert runoff from the sheet onto the field to increase soil moisture content. The film has a thickness of 0.01 mm and has a service life less than half year and is abandoned after harvest.

In the second way, plastic film is buried with straw mud, coarse sand, or low-grade cement mortar with a thickness of 3–4 cm to avoid sun damage. It is estimated that the service life of the plastic film can be extended to more than five years but the RCE is lowered significantly. According to tests, the RCE of buried plastic film is around 0.3–0.4 in Gansu.

3.1.4.4 Consolidated Soil

Since the 2000s, the use of consolidated soil for rainwater catchment has been studied by the Inner Mongolian Research Institute for Water Resources and the Northwestern Agriculture and Forestry University (Cheng et al. 2009). The agent for soil consolidation is a high-strength and water stability earth consolidator called HEC. The consolidator is a gypsum/cement mixture with proportions of HE C:gypsum:cement = 1:4.19:129.2. The proportion between consolidator and earth is 1:8 (by weight). The thickness of the consolidated earth for catchment is 6 cm.

According to tests, the consolidated earth can have RCE up to 0.75, about the same as concrete but at a lower cost. However, some tests showed that the consolidated soil is more easily cracked and eroded. It is still at an early stage of research and development. So far there are no reports on using this kind of material for catchment in practical RWH projects.

3.1.4.5 Other Kinds of Earthen Material for Catchment

Several kinds of earthen material can be used as catchment: lime soil, 3-component mixed soil, and compacted soil. Lime soil is a compacted mixture of soil and slaked lime (powder). The proportion of the two components is lime to soil = 3:7 or 2:8 (by volume). The water content is the same as for compacted soil. The 3-component mixed soil is a mixture of soil, clay, and slaked lime. It has also to be compacted. The proportion usually taken is lime:clay:loam/sandy loam = 1:2:4.

Compacted soil is compacted to a depth of 30 cm. Tests show that the RCE can be raised significantly by compaction. First, the surface layer of soil with depth of 30 cm is loosened. Then loose soil is compacted layer by layer with each layer of 15 cm to a dry soil density of 1.55 t/m^3 by manual or by portable compaction machine. It is important to have suitable soil moisture during compaction. For loam, the best soil water content is in the range of 13–15 % by weight. For sandy soil and clay soil, the moisture content is to be reduced and added by 2–3 %, respectively. According to tests, RCE of compacted soil can be 0.15–0.25 in semi-arid climates. The problem of compacted soil for catchments is they are easily eroded during rain storms. Another problem is that under weathering action such as freeze/thaw cycles, the compacted soil will be loosened and thus the RCE lowered.

The earthen catchment is widely used in south China where the climate is humid and the soil moisture content is often high. RCE of earthen catchment in humid areas is much higher than that in the semi-arid regions.

3.1.4.6 Economic Comparison Among Different Materials for Building Catchment

GRIWAC has undertaken economic comparisons among different kinds of materials for the purpose-built catchments. The results are shown in Table 3.2.

Table 3.2 Economic comparison among different materials for purpose-built catchments

Item	Concrete	Cement tile	Cement soil	Buried plastic	Factory made clay tile	Village workshop clay tile	Compact soil
Annual RCE	0.75	0.72	0.48	0.42	0.44	0.36	0.23
Annual collected water (m^3/m^2)	0.181	0.168	0.112	0.098	0.104	0.085	0.053
Unit initial cost (RMB/m^2)	4.82	4.96	3.64	1.94	3.98	3.0	0.25
Service life (years)	20	25	15	10	25	20	4
Total collected water in service life (m^3/m^2)	3.62	4.2	1.68	0.98	2.6	1.7	0.21
Annual O&M cost (RMB/m^2)	0.145	0.099	0.146	0.155	0.08	0.09	0.05
Total initial and O&M cost in service life (RMB/m^2)	7.71	7.44	5.82	3.49	5.97	4.8	0.45
Unit cost of water (RMB/m^3)	2.13	1.77	3.47	3.56	2.3	2.4	2.14
Land occupation for collecting 1 m^3 of water (m^2)	5.52	5.95	8.92	10.2	9.62	11.8	18.9

Note (1) Cost based on 1990 price; (2) Annual rainfall = 350 mm; (3) Annual O&M cost is estimated by percentage of initial cost based on practical experience

From Table 3.2, it can be seen that among the different materials, the ranking of cost for collecting 1 m^3 of rainwater (from small to large) is cement tile, concrete, compacted soil, tile made in factory, tile made in village workshop, cement soil, and buried plastic film. From an economic point of view, the cement tile and concrete-lined surface should be the first choices. The cost of compacted soil is lower than that of clay tiles, but the service life is short and land requirement is high.

3.1.4.7 Ground Preparation

For purpose-built catchments, the ground should be compacted first to avoid any damage owing to subsidence. The procedure for ground compaction is same with that for compacted soil (see Sect. 3.1.4.5).

3.2 Water Storage Subsystem

3.2.1 Classification of the Water Storage Subsystem in China

The purpose of the rainwater storage subsystem is to store excessive rainwater during the rainy season to meet water demand in the dry season. To fulfill this function, the rainwater storage subsystem is composed of a storage tank and auxiliary structures including: a settling basin and filtration equipment. In China, owing to the wide range of climatic, topographic, and geologic conditions, there are different types of rainwater storage tanks in different regions. Basically, five types exist, namely, underground tanks, surface tanks, ponds, prefabricated containers, and channel networks. The composition of rainwater storage subsystem is illustrated in Fig. 3.10.

Underground tanks include the water cellar and water cave. These used to be very popular in the north and northwest China. Its advantages and disadvantages are shown in the following.

1. Advantages

   - Being able to store ground runoff that is necessary to supplement to the roof runoff for domestic supply in the semi-arid area as well as for the irrigation and other production water supply;
   - Keeping stored water cool: when the water table is located at 1.5 m below the ground then the water temperature is close to the average annual air temperature. In Gansu, the water temperature in the water cellar is generally under 10 °C, which is favorable for keeping good water quality;

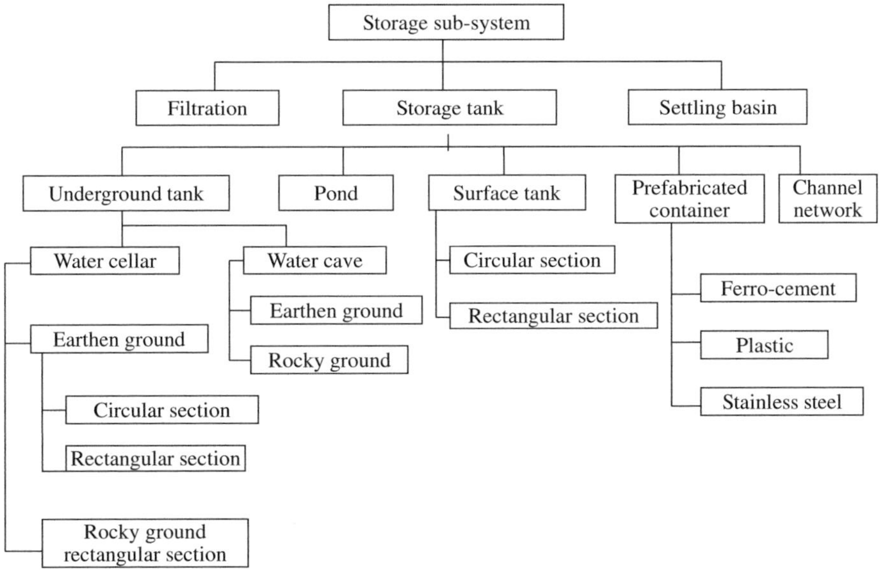

Fig. 3.10 Composition of storage subsystem

- The underground tank is usually covered keeping out dust, insects, and animals;
- In semi-arid areas, the annual evaporation is always greater than the annual precipitation. In underground tanks evaporation can be reduced or even avoided;
- Avoidance of freezing of the stored water in winter in the colder regions; and
- Under certain soil conditions, the ground can help the stability of the structure thus saving construction materials and lowering the cost.

2. Disadvantages

- Large excavation requirement increases the cost especially in rocky areas;
- Cannot be used in areas where the groundwater table is high; and
- If soil has low shearing strength and/or is composed of fine sand, the underground structure would be subjected to higher load than a structure constructed above the surface.

3.2.2 Underground Rainwater Storage Tank

The underground tanks in China can be divided into different types, for which the structure is not the same. Design of the tank for earthen and rocky ground conditions is also quite different. According to the shape of the tank, the underground tank can be divided into circular or rectangular ones. Figure 3.11 shows the classification of the underground tank.

3.2.2.1 Water Cellar with Circular Section

A water cellar is an underground tank and is popularly used in the northwest and north China. It has a local name of "*Shuijiao*" and in some areas also "*Hanjing*" (dry well). In earthen ground, water cellars mostly have a circular section due to the higher resistance to the surrounding earth pressure and to minimize

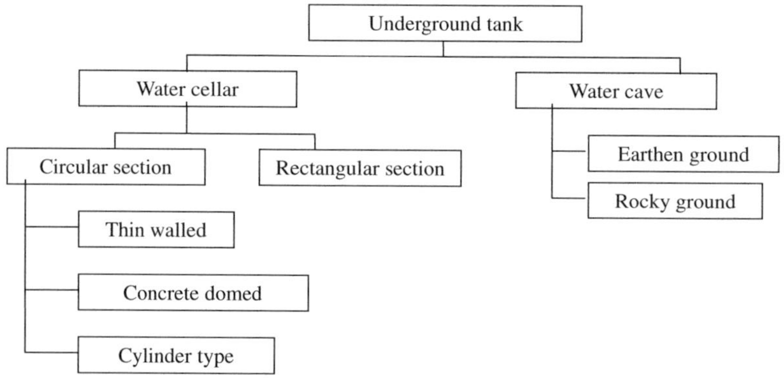

Fig. 3.11 Classification of underground rainwater tank

construction material requirements per unit of storage volume. Usually, it has a bottle-like shape with a ratio of 1:1.5–2 between the height and the diameter. Water cellars are especially popular in the loess plateau area where the loess soil is predominant. The property of loess soil has a large impact on the design and construction of water cellar. According to different soil conditions, the structural design of water cellar can be divided into three types, namely, the thin walled, the concrete domed, and the cylindrical (Zhu et al. 2012).

(1) Thin-walled water cellar

The section of thin-walled water cellar is illustrated in Fig. 3.12. The thin-walled water cellar is suitable for a good soil condition with: a soil dry density larger than 1.5 t/m^3 and clay content higher than 15 %. Besides, the loess soil is dry and has good structure. In this case, the soil can keep stable in a vertical cut up to 10 m without any support. It means a cut configuration of a water cellar can be stable just by the soil itself. The only need for stability is to keep the soil dry. So a thin layer of cement mortar with nice anti-seepage performance is enough to keep the structure stable. Another role of the layer of cement mortar is to avoid slumping, which would damage the integration of the top soil body and might induce collapse of the tank.

In the thin-walled water cellar, the wall and the bottom are all plastered with cement mortar of grade M10. Here M10 means the mortar has compressive strength of 10 MPa (100 kg/cm^2). The upper part of the tank where the wall has an inverted slope is called the "dry cellar." To avoid collapse of this part, water is not allowed to be stored here. In the bottom of the tank, where the water pressure

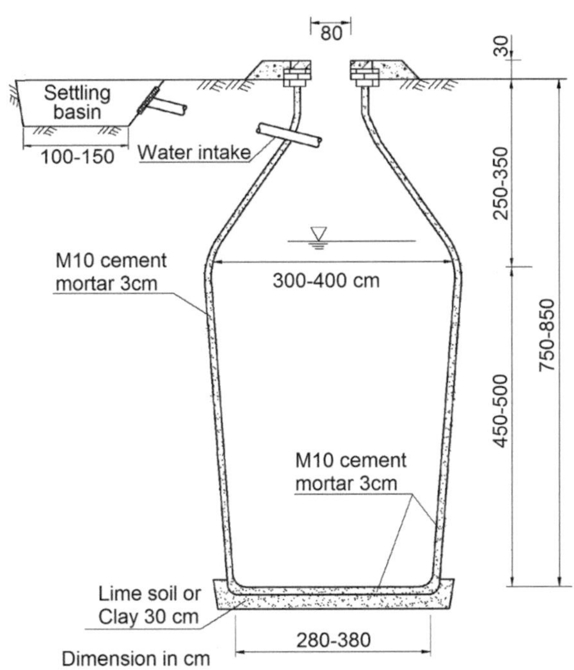

Fig. 3.12 Thin-walled water cellar

Fig. 3.13 Concrete-domed water cellar

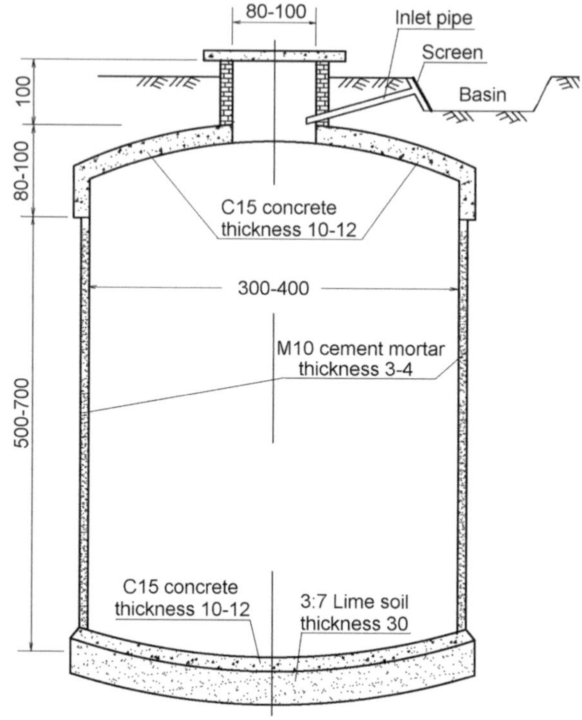

is the highest, it is preferable to put a layer of lime soil (lime to soil = 3:7 by volume) before plastering with the mortar.

(2) Concrete-domed water cellar

When soil is not as strong as in the above case, to sustain the weight of the top soil and the upper load, a concrete dome with thickness of 10–12 cm is used for the top structure. The concrete dome section is shown in Fig. 3.13.

The dome is built with plain concrete. Sometimes, a glass fiber mesh is put in the concrete to increase its integration. The concrete is grade C15, which denotes the compressive strength of the concrete is 15 MPa (150 kg/cm^2). Ratio between the dome height and the diameter is about 0.3–0.4. The side wall of the tank is built of cement mortar with thickness of 3–4 cm. The bottom of the tank is also built of plain concrete in a basin-like shape or a slab.

The conventional way for casting the concrete dome structure needs a mold (made of either timber or steel), a framework to support the mold, and a platform for the workers. Since this method is expensive and the construction procedure complicated, local people have invented a simple and cost-effective method for casting the dome concrete using an earth mold. This will be described detail in Chap. 4. The dome structure can also be built with bricks, in which case, only a platform is needed, and material for the mold and framework can be saved. However, this method requires skilled labor to do the work, also see Chap. 4.

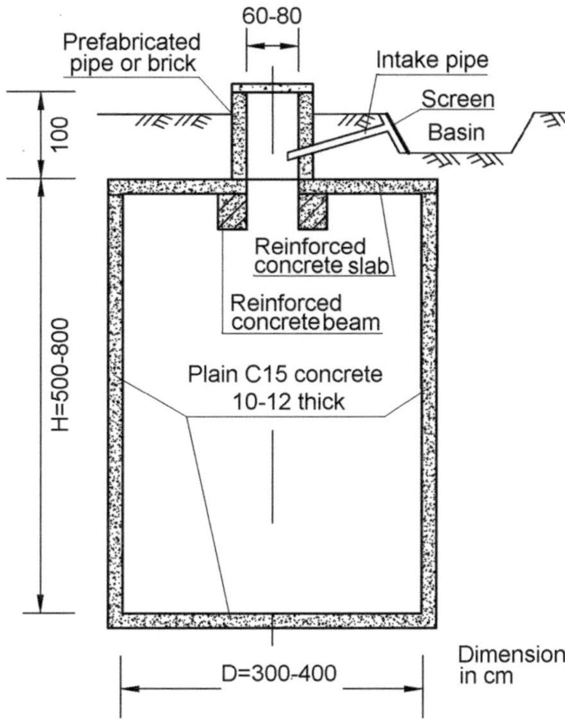

Fig. 3.14 Cylindrical water cellar

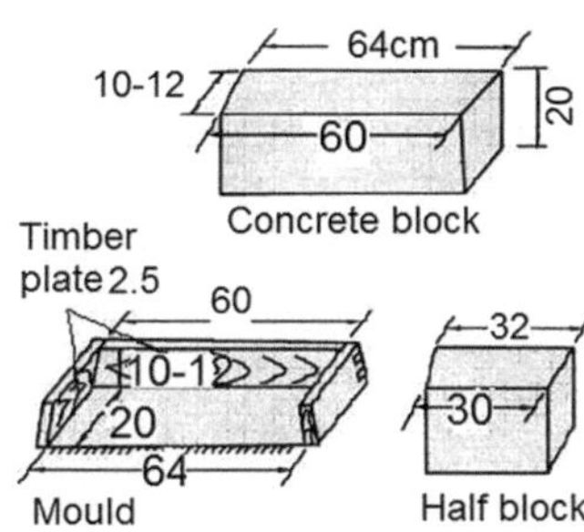

Fig. 3.15 Prefabricated concrete blocks for cylindrical water cellar

(3) Cylindrical water cellar

When the ground is composed of fine sandy soil with low cohesive strength and/or with a loose structure and low dry density, a thin layer of cement mortar cannot stand the surrounding soil pressure. In this case, the side wall should be built like a cylinder with concrete 10–12 cm thick. Plain concrete with Grade C15 is used. For the top structure, there can be two alternatives: the concrete or brick dome and the reinforced concrete slab. Figure 3.14 shows a cylindrical water cellar with a reinforced concrete slab for the top cover.

To cast the concrete wall, a mold and a supporting frame work are needed, which are costly. To avoid using a mold, the wall can be built with prefabricated concrete blocks. The blocks have a trapezoidal section to form the circular shape, see Fig. 3.15.

The grade of concrete is C15. For the top reinforced concrete slab, a Grade of C20 is adopted. The slab is prefabricated on the ground and then put in place. If the length of the slab is 3 m or less, it can be installed manually. But if the length (the diameter of the tank) exceeds 3 m, it is too heavy for handling manually. Then a crane is necessary. Another alternative is to fabricate shorter slabs and put them on two reinforced concrete beams that are supported on the edge of the cylinder wall.

(4) Economic comparison between the three types of water cellar with circular section

Table 3.3 shows the cost per storage volume for the 3 types of water cellar. It is apparent that the cost of the thin-walled water cellar is the cheapest among the three types of water cellar. This is also illustrated in Fig. 3.16. If the soil conditions allow, the thin-walled water cellar should be adopted whenever possible. However, for safety reasons, the concrete-domed type is the most common. It can

Table 3.3 Materials and unit cost for different types of water cellars

Type	Volume (m^3)	Cement (kg)	Gravel (m^3)	Sand (m^3)	Labor day	Total unit cost (CNY/m^3)
Thin walled	15	345		1.29	10	52.4
	20	375		1.41	11	43.1
	25	402		1.51	12	37.4
	30	433		1.62	14	35.6
Concrete domed	15	630	0.78	1.6	9	61.6
	20	750	0.9	1.89	11	55.6
	25	850	1.03	2.16	14	53.9
	30	930	1.19	2.27	16	50.3
Cylindrical	15	888	2.97	2.02	11	86.8
	20	1018	3.41	2.32	13	75.8
	25	1151	3.85	2.62	15	69.3
	30	1284	4.3	2.92	17	64.9

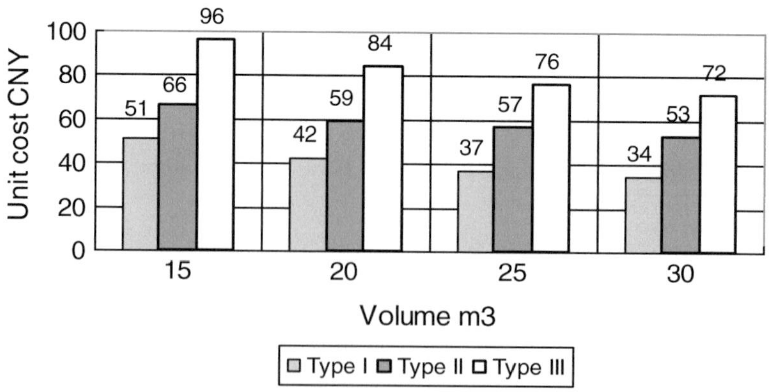

Fig. 3.16 Unit cost versus volume for different types of cylindrical water cellar (*Type I*—Thin walled, *Type II*—Concrete Dome, *Type III*—Cylindrical)

also be seen from Fig. 3.16 that the unit cost of construction decreases as the volume increases. So to build a single tank costs less than to build two or more tanks with the same volume.

3.2.2.2 Rectangular Shaped Underground Tank

The rectangular shaped underground tank has the advantages of avoiding deep excavation and its length can be extended to have a large volume. In China, the maximum volume of rectangular RWH tank is more than 1000 m^3. The disadvantage is that the sidewall of the rectangular tank is subjected to pressure from the surrounding soil, resulting in bending stress in the wall, which cannot be sustained by the plain concrete or masonry structure. In this case, either the reinforced concrete wall or gravity retaining wall with large thickness built of plain concrete or masonry structure is used. To improve the loading condition of the rectangular tank, lateral walls at an interval of 5 m need to be built inside the tank as a support to the side wall so the bending load on the side wall can be reduced. Figure 3.17 shows the plan and vertical sections of the rectangular underground tank. The section B—B shows a lateral wall supporting the side wall. With the hole in the lower middle part of the lateral wall, water in the compartments of the tank is connected together.

The gravity retaining wall is usually adopted for the structure of the side wall and can be built with concrete or masonry. The top structure of the tank is either a reinforced concrete slab or an arch structure, see Section C—C of Fig. 3.17. In the latter case, masonry is commonly used.

The reinforced concrete slab for the top structure is better prefabricated on the ground so as to avoid using the framework for casting the concrete and the complicated construction procedure. If the prefabricated slab is installed manually, then it should not be too heavy. The length of the slab should preferably be not greater than 3 m. The width and the thickness are 60 and 10 cm, respectively. For the bottom, a plain concrete slab with thickness of 10–20 cm is commonly used. It is very

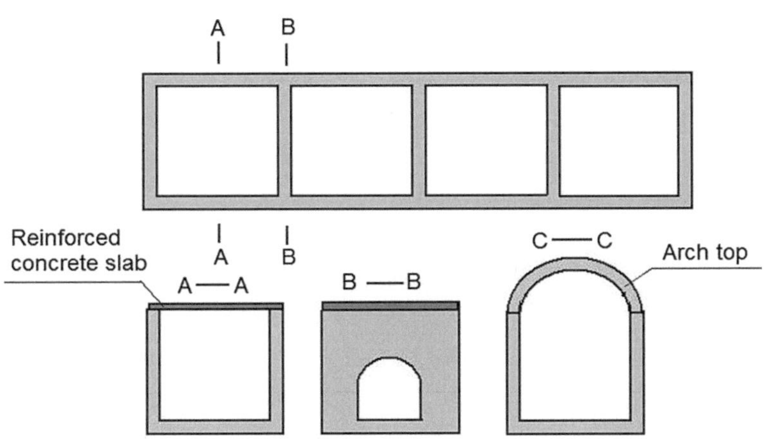

Fig. 3.17 Plan and section of rectangular water cellar

important to carefully seal the joint between the wall and the bottom slab to avoid any seepage, which could erode the subsoil and damage the structure. The most commonly used sealing materials for this purpose are asphalt mortar, clay, PVC-tar mixture, and rubber.

When the RWH project is built on rocky ground, a rectangular tank design is normally adopted. To avoid deep excavation, sometimes the tank is built with half cut and half fill. Figure 3.18 shows an underground tank built in a rocky hill side in the southwest Guizhou Province.

In this example, half of the tank was cut in the rock and wall of the upper half of the tank is built of masonry using stones excavated from the pit. The remained stones were piled around the wall and on the top of the slab as insulation.

3.2.2.3 Cave-Type Underground Tank

The water cave is another kind of underground tank and has a local name of *Shuiyao*. It is like the cave dwelling (cave in a cliff), which was common in the past in north and northwest rural China and is still seen even today in some areas. The cave-type underground tank is built inside a loess cliff.

When the ground is earthen, the water storage is inside the cave and located below the ground level. People go into the cave and fetch water either by hand or using an electric pump. See Fig. 3.19. The loess soil inside the cliff usually has dense structure and high strength. The top structure of the cave is usually an

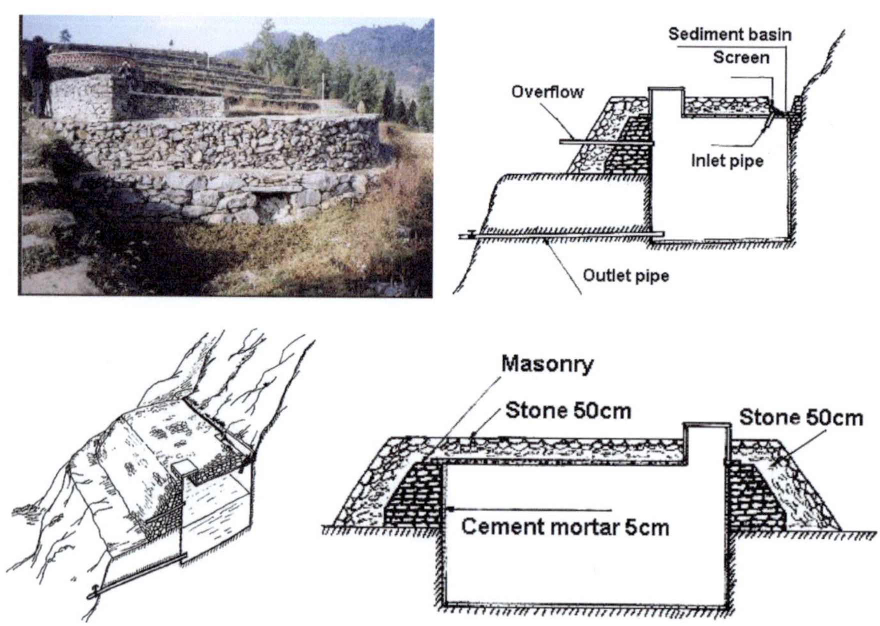

Fig. 3.18 Underground tank built at the rocky hill side. *Source* IDMWR and RDMOF (2001)

arch which is plastered with cement mortar to avoid collapse of loose soil particles, which could result in instability of the cave. The wall and the bottom are also paved with cement mortar to prevent water seepage. A door is installed in the entrance of the cave. Figure 3.20 shows a view of two sets of water caves, an old traditional type and a modern example.

In the southwest China, where rocky hills are common, water caves are shaped like a tunnel. Figure 3.21 shows a design of this kind of water cave by Guizhou Provincial Design Institute of Water Resources.

The water level stored in the cave is higher than the ground. A dam is built at the entrance of the cave to store water inside the cave. The outlet and drainage pipes are installed with control valve. If there are no significant fissures or cracks on the rock, then the wall and bottom are pasted with cement mortar of 3 cm in thickness.

3.2.2.4 Sizing the Underground Tank

The recommended dimensions for circular water cellars of different volumes are shown in Table 3.4.

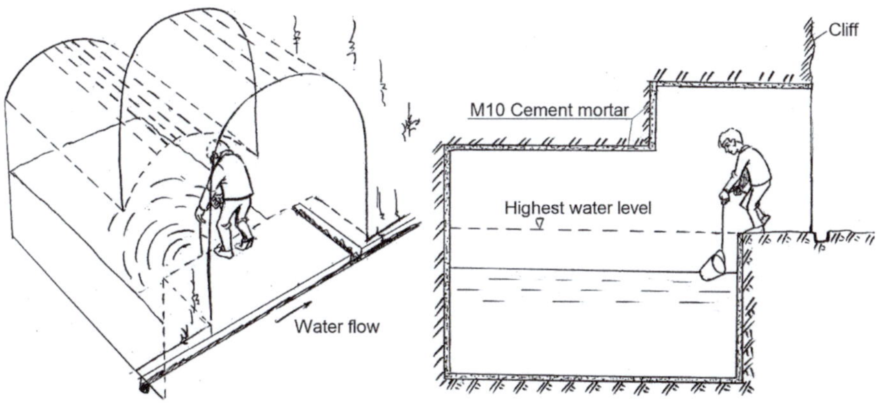

Fig. 3.19 Cave-type underground tank (Shuiyao). *Source* IDMWR and RDMOF (2001)

Fig. 3.20 Water caves: old (*left*) and new (*right*)

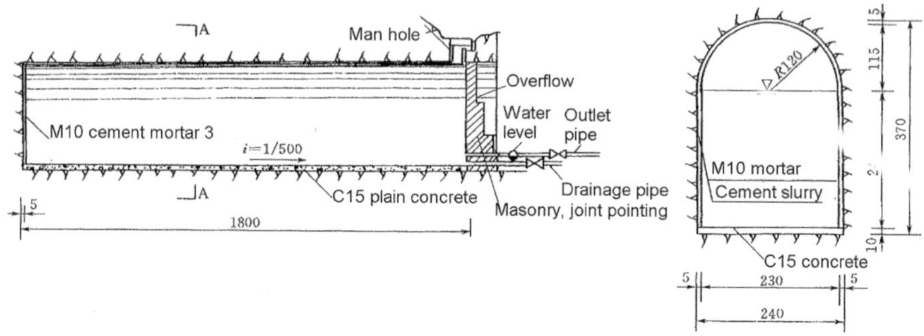

Fig. 3.21 Tunnel-like water cave (dimensions in cm). *Source* IDMWR and RDMOF (2001)

Table 3.4 Sizes and dimensions of circular water cellar

Volume (m^3)	Diameter (m)	Water depth (m)	Thin wall cellar		Domed wall cellar		Total height of cylinder type cellar (m)
			Total height	Dry cellar height	Total height	Dry cellar height	
20	2.5	4.1	5.4	1.2	5.2	1	5.0
30	3.0	4.3	5.8	1.5	5.6	1.2	5.2
40	3.5	4.2	6.2	1.9	5.6	1.4	5.2
50	3.8	4.4	6.6	2.1	6.0	1.4	5.3
60	4.0	4.8	7.0	2.2	6.5	1.6	5.7

Table 3.5 Recommended size of rectangular tank

Volume (m^3)	20	30	40	50	60	80	100
Depth (m)	2.1	2.3	2.6	2.9	3.1	3.2	3.4
Width (m)	3.2	3.5	3.5	3.5	3.5	3.5	3.5
Length (m)	3.3	4.1	4.8	5.6	6.1	7.9	9.2

The recommended dimensions for rectangular tanks are shown in Table 3.5. In the table, the maximum width of the rectangular tank is taken as 3.5 m. This is an upper limit of dimension for the slab that can be prefabricated and handled manually. Using prefabricated slabs to cover the tank avoids the need for forms and a frame when casting in situ. If crane is available for construction, then the width of the rectangular tank can be larger.

3.2.3 Design of Surface Tanks

Surface tanks are those located above or at ground level. These are popular in the humid and sub-humid areas in mid-, southwest, and south China, where the precipitation is high and the evaporation is low.

3.2.3.1 Classification of Surface Tanks

These can be divided into two types: the open tank and sealed tank. Tanks for domestic water supply have to be covered. This is not only to avoid debris falling into the tank but also to exclude direct sunlight which would result in stagnant water with poor quality (Thomas and Martinson 2007).

The tank types can also be divided into circular and rectangular in shape. The circular tank is most commonly used because of its higher structural resistance to the soil pressure.

Tanks can further be classified according to the construction material used, namely concrete, masonry, and brick. Usually, for the same storage volume, the concrete tank requires less construction materials than the masonry one. However, cast-in-place concrete needs a mold and framework to support it, which are costly and time-consuming to erect. Besides, in areas where the rocks are available and easy to extract, masonry can be an economic solution. Brick structures are cheaper than masonry and concrete not only because the material is cheaper but also because no mold and complicated framework are needed. But bricks are only suitable in areas with mild climates.

The tanks can also be divided into those with an integrated and separate structure. In the integrated structure, the sidewall and the bottom are linked to each other as a whole without joints between bottom and the wall and even the top slab is sometimes integrated with the wall. Tanks with integrated structures are usually built of reinforced concrete. The construction procedure is complicated, and requires a professional construction team. Generally, in rural areas in China, the integrated structure tank is seldom adopted. For RWH tank, usually the separated structure is used, where the wall, roof, and the bottom are all separately built with water proof joints between them.

3.2.3.2 Design of the Circular Surface Tank

(1) Concrete circular surface tank

Compared to the masonry tank, the main advantages of the concrete circular tank are

- Less construction material required.
- Cast in situ concrete has better integration to adapt to any deformation of the ground.

The disadvantages are

- Complicated construction procedure.
- For casting concrete in situ, mold and the framework are costly.

Figure 3.22 shows design for the concrete surface tank by the Design Institute of Guangxi Autonomous Region. In the design, only the earth pressure from outside of the wall and water pressure from inside the wall are taken into consideration, while the load from earth quake and uplift pressure from groundwater flow are not

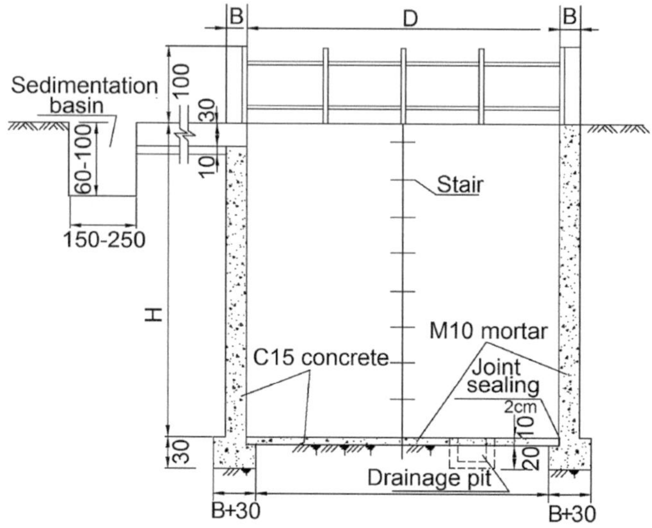

Fig. 3.22 Concrete circular surface tank

Table 3.6 Size and amount of major work of the concrete circular tank. *Source* Gu et al. (2001)

Storage volume (m^3)	Inner diameter (cm)	Height of tank (cm)	Thickness of wall (cm)	Thickness of bottom (cm)	Amount of concrete (m^3)	Amount of cement (t)	Amount of earth cut (m^3)	Amount of earth fill (m^3)
30	400	240	13	10	7.7	2.7	59	16
40	400	320	17	10	11.2	3.8	85	26
50	500	260	16	10	12.1	4.2	95	23
60	500	310	19	10	15.4	5.2	118	32
70	500	360	21	10	18.5	6.2	141	41
80	550	340	21	10	19.8	6.6	153	39
90	550	380	23	10	23.1	7.6	176	48
100	550	420	25	10	26.7	8.7	201	58

taken into account. If the groundwater table is high, then drainage measure should be taken to divert the groundwater away.

When the wall is mostly above the ground and bottom of the tank is close to the ground level, the wall is only subjected to the internal water pressure as there is no external soil pressure applying on it. In this case, the wall has to be strong enough to resist the load from the internal water pressure. If the tank has its top located at ground level and the wall is located completely below ground, the wall has to resist only the external soil pressure applying on the wall when the tank is empty. When the tank is filled, the water pressure from the inside can be partly balanced by the earth pressure from the outside.

Table 3.6 shows the size of the tank with top at the ground level for different storage volumes ranging from 30 to 100 m^3. The thickness of the wall and bottom as well as the quantity of concrete and cement for the tank built either inside or above the ground are the same. Only when the tank wall is built under the ground surface, there are additional work of earth cut and fill.

(2) Masonry circular surface tank

The masonry tank usually requires more work and materials than the concrete one. However, when stone is available on-site and the local people are familiar with masonry work then it can be an economic option. Besides, construction of masonry tank does not need the mold and supporting framework as construction of concrete tank needs.

Figures 3.23 and 3.24 show two designs for masonry tanks. One has the bottom at the ground level and another has the top at ground level. The loads on these two

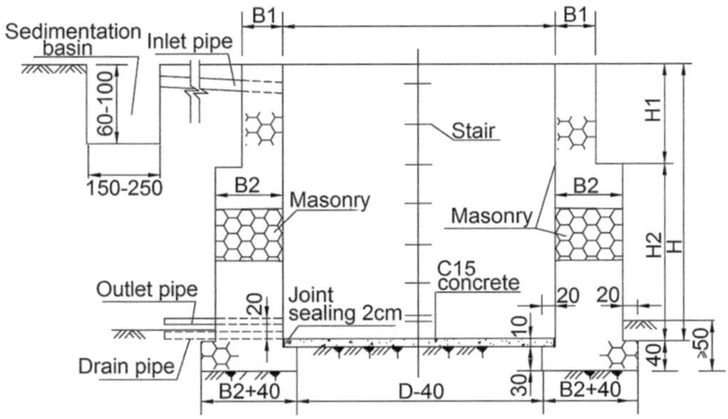

Fig. 3.23 Circular masonry surface tank with bottom at the ground level. *Source* Gu et al. (2001)

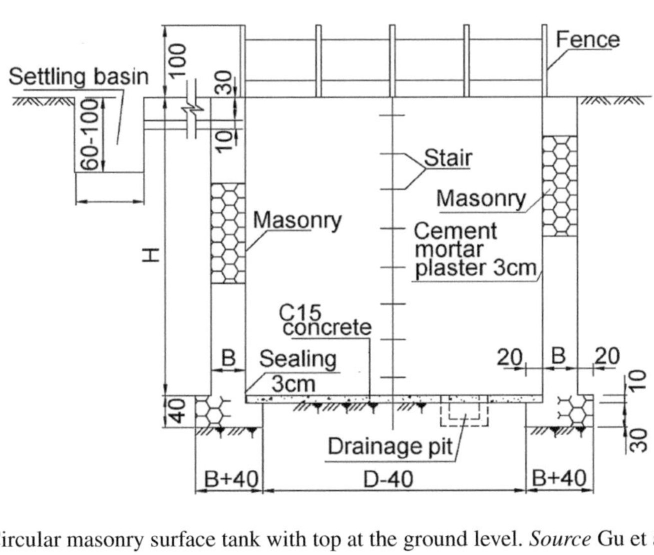

Fig. 3.24 Circular masonry surface tank with top at the ground level. *Source* Gu et al. (2001)

Table 3.7 Size and major work amount of the circular masonry surface tank with bottom at the ground level. *Source* Gu et al. (2001)

Storage volume (m^3)	Inner diameter (cm)	H_1 (cm)	H_1 (cm)	Total height H (cm)	B_1 (cm)	B_2 (cm)	Thickness of bottom (cm)	Amount of concrete (m^3)	Amount of masonry (m^3)	Amount of cement (t)
30	400	240	0	240	50	50	10	1.4	24.3	3.5
40	440	265	0	265	60	60	10	1.7	34.4	4.7
50	500	260	0	260	65	65	10	2.2	41.2	5.7
60	500	90	220	310	65	80	10	2.2	56.3	7.4
70	550	80	220	300	65	80	10	2.6	59.8	7.9
80	550	70	270	340	65	95	10	2.6	78.9	10.0
90	600	70	250	320	70	90	10	3.1	77.4	10.1
100	600	80	280	360	80	105	10	3.1	100.8	12.6

Table 3.8 Size and major work amount of the circular masonry surface tank with top at ground level. *Source* Gu et al. (2001)

Storage volume (m^3)	Inner diameter (cm)	Height H (cm)	Wall thickness B (cm)	Thickness of bottom (cm)	Amount concrete (m^3)	Amount masonry (m^3)	Amount cement (t)	Amount of earth cut (m^3)	Amount of earth fill (m^3)
30	400	240	50	10	1.4	20.0	2.8	71	15
40	440	320	60	10	1.4	24.0	3.4	95	23
50	500	260	65	10	2.2	32.0	4.5	116	20
60	500	310	65	10	2.2	36.0	5.0	137	25
70	500	360	65	10	2.2	41.0	5.5	160	33
80	550	340	65	10	2.6	52.0	6.9	183	33
90	550	380	70	10	2.6	56.0	7.3	205	38
100	550	420	80	10	2.6	62.0	8.0	228	45

kinds of masonry tank are the same with that on the concrete tank. Tables 3.7 and 3.8 show the sizes and labor requirements for these two kinds of masonry tank. To enhance the water proofing performance of the masonry work, it is better to plaster cement mortar with thickness of 3 cm on the surface of masonry. The mortar is to be applied in 3 layers each 1 cm thick.

From the above examples, it can be seen usually that construction material consumption for concrete tank is less than that for the masonry ones. Furthermore, since the soil pressure on the tank wall mainly causes compression stress, this can be well resisted by the masonry and concrete materials, while the internal water pressure causes tension forces in the wall that are harder for the concrete, and especially the masonry to resist. So theoretically, when tank is put inside the ground, the load condition would be more favorable because the soil can provide some help to resist the internal water pressure. However, there is no difference of wall thickness for concrete tank in these two situations but for a masonry tank built completely underground, less materials and work are required.

For the RWH tank, the grade C15 or C20 concrete is commonly used, and for the masonry grade M10 cement mortar is recommended.

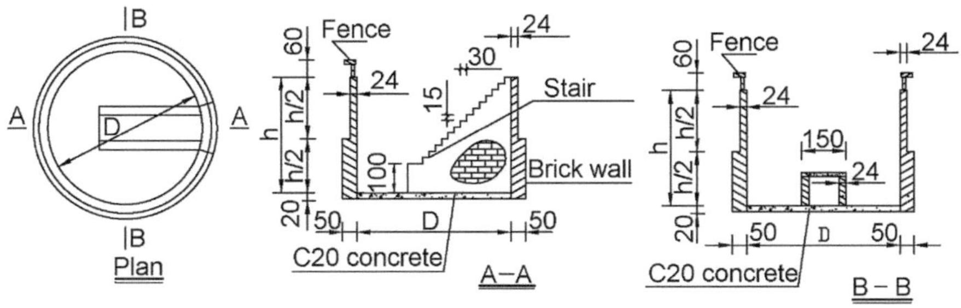

Fig. 3.25 Brick-built circular surface tank. *Source* Gu et al. (2001)

(3) Brick circular tank

In areas with mild climates, surface tanks can also be built with bricks. Figure 3.25 shows a design for a brick-built surface tank with a volume ranging from 50 to 200 m^3. Table 3.9 is the size and amount of main work for the tank with different volumes.

For the brick-built tank, it is necessary to plaster cement mortar on the brick surface to enhance its waterproofing performance. The cement mortar needs to be 2–3 cm in thick and applied in 2–3 layers with 1 cm each.

(4) Piping system in the surface tank

The inlet and outlet pipe as well as the drainage pipe should be installed in the tank. The inlet pipe is put above the highest water level by 5–10 cm. The outlet pipe is located 20–30 cm above the tank bottom. The drainage pipe is to be placed in a collection pit lower than the bottom by 30–50 cm, depending on the storage volume. For safety reason, an overflow pipe (spill way) installed at 5–10 cm above the highest water level is necessary. However, for underground tanks, the outlet and drainage pipes are difficult to place unless on a steep slope where an outlet can be found. In most cases, the water delivery from this kind of tank can only be by pump or sometimes by siphon.

Figure 3.26 illustrates a partially buried circular tank built at the hill side. In this case, the water supply pipe can be fitted on the down slope side.

3.2.3.3 Rectangular Surface Tank

The rectangular surface tank has the advantage of being easy to build. The disadvantage is the unfavorable loading applied on it when compared to the circular tank. Usually, a rectangular tank is built below ground with the top at ground level. This way the internal water pressure is partly counter-balance by the surrounding soil pressure and only part of the load is applied on the wall. In this case, the critical load is that of the external soil pressure when the tank is empty. Since the load applied on a longitudinal wall is larger with its length increases, building lateral

Table 3.9 Size, dimensions, and material requirement for brick-built circular surface tanks. *Source* Gu et al. (2001)

Vol.	Diameter of tank (m) with different heights (m)				Amount of brick and concrete for different heights (m)							
					1.5		2.0		2.5		3.0	
	1.5	2.0	2.5	3.0	Brick (m^3)	Concrete (m^3)	Brick (m^3)	Concrete (m^3)	Brick (m^3)	Concrete (m^3)	Brick (m^3)	Concrete (m^3)
50	6.5	5.6	5.0		18.3	6.7	19.6	5.0	20.9	4.0		
100	9.2	8.0	7.2	6.5	25.1	13.4	28.1	10.0	28.7	8.0	30.4	6.7
150	11.3	19	9.0	8.0	30.3	20	32.4	15.0	34.5	12.0	36.6	10.0
200	13.0	11.3	10.1	9.2	34.6	26.6	37.1	20	39.5	16.0	41.9	13.4

Note Figs. 3.23, 3.24, 3.25, 3.26, 3.27 and Tables 3.6, 3.7, 3.8, 3.9, 3.10 are sourced from Gu et al. (2001)

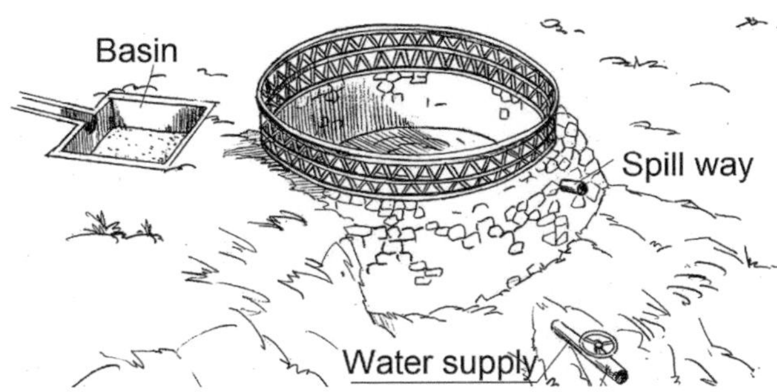

Fig. 3.26 Illustration of a circular surface tank popular in southern China

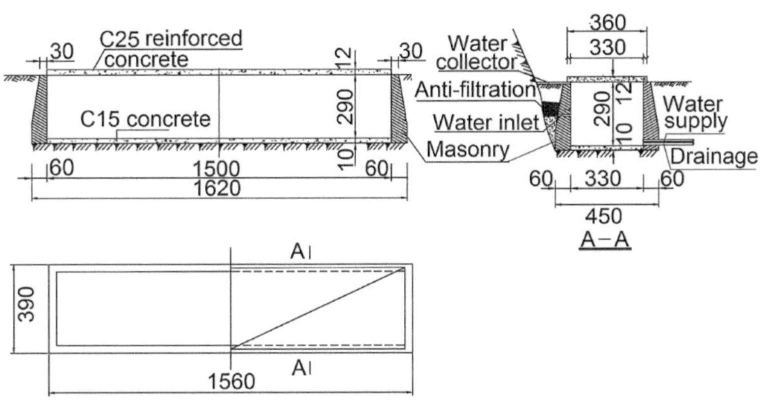

Fig. 3.27 Example of rectangular surface tank by hillside. *Source* Gu et al. (2001)

walls with given intervals can significantly reduce the moment load, as shown in Fig. 3.17.

Figure 3.27 is an example of a rectangular tank built by the hill side with a storage volume of about 120 m^3 in southwest China (Gu et al. 2001).

In this example, the length of the tank amounts to 15 m. The moment load is high on the longitudinal wall. If adding two lateral walls to provide supplemental support to the wall, a rough estimation shows that the maximum moment load applied on the wall can be reduced by 130 %. Thus, the safety of the structure can be greatly enhanced.

3.2.3.4 Cover of the Surface Tank

Tanks for domestic water supply should be covered and sealed. The cover or top structure is usually built of reinforced concrete. Design of the cover for a surface tank and underground tank uses the same design principle, and only the load

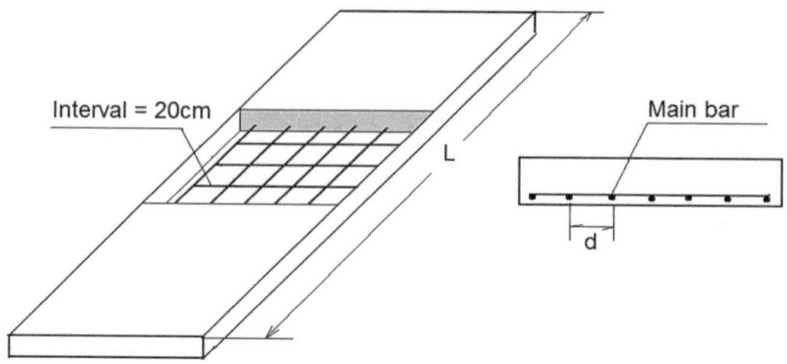

Fig. 3.28 Top cover slab of surface and underground tanks

Table 3.10 Size of top cover slab and reinforcement. *Source* IDMWR and RDMOF (2001)

Depth of earth overlying (m)	Thickness and reinforcement	Span of the slab (m)		
		1.5	2.0	2.5
0	Slab thickness (cm)	10	10	10
	Interval of main bar (cm)	20	20	20
	Diameter of main bar (mm)	10	10	10
0.5	Slab thickness (cm)	10	10	10
	Interval of main bar (cm)	20	12	10
	Diameter of main bar (mm)	10	10	10
1.0	Slab thickness (cm)	10	10	12
	Interval of main bar (cm)	12	10	10
	Diameter of main bar (mm)	10	10	14

conditions are different. In the case of the surface tank, there is no earth on the top. The load for the cover structure is only from the weight of human stepping on the cover. While for the underground tank, the load includes the weight of the earth and human together.

For rectangular tanks, the cover slab has a length equal to the width of the tank. To avoid using costly molds and a supporting frame, the slab is usually prefabricated on the ground and then moved on to the top of the tank. Handling of the slab is usually by hand so the slab should not be too heavy and the width of the slab has is a limited range.

Figure 3.28 shows the top slab and the suggested reinforcement under different loads. Table 3.10 shows the size of the slab and the reinforcement with different depths of overlying soil. In the table, the slab is supported on the tank wall so the length of the slab should be longer than the span by 10–15 cm. The width of the slab can be decided depending on the suitable weight of the slab for handling. For example, if the slab is weighed less than 300 kg, then the width of a 2.5-m long slab with thickness of 10 cm is about 50 cm.

For a large tank, when the tank is too wide to have the slab handled manually, the length of the slab can be shortened using a slab-beam system.

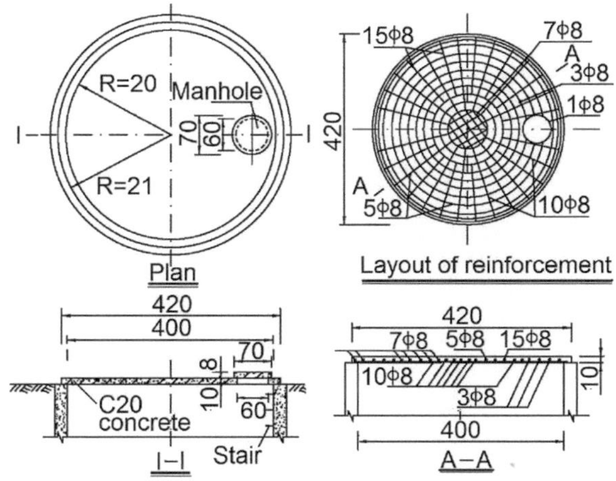

Fig. 3.29 Cover for circular surface tank and reinforcement arrangement. *Source* Gu et al. (2001)

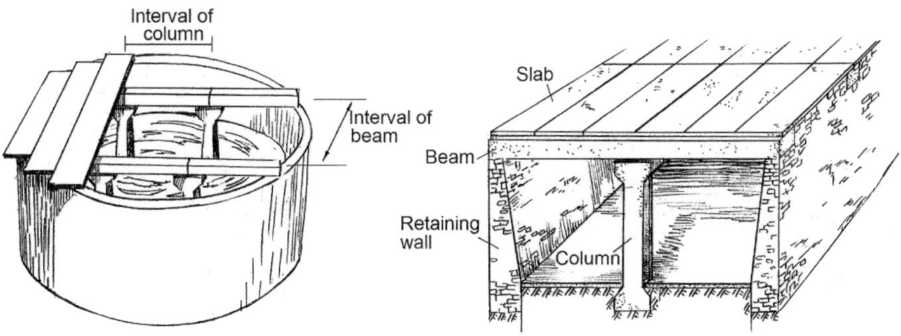

Fig. 3.30 Slab-beam system for the top structure of a large tank. *Source* IDMWR and RDMOF (2001)

For a circular tank, the top cover slab is also circular in shape and usually cast in situ. Figure 3.29 shows the design of reinforcement for a circular cover with diameter of 4 m. The thickness of the top cover slab is 10 cm. The quantity of concrete is 1.5 m^3 and the total weight of reinforcement bar is 55.5 kg. The rectangular slab can also be used for a circular tank. Slabs with different lengths are prefabricated and supported on the wall as illustrated in Fig. 3.30.

3.2.4 Prefabricated Water Tanks

Water storage tanks are now increasingly being manufactured in large factories. These tanks are currently used for rainwater storage mainly for domestic use especially in the southeast China. These tanks are made from reinforced high-strength cement mortar (Ferro-cement tank), plastic, or stainless steel. The capacity of the

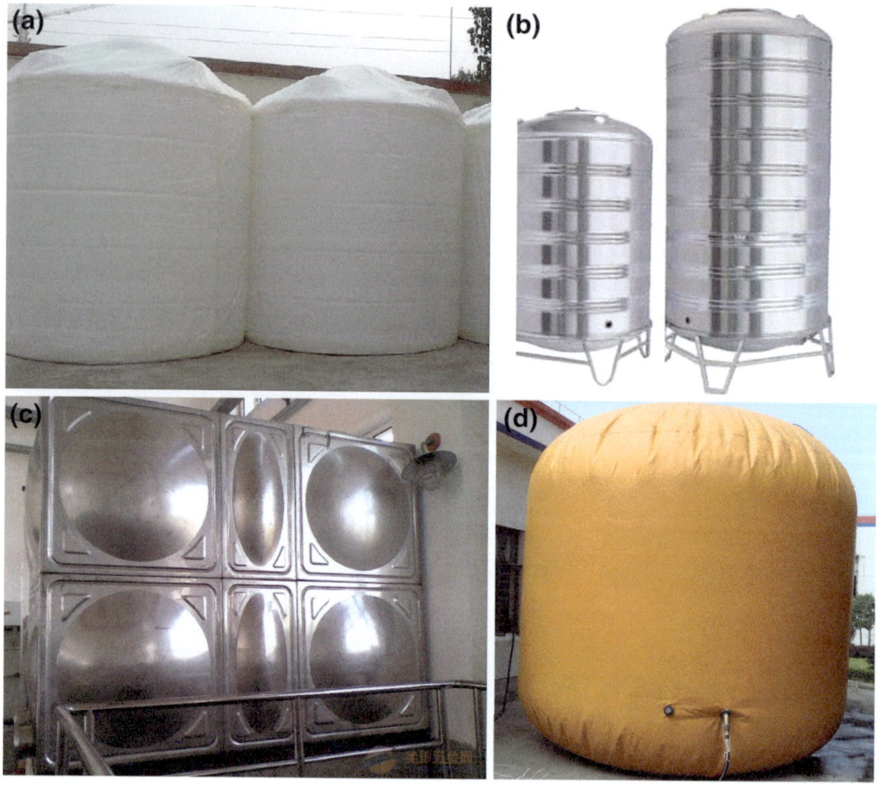

Fig. 3.31 Plastic and stainless steel water tanks. **a** PE tank. **b** Stainless steel tank. **c** Composed type stainless steel tank. **d** Flexible polymer tank

cement mortar tanks range between 0.6 and 2 m^3. The plastic polymers for making these tanks include Polyethylene (PE), High-Density Polyethylene (HDPE), and Linear High-Density Polyethylene (LHDPE). The volume of plastic tanks ranges between a 100 and 50,000 l. There are also different types of stainless steel water tanks. The cylinder-type tank is an integrated one. It has volume from 100 to 10,000 l. The composed type stainless steel container can have volume from 1 to 500 m^3. These are composed of standardized stainless steel plate, each of which is 1 by 1 m. The plate is welded and fastened with steel rods by screwing or welding. The container can be composed into different volumes and with different sets of length, width, and height according to the customer's order. The plastic polymer and stainless steel water containers are widely available in the Chinese market and can be easily ordered. Figure 3.31 shows some examples of plastic polymer and stainless steel container. Besides, there is also tank made of fabric glass, mainly used for septic tank.

Figure 3.32 shows a rubber bladder tank that has been used for greenhouse irrigation in the Inner Mongolia Autonomous Region, a cold region in north China. To avoid freezing of the water in the bladder, it was buried in a pit with earth cover

Fig. 3.32 Rubble made water container in the Inner Mongolia Autonomous Region

on it. The 40 m^3 rubber bladder cost $815 in 2004. Cost per unit volume is about $20.5/$m^3$, a little higher than that of the cylinder-type water cellar.

The cement mortar tanks are manufactured in small village workshops and are widely available in the southeast China. These are made of high-strength cement mortar and steel mesh. In Yuyao Municipality, Zhejiang Province, the cement mortar is 2 cm thick and has proportion of 1:1.5 (cement to sand). The steel mesh is made of steel wire with diameter of 1 mm and welded at space of 2 cm × 2 cm. The bottom of the jar is integrated with the sidewall. The cover is prefabricated separately and is made of the same material. Figure 3.33 shows the structure of the cement mortar container.

The manufacture procedure is as follows:

(1) First make the bottom: put the round wire mesh at 1 cm higher than the bottom mold, on which lubricate oil is applied on. Then paste and press the cement mortar to make it firm. The steel mesh at the bottom edge should be bended and extended upward to connect with the mesh of the wall.

(2) Assemble and put on inner mold, which can be made of steel or timber. Then install the wire mesh around the mold and firmly connect (better weld) it with steel mesh extended from the bottom. Make sure the space of 1 cm between the mesh and the mold.

(3) Then plaster and press the cement mortar onto the mold. To avoid slippage of the mortar, the water content should be controlled strictly with zero slump (ratio of water and cement is less than 0.4). To make removal of the inner mold easier, the mold should be assembled in 2 or 3 pieces and lubricant oil applied on the surface of mold before pasting on the cement mortar.

(4) Depending on the temperature, the mold can be removed after 2 or 3 days of curing the mortar.

When manufacturing the container, holes for installing drainage and water supply pipes near the bottom and ventilation hole on the upper part should be installed.

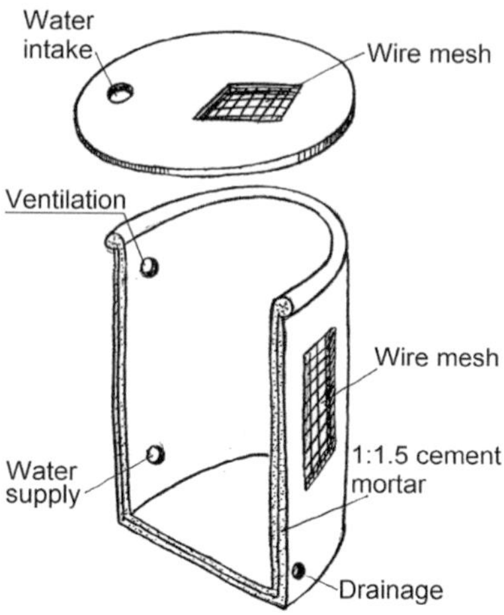

Fig. 3.33 Structure of cement mortar tank. *Source* IDMWR and RDMOF (2001)

3.2.5 Pond and Channel Networks

3.2.5.1 Pond

Ponds are water storage reservoirs formed in natural depressions. These are common in the mountainous areas of southern China. The ponds store not only rainwater but also spring water. It was the main water source for domestic use including water for drinking and cooking. Nowadays, owing to pollution, in some areas, pond water can no longer be used for drinking.

Ponds are usually located in natural depressions with water levels lower than the surrounding ground. In many cases, no structure is needed. People fetch water by using buckets or sometimes using an electric pump. In the latter case, a simple intake structure has to be built.

To enlarge the volume of the pond, a dam or a dike in the lowest position can be built to raise the water level. The most common type of dam/dike is earth one, sometimes when the height of the dam is large, dry pitching (dry rubble masonry) is placed on the slope to prevent from erosion. To avoid spillage over the dam wall during storms, a sluice by the side of the dam or a culvert under the dam to drain excessive water is necessary.

One of the problems in pond design and operation is the seepage control of the pond. To avoid or reduce seepage loss, the bank and sometimes also the bottom of the pond have to be lined with impervious material. Two types of lining can be used: the rigid lining and the flexible lining. The rigid lining can be built with a concrete slab or masonry. The flexible lining mainly includes different kinds of membrane made of plastic polymers and asphalt-plated grass fabric. The rigid lining is more easily cracked causing seepage when the ground experiences subsidence. So the flexible lining is more commonly used.

Aging of the membrane due to solar radiation shortens its service life. The composite geomembrane is made of one or two layers of polymer and one layer of geotextile. It has better performance than a single plastic sheet. The composite geomembrane has higher resistance to radiation. It is claimed that some products can have a useful service life of 10 years. Nevertheless, it is strongly recommended to have cover materials to protect them from direct solar radiation. The plastic polymers used include Polyvinylchloride (PVC), Polyethylene (PE), High-Density Polyethylene (HDPE), Low-Density Polyethylene (LDPE), Linear Low-Density Polyethylene (LLDPE), and Ethylene Vinyl Acetate (EVA). The textile is made of woven or non-woven polymer fabric (punctured or non-punctured).

The covering material on the membrane to protect it from solar radiation include coarse sand, mud soil (clay or loam), and low-grade cement mortar.

3.2.5.2 Channel Network

In the plains region of southeast China, numerous channels were built and formed a network in ancient China and were originally built mainly for transportation. The density of the channels can be 1–2 km/km^2. Later these were also used for flood water mitigation as well as for storing rainwater. The channel network is linked with the river system. To control the water level, sluices were built between the river and the network. When there is big flood flow or high tide in the external river, the sluice will stop the inflow to the network to prevent water flooding in the network area. When the water level is too low to meet the demand of irrigation or water supply as well as the transportation, the river water will be diverted or pumped into the network. In the past, the water quality was good and suitable for domestic use. People cleaned rice and vegetables in the channels. However, due to the pollution from municipal, industrial, and agricultural sources (due to over use of chemicals for fertilizer, pest and weed control), water quality has deteriorated

Fig. 3.34 An updated channel in Zhejiang Province

and can no longer be used for domestic purposes, but only for irrigation. Due to urbanization, many channels have also been filled to create land. The role of the channel networks has been gradually reduced. However, measures have been taken recently to preserve the channel network. To prevent scouring of the banks during storms, the bank is to be lined with concrete or masonry. Figure 3.34 shows an updated channel in Zhejiang Province.

3.2.6 Auxiliary Facilities of Storage Subsystem

3.2.6.1 Settling Basin

When runoff passes over a bare earth catchment, the flow may bring a lot of silts causing loss of storage volume and poor water quality in the tank. A settling basin (sedimentation chamber) can help settle down most of the silt and dirt before it enters the storage tank. The simplest settling basin is a rectangular tank, with dimensions of 1–1.5, 0.8–1.0, and 0.5–0.6 m for length, width, and depth, respectively. To enhance the settling efficiency, two new designs are shown in Fig. 3.35. The idea of them is just to extend the flow path while reducing the size of the basin. In the first design (Fig. 3.35b), water flows in horizontal zigzags, while in the second design (Fig. 3.35c) water flows in vertical zigzags.

The structure of the settling basin is similar to that of the rectangular surface tank. It can be built with concrete, masonry, or brick. After each rain event, there is water remained in the settling basin, which will be lost by evaporation or seepage

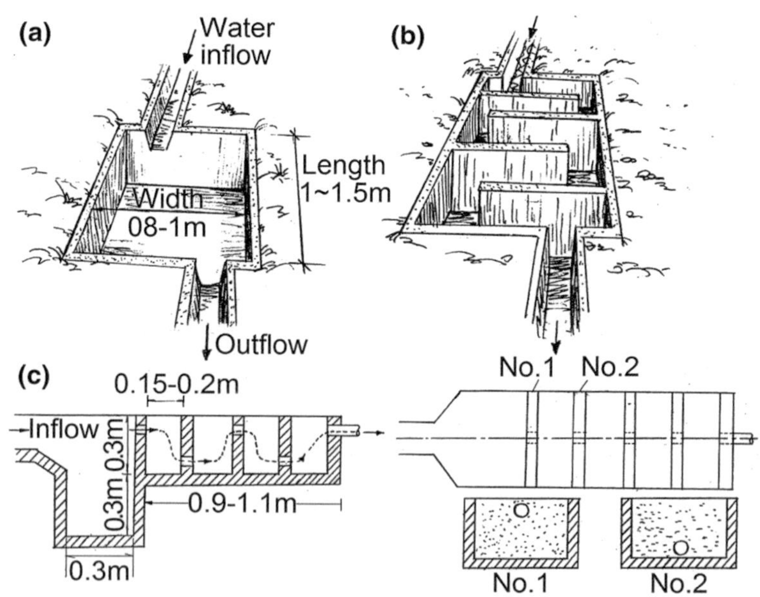

Fig. 3.35 Updated design of settling basin. *Source* IDMWR and RDMOF (2001)

which is valuable in the dry areas. It is suggested to install a pipe above the bottom of the basin by 10 cm to divert out the water after the storm is over for irrigation or husbandry.

3.2.6.2 Filtration Equipment

When runoff is from a large, earth catchment and the water is for domestic water supply or for mini-irrigation (drip, micro-spray, and bubble irrigation), it needs to get rid of coarse sand and silt particles, and sand–gravel filter should be installed before water enters the tank. The filtration system consists of layers of fine to course sand, gravel, and rubble put along the flow path. Water flows from the upper ditch into the filter and flows out through pipe at the bottom. Figure 3.36 shows two examples of the sand–gravel filtration equipment (Gu et al. 2001). Figure 3.36a is used in Longquanyi District in Sichuan Province for an earth catchment. The tank is built of masonry. Figure 3.36b is used in the Changdao County in Shangdong Province for filtering runoff from a road catchment.

3.2.6.3 Screening

To avoid debris entering the tank, a screen made of mesh is put at the intake of the tank. It is very simple so that every household can make it by themselves. The screen is shown in Fig. 3.37.

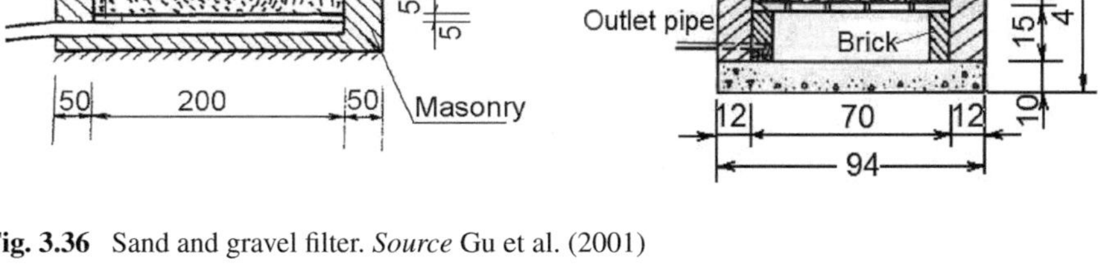

Fig. 3.36 Sand and gravel filter. *Source* Gu et al. (2001)

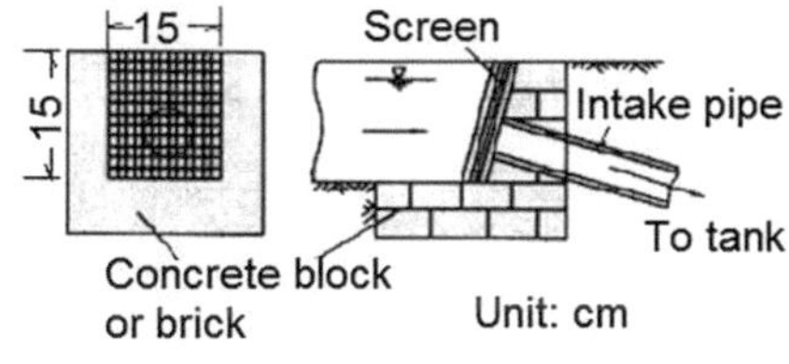

Fig. 3.37 Screen before the tank. *Source* IDMWR and RDMOF (2001)

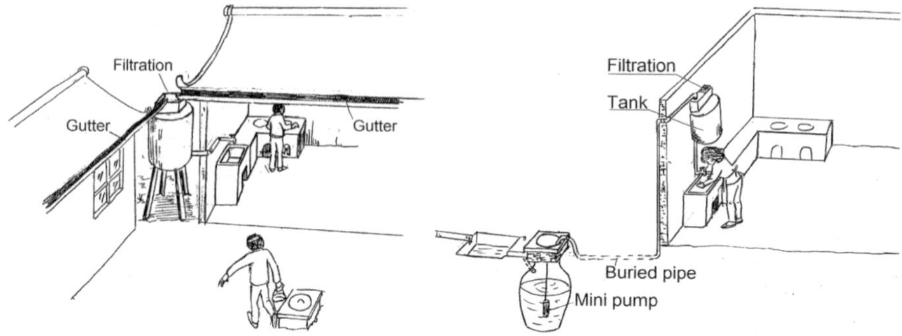

Fig. 3.38 Simple household tap water supply using roof catchment

3.3 Water Supply Facility

3.3.1 Water Delivery for Domestic Supply

The simplest way of fetching water from an underground tank is to get water using a bucket. However, as the bucket is often put on the ground, pollution of the tank water is inevitable.

A number of ingenious arrangements have been developed to get household tap water connections. See Fig. 3.38. In Fig. 3.38a, runoff collected from the roof-gutter system is flowing into an elevated tank and then through a filtration facility to provide water for cooking and drinking. Water collected from ground catchment is used for other domestic use. In Fig. 3.38b, an electric mini-pump lifts water from water cellar to the tank and then through the filtration system. In this case, water from roof and ground catchment is mixed and used for drinking and other purposes.

3.3.2 Water Delivery for Irrigation

Water delivery for irrigation can be done in 3 ways: by gravity flow through a siphon, by hand pump, and by electric pump. The first case is only suitable where the water level in the tank is higher than the field to be irrigated. The hand pump can be used only for small-scale drip or micro-spray system. In China, hand pumps used for irrigation has been developed. The specification of the hand pump is shown in Table 3.11.

When the flow rate for irrigation is too large or the lift is too high for a hand pump, an electrical pump needs to be used. Usually, for RWH irrigation systems, a

Table 3.11 Specification of large capacity hand pump

Type	Discharge (m^3/h)	Lift head (m)	Suction head (m)	Diameter of piston (mm)	Diameter of I/O pipe (mm)	Weight (kg)
RB1.5	1.5	15	7	86	25	15

submerged pump (diving pump) is used. Table 3.12 lists the specifications of small capacity submerged pump.

Figure 3.39 is photo of hand pump for irrigation.

Table 3.12 Specification of mini-diving pump

Model	Discharge (m^3/h)	Lift (m)	Motor power (W)	Weight (kg)
QDX3-8	3	8	180	10
QDX3.6-5	3.6	6		
QDX1.5-12	1.5	12	250	17
QDX3-10	3	10		
QDX5-7	5	7		
QDX3-12	3	12		
QDX3-10	3	10		
QDX3-15	3	15	370	
QDX3-20	3	20	550	23
QDX3-22	3	22		
QDX4-20	4	20		
QDX1.5-32	1.5	32	750	19
QDX3-30	3	30		

Note The model denotes: diving pump operated in wet condition discharge (m^3/h)-lift (m). Motor is single phase, working at 50 Hz, 220 V. The diving pump is put under water 0.5–5 m

Fig. 3.39 Hand pump for irrigation

References

Cheng M, et al. Rainwater harvesting water saving irrigation and dry farming technology in the semi-arid loess hilly area of North China. Zhengzhou: Yellow River Water Resources Publications; 2009 (in Chinese).

Gu B, et al., editors. Rainwater harvesting and utilization technology and practice. Beijing: China Water Power Press; 2001 (in Chinese).

Irrigation and Drainage Department of Ministry of Water Resources (IDMWR) and Rural Department of Ministry of Finance (RDMOF) (ed). Concise textbook of rural rainwater harvesting project. Beijing: China water Resources and Power Publications; 2001 (in Chinese).

Thomas TM, Martinson DB. Roofwater harvesting. Delft: International Water and Sanitation Centre; 2007.

Zhu Q, Li Y, John G. Rainwater harvesting and sustainable technologies in rural China. Rugby: Practical Action Publications; 2012.

Chapter 4
Construction and Operation and Maintenance of Rainwater Harvesting Project

Chengxiang Ma

Keywords Construction of rainwater harvesting project · Operation and maintenance

As discussed in the previous chapters, the rainwater harvesting system consists of three components, namely the rainwater collection, rainwater storage, and water supply/irrigation subsystems. In this chapter before discussing the construction and operation and maintenance (O&M) of these 3 subsystems, we will start with a general consideration of the building materials for the RWH project.

4.1 Building Materials

4.1.1 Concrete

4.1.1.1 Component Materials of Concrete

The raw materials for concrete include cement, sand, gravel or crushed stones, and water.

- Cement

There are many types of cement for different uses. For rainwater harvesting projects, normal Portland cement is commonly used. Owing to the mini-size of RWH project, it is recommended to use low-grade cement such as Grade 32.5 or 42.5

C. Ma (✉)
Gansu Research Institute for Water Conservancy, Lanzhou, China
e-mail: machengxiang@hotmail.com

Q. Zhu et al. (eds.), *Rainwater Harvesting for Agriculture and Water Supply*,
DOI 10.1007/978-981-287-964-6_4

slag cement and Grade 42.5 ordinary Portland cement (ISO standard). It is important to store cement in a dry environment to prevent it from becoming moist. The moistened or caked cement should not be used.

- Sand

Sand prepared for RWH projects must be rigid, free of weak, pin-type, and/or flake-shaped particles. A suitable diameter of grains ranges between 0.35 and 0.5 mm. Moreover, it should be clean and the content of silt ($d = 0.005$–0.05 mm) and clay ($d < 0.005$ mm) should not be more than 4 %.

- Gravel or crushed stone

The gravel or crushed stones should be solid and free of weak or weathered materials. To ensure the rigidity of concrete, the gravel or crushed stone should be completely enveloped by cement mortar. For this purpose, the largest diameter of gravel or crushed stone should be smaller than 1/3 of the smallest dimension of the concrete slab and 1/4 of the smallest dimension of any other concrete component. For example, when building a concrete slab for lining the surface, the largest diameter of the gravel or crushed stones should not be more than 1/3 of the thickness of the slab. In the case of reinforced concrete, the diameter of the gravel or crushed stones should be smaller than 2/3 of the space between the steel bars.

- Water

Industrial and household wastewater as well as turbid water containing high silt load are not permitted for use in the concrete mixture. It is also advised not to use any swamp water, which might be highly mineralized and is detrimental to the concrete.

4.1.1.2 Proportion of Concrete Mixture

The proportion of concrete mixture is determined according to the concrete strength and the workability of concrete. In general, the water–cement ratio is one of the most important factors that affect the strength of concrete. The workability of concrete indicates the plasticity of the concrete mixture during casting. It can be measured by the slump or degree of consistency. The larger the slump is, the more plastic the mixture and the easier its casting will be.

Once the water–cement ratio (W/C), slump, grade of cement, and the properties of gravel and sand are determined, the proportion of various components of the concrete mixture with a certain concrete grade can be determined through laboratory tests and analysis. However, since the RWH project involves the construction of a large number of separate systems, it is not feasible to conduct tests at each project site. Instead, the technical extension service at county level should recommend the proportion of concrete mixture based on their experience of existing water resource projects and previous test results for the materials in their area. The builder of the RWH project, usually the farmers, may seek assistance from the local extension service.

4.1.2 Cement Mortar

Cement mortar is composed of cement, sand, and water. Demands of cement mortar for cement and water are the same with that of concrete. The suitable diameter of sand depends on the purpose of cement mortar. For the inner layer of the mortar, sand with medium diameter of 0.35–0.5 mm is used while for the layer to finish the surface, fine sand with diameter of 0.2–0.35 mm is used.

4.1.3 Other Materials

In the RWH project, in addition to the materials for concrete and cement mortar, other materials include rubble stones, big pebbles, and purchased materials. The rubble stone and big pebbles should be rigid, and with high strength which is desirable to avoid weaknesses and cracking when subjected to weathering.

Quality demands on lime are mainly on its content of calcium dioxide. This can be identified through testing. Besides, caked lime should be avoided through careful inspection. There are different grades of brick. For the RWH project, bricks with a grade higher than #80 are used.

For other purchased materials like plastic film, pipes, etc., it is important to check the appearance and test the properties of these materials to see if demands stipulated by relevant standards are met.

4.2 Method for Constructing Purpose-Built Catchment

4.2.1 How to Build a Catchment with Concrete Pavement

To avoid cracks on the concrete surface due to ground subsidence, the earth under the concrete slab should be compacted firmly. The thickness of compaction should not be less than 30 cm. The 30 cm topsoil is loosened first and then it is compacted to a density of no less than 1550 kg/m^3. To achieve this, the compaction is done in two layers, each 15 cm thick.

The concrete is then cast in a form made with timber or metal (Fig. 4.1). The concrete is cast into square blocks each with dimensions of 1.5 m $\times$ 1.5 m, or 2 m $\times$ 2 m. This is to prevent the concrete from cracking due to shrinkage or expansion in the drying process and/or the effect of temperature change. Casting is done using the concreting sequence method shown in Fig. 4.2a). Concrete is cast in two sets of blocks in sequence. After the first set of concrete is cast and the shrinkage process of concrete is concluded, then the mold is removed and prefabricated sealing material is placed at the edge of the cast concrete block (Fig. 4.2b). The other blocks are then cast. The sealing material is usually made of three layers of asphalt felt with four asphalt coatings.

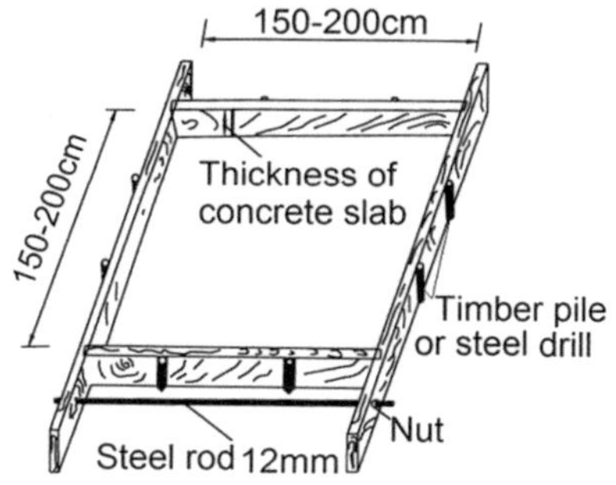

Fig. 4.1 Mold planks for casting concrete catchment. *Source* IDMWR and RDMOF (2001)

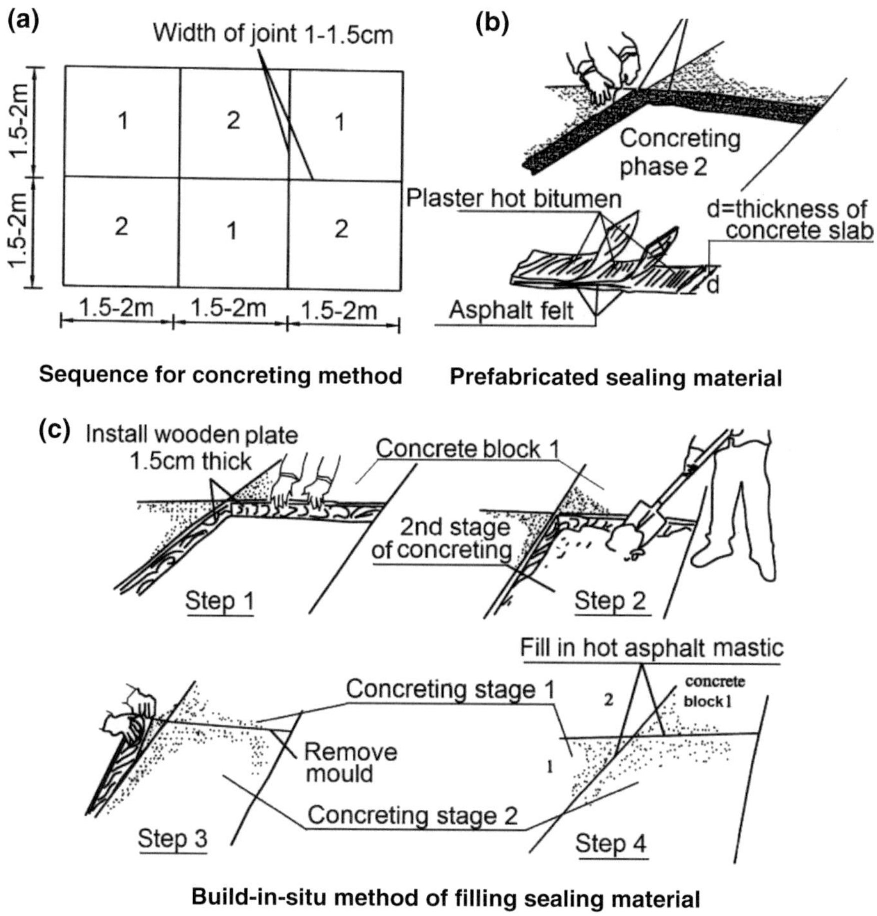

Fig. 4.2 Illustration of building concrete catchment with sequence concreting method and filling the joints with sealing materials. **a** Sequence for concreting method. **b** Prefabricated sealing material. **c** Build-in situ method of filling sealing material. *Source* IDMWR and RDMOF (2001)

The sealing material can also be filled in the joints after the two sets of concrete blocks are finished concreting (Fig. 4.2c). In this case, the commonly used material to fill the joints is asphalt–sand mortar or plastic tar. The asphalt–sand mixture is a mastic product of asphalt and fine-medium sand with proportions of 1:3–4 applied at a temperature of 180 °C. The plastic tar mortar is a product of tar, waste plastic film, fly ash, and a chemical called T50 with proportions of 1:0.15–0.2:0.3:0.04 and at a temperature of 110–120 °C. The asphalt–sand mix is poured into the joints at a high temperature. Recently with certain solution, it can also be applied under normal temperatures. In the latter case, some volatile solvent is added. The tar plastic mix is commercially available and has better performance vis-à-vis flexibility and waterproofing under low-temperature conditions but its cost is higher. If both materials are not available, clay can be used for sealing the joints but its service life is short and frequent maintenance is necessary.

The thickness of mold planks is around 2 cm. To remove the frame easily after concrete setting, surface of plank should be smooth and be pasted with used machine oil (Fig. 4.1).

To ensure high quality concrete, the concrete mixture should have the correct proportions. During mixing, components of the concrete can be weighed according to the required proportions. However, to simplify the procedure, the amount of various components can also be measured by volume based on tests or practical experience.

When the mixing is done manually, the dry components of the concrete mixture, namely sand, gravel, and cement should first be mixed three times. Then water is added and an additional three mixings follow. Mixing can be done on a metal plate or on a platform made of bricks and with 1 cm cement mortar layer pasted on top of it. The brick–mortar platform should be cured with water for 7 days before use.

During casting, the fresh concrete mixture should be vibrated to obtain a higher density. Ideally, a flat-plate vibrator should be used for vibrating a concrete slab such as the catchment pavement. However, for many of the RWH project sites electricity and/or the vibrator are not available, in this case, the concrete mixture can be vibrated using a shovel and trowel. Particularly at the edges of the mold, vibrating should be carried out carefully. After the concrete gets dense and leveled, the surface should be finished twice: once after leveling and another fine finishing after the initial setting.

The finished concrete should be cured by spraying with water 3–4 times a day over a two-week period. Straw bags are put on the surface of the concrete to keep the concrete moist. Another method is to spread a plastic-based agent on the surface instead of spraying water. However, in this case, the surface should be covered with straw bags or cement paper bags to avoid damage by strong sunshine.

4.2.2 How to Build a Catchment with Masonry or a Brick Pavement

One layer of high quality rubble-stone pavement with thickness of 15–20 cm is enough for controlling seepage. Flat stones are preferable and should be free of cracks. The key to getting a nice waterproof pavement is to ensure the cement mortar fills all the voids between the stones. The following steps are suggested:

- Pave a layer of cement mortar to a thickness of 2–3 cm on the ground;
- Wash the dirt away from the stone surface;
- Place the stones onto the mortar layer and press firmly with the larger surface of the stone facing downward (Fig. 4.3a);
- Before placing the next block, plaster 2–3 cm of mortar on the side of the block already laid, then press the new block to the previous one as close as possible to make the two stones stick together tightly;
- Place the stone in an indented way (see Fig. 4.3c), so any through joints are avoided;
- During the laying of the stones, concave or convex surfaces should be avoided as far as possible to ensure a smooth surface. After laying, press mortar into all the voids and point all the joints;
- Sometimes people use a layer of cement mortar over the masonry to prevent seepage loss. However, if the above steps are carefully followed, this should not be necessary and just waste materials.

The cement mortar to build masonry catchment is Grade M10 (compression strength of 10 MPa). The standard of the cement mortar depends on proportion of the components, particularly the water–cement ratio, the grade of cement, and the construction quality.

For a brick pavement, the construction procedure is similar to that of a masonry pavement but is easier because of the regular shape of the bricks. However, since the bricks are permeable, after brick laying a 2-cm-thick layer of M10 mortar should be plastered over the surface. To ensure the bricks and mortar stick together strongly, the bricks must be wetted before use.

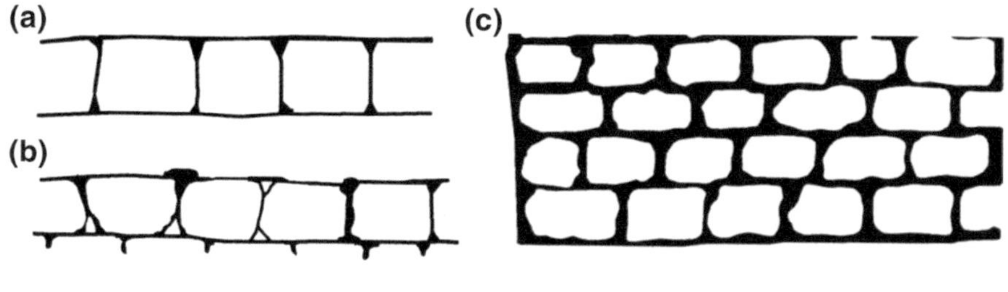

Fig. 4.3 Plan for building masonry catchment: **a** correct laying method; **b** incorrect laying method; **c** lay the stone in indented way

4.2.3 How to Build a Catchment with Plastic Sheeting

There are two ways to install plastic sheeting as a catchment. One is lay plastic film directly on the surface without any cover on it. Plastic film with thickness of 0.01 mm is used in this case. This has the advantage of having high collection efficiency up to 0.9–0.95. However, due to the material aging caused by solar radiation and mechanical damage, the service life is only one crop growing season or less. Another way is to cover the film with a 3–4 cm layer of coarse sand/straw-mud or low-grade cement mortar to prevent aging and mechanical damage. A PVC or PE membrane with thickness of 0.1–0.2 mm is usually used with this method and can last for more than 5 years. However, due to water absorption by the covering layer, the rainwater collection efficiency is much lower than that without a covering. Typically in semi-arid areas, the efficiency is only half of the exposed one.

There are two ways to join the plastic sheets. Welding is the most reliable and waterproof way. Welding can be carried out either with a specially designed machine or manually. The manual operation is conducted using the following steps: two sheets are overlapped by a width of 10 cm; a wooden plank is put under the overlapped sheets; and an iron with a constant temperature is pressed on the joint to make the seal. The temperature and passing speed of the iron has to be determined by trial and error. To simplify the process, the joint can also be made by overlapping and folding two sheets. To reduce the number of joints, it is preferable to use a large plastic sheet.

It is necessary to prepare the ground properly before installing the membrane. The ground should be leveled and compressed. All kinds of debris like stones, roots, etc. should be removed.

4.2.4 How to Build a Pavement of Cement Soil and Lime Soil

A mixture of cement soil is hardened by compaction. There are two kinds of compaction forming the rigid and plastic cement soil. The rigid cement soil is made by heavy compaction with a machine while the plastic one uses only a manual or light machine compaction method. The water content in the rigid cement soil is less than that in the plastic one. The mixture is placed in the mold and the press applies a compression force to the mixture, creating a product of high density, high strength, and high performance for seepage control. However, the machine is not often available in countryside. Besides, the press can only produce small-sized slabs of cement soil, around 50 cm × 50 cm. There will be a lot of joints on the pavement, which would cause high seepage loss. Since the plastic cement soil can be compacted in a continuous manner and is a simpler technique it is the one most commonly adopted for RWH projects.

The plastic cement soil is placed in situ. The construction procedure for the pavement is similar with that of soil compaction. The content of cement in the cement soil mixture is about 8–12 % of the total weight of the mixture. Under normal conditions, the density of cement soil amounts to 1500–1600 kg/m^3. So for each cubic meter of plastic cement soil, 120–190 kg of cement is needed. The water content of cement soil can be taken as the plastic limit of the soil. A simple way can be used to judge if the moisture content of cement soil is suitable. If a handful of kneaded cement soil can be agglomerated and becomes loose when it drops to the ground then the moisture content is fine. If the cement soil cannot be agglomerated then it is too dry and if after dropping the soil, the lump does not break up, and then the mixture is too wet.

Lime soil is a traditional material for water sealing. The most commonly used proportion of lime to soil is 3–7. Another kind of mixture that can be used for seepage control for the catchment is a mixture of lime, clay, and local soil with a proportion of 1:2:4. These two mixtures are compacted to form a less permeable surface.

4.3 Construction of Water Storage Tank

The standard method to construct an underground or surface tank with concrete and masonry materials can be found in many textbooks and construction manuals. In this chapter, we will introduce some unconventional ways to build a tank, which can either avoid using a costly mold and framework during casting the concrete structure or using a wise method to reduce excavation amount. This can save on construction materials and the cost. This simple method is also widely accepted by the local farmers who are the main builders of the RWH systems in China.

4.3.1 Method to Build the Water Cellars

4.3.1.1 How to Construct a Circular-Shaped Cellar with a Thin Wall

A circular-shaped cellar with a thin wall is the economic design when the subsoil is firm and dense enough to keep the excavated earth pit stable. The cement mortar surrounding the wall does not act as a support but is mainly for seepage control. A bottle-like shape is most commonly used and the local people in the northwest China have developed the technique for the excavation work. It may be called the "hollowing-out" method. The construction procedure is described as follows:

(1) Firstly, locate the center point for the cellar to be built on the ground, and mark out a circle on the ground for the outer-diameter of cellar.

(2) Dig a hole with the diameter of the opening of the cellar around the center point to a depth of a person. Earth within the cellar is dug and thrown out to the ground. When the excavation exceeds a depth at which the earth cannot be thrown out manually, a tripod with a winch is installed above the opening and the earth is hoisted up in a bucket either using manpower or animal power.
(3) The excavation proceeds gradually from top to bottom and from center outwards. It is important to keep the excavation just surrounding the centerline. The earth for the last 3–4 cm at the edge of the outer diameter of the wall is left uncut so this can be compacted to the designed size using a wooden hammer to increase the density of the soil. This is good for improving seepage control and better integrating the cement mortar with the soil.
(4) When the excavation is finished, plaster the cement mortar on the wall. For a 3-cm layer of mortar, plastering takes place three times, each adding a 1 cm layer. Mortar mixed with finer sand is used for the last plaster layer. Finally, a layer of pure cement grout is washed on the surface for improving its waterproofing performance.
(5) The bottom concrete slab of the water cellar is usually cast in situ. The procedure is similar to the casting of the catchments concrete slabs except no joints are needed because of a relatively stable climate inside the cellar. However, the joint between the bottom slab and the wall should be carefully filled with cement mortar to prevent any leakage. To further improve anti-seepage performance, a layer of compacted lime soil (lime to soil ratio of 3:7) 30 cm thick should be added before the concrete is cast. Alternatively, the subsoil should be compacted to a depth of 30 cm. Compaction should be done in two steps: first to loosen soil to a depth of 30 cm and then to compact it in two layers, each 15 cm thick.
(6) The final step is to finish installing the upper structure of the cellar. A prefabricated concrete ring pipe and bricks are often used.

4.3.1.2 How to Build a Circular Water Cellar with a Concrete Dome Structure

To build a dome structure in the traditional way, a timber or steel mold and framework to support the mold is necessary. It needs a lot of materials and labor. In the northwest China, the local people have developed a cost-effective method for building the circular domed water cellar. The procedure is described as follows:

(1) Firstly, locate the center point of the cellar on the ground, and mark a circle showing the outer diameter of the dome structure.
(2) Excavation is carried out within the outer diameter of the concrete dome. The earth above the contour of the lower surface of the concrete dome is excavated. An allowance of 2–3 cm of the soil is left to be compacted to the design size of the surface on which the concrete dome will be cast (Figs. 4.4a and 4.5).

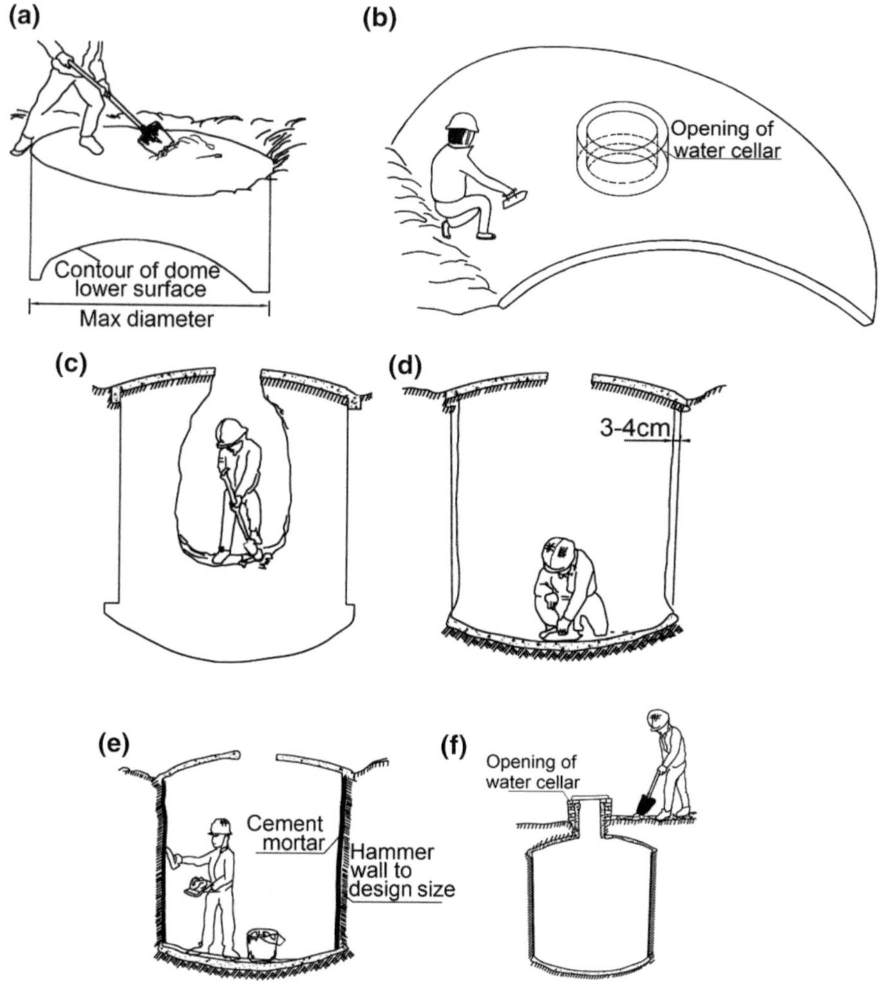

Fig. 4.4 Construction procedure for a circular-shaped cellar with concrete dome. *Source* Zhu et al. (2012)

(3) Then the concrete dome is cast on the prepared ground using the earth as a mold. A hole for the opening of the water cellar is left (Fig. 4.4b).
(4) After 7–10 days of hardening of the concrete dome, excavation is restarted from the opening of the cellar (Fig. 4.4c). Again, excavation proceeds from top to bottom and from the center outwards. An earth layer 2–3 cm thick on the outer diameter of the wall is left and compacted to the design size by wooden hammer.
(5) Casting of concrete bottom and plastering of the cement mortar on the wall are then carried out (Fig. 4.4d, e).
(6) Build the opening structure for the water cellar (Fig. 4.4f).

Fig. 4.5 Earth mold for casting concrete dome of water cellar

The top dome of the water cellar can also be built with bricks. The traditional way to build the brick dome needs a timber or steel framework that is costly and time-consuming. When implementing RWH projects in the northwest China, the job is done in a cost-effective way. The bricks are laid in a circle, layer by layer, with joints between bricks arranged in an alternate way. When 2–3 circular layers of bricks are in place, cement mortar is plastered on the inside and outside of the bricks for stability of the structure. During construction, a platform supported on the wall is needed for the workers to stand on, see Fig. 4.6.

4.3.1.3 How to Build the Concrete Wall of the Cylindrical Water Cellar Using Brick as Mold

To increase the strength and stability of the water cellar and to prolong the service life especially when the subsoil is weak and/or sandy, a cylindrical shape water cellar with a thick wall can be constructed. Since the water cellar is underground, the earth wall can act as an outer mold for casting the concrete wall. But in this kind of structure an inner timber or steel mold for casting the concrete wall is needed. This has a high cost and is not affordable by most farmers. Besides, the technique for preparing the mold is complicated and hard for local people to manage.

A simplified method using bricks instead of timber or a steel mold has been developed in Gansu and Inner Mongolia Autonomous Region in China and later

Fig. 4.6 Cost effective way for construction of brick dome of a water cellar. *Source* Zhu et al. (2012)

adopted in RWH projects in some African countries. The procedure is described below and shown in Fig. 4.6.

(1) Excavate a circular pit with the design depth and outer diameter of the water cellar.
(2) Prepare standard bricks with a size of 24 × 12 × 5.5 cm. This size brick is common in China and available everywhere in the country but in other countries the size may be different. For instance in Africa the brick size is 25 × 12 × 7 cm. To build one water cellar with volume of 25–40 m^3 needs 400–500 bricks for the mold. The bricks should be submerged in water and wetted before use.
(3) Firstly, cast the concrete floor and then lay bricks on the floor one by one tightly in the form of circle. The brick should be placed with the side of size 24 × 5.5 cm downward and kept equally spaced to ensure a consistent concrete wall thickness between brick circle and the earth wall. To make laying of the bricks in a tight manner, it is necessary to cut the last brick into size to fit the space between the last bricks as shown in Fig. 4.7.
(4) Fill the concrete mixture into the space between the earth wall and brick mold. The concrete should be filled and compacted to avoid any voids. Tools including a steel rod and trowel, etc. can be used.
(5) Lay another layer of the brick circle in the same way, and then fill the concrete again. The operation is repeated. After the first layer of concrete has been cast for 8–10 h the bricks for the mold used for the bottom section can be removed and reused as a mold for the next section. The bricks are recycled for the mold until the completion of casting the concrete wall. The bricks can then be reused for building other cellars.

Fig. 4.7 Use bricks as mold for casting the concrete wall of water cellar

(6) When the casting work is finished, the M10 cement mortar is plastered on the concrete wall to a thickness of 2–3 cm, the plastering takes place 2–3 times, each time a 1 cm layer is added as shown in Fig. 4.8. Finally a finish of mortar, i.e., mortar mixed with fine sand is used.

4.3.1.4 How to Build a Circular-Section Water Cellar with a Thick Wall on Unstable Subsoil

When the soil is weak and/or sandy it is difficult to excavate a cylindrical pit with a vertical wall for building a water cellar. The so-called "hollowing out method" for excavation cannot be worked under these conditions so instead an "open excavation method" is used. To ensure the stability of the side wall, it is necessary to have sloping sides to the pit for safety reasons. In this case, the earth itself cannot be used as the outer mold for casting the concrete wall and an outer mold made with timber or steel would normally be installed. This will increase the cost and make the construction procedure more complicated. A better solution is to adopt the surface tank instead of an underground one to reduce the depth of excavation. If due to some reason an underground tank has to be adopted then it is suggested to use prefabricated components for building the side wall and the top cover structure.

However, in case the side wall can be kept vertical when excavating the pit to a limited depth of 1 m, for example, then the above-mentioned use of bricks as a mold can still be used. During their assignments in Nigeria and Kenya for implementation of the RWH demonstration projects, the technical assistants of

Fig. 4.8 Plastering the wall of the cellar with cement mortar

GRIWAC developed a cost-effective method for building the water cellar. This method has led to the construction of a circular-section water cellar in unstable subsoil becoming easier, simpler, and with lower cost.

With this method, the concrete side wall can be cast using bricks as an inner mold and earth as outer mold section by section to avoid collapse of surrounding soil of the pit. The construction procedure is described as below:

(1) Prepare 400–500 bricks of the same standard and size as described above.
(2) Determine the center point of the cellar and draw the circle on the earth to indicate the outer diameter of the side wall of the water cellar.
(3) Excavate this circular pit with a vertical wall to a depth of 1 m or a little more. Keep a careful observation of the pit slope to make sure it remains stable without any risk of collapse.
(4) Use the bricks as the inner mold to cast the concrete wall with designed thickness from bottom to top using the same procedure mentioned above.
(5) After 8–10 h following casting of the first section, remove the brick mold. Continue to excavate the pit for another section of same depth and diameter. The bricks are again used for the mold and the concrete wall is cast in the same way. Special care should be taken in making proper joint of the concrete wall between the two sections.
(6) Repeat the same procedures until the concrete wall reaches the designed depth. The bottom soil is to be compacted with depth of 30 cm. First the soil layer of 30 cm is loosened and then compacted by 2 layers with 15 cm each. Then a concrete bottom 10–12 cm thick is cast.

(7) Plaster 3 cm of cement mortar of grade M10 on the wall and the bottom. The plastering is done in 3 layers with each 1 cm thick. At the joints between sections of concrete wall special care should be taken in plastering. All the voids should be filled in with mortar and the concave surfaces carefully leveled. It is preferable that anti-seepage glue is mixed in with the mortar to improve waterproof performance. Finally, a layer of cement grout (cement to water ratio of 1:1) is washed on the surface.

(8) For this kind structure, a reinforced concrete slab is prefabricated and installed for the top structure of the water cellar.

This method of construction has proved the reliability for building water cellars with volume no greater than 50 m^3 (see Fig. 4.9). So far there is no experience using this method for building water cellars having volumes larger than this. It is not recommended to apply this methodology for construction of cellars larger than 50 m^3 in unstable subsoil until any further new development of this technology.

Another approach to building the concrete wall of the water cellar in a pit with sloping sides is to use prefabricated concrete blocks again avoiding a costly mold and its supporting framework. Figure 4.10 shows the prefabricated blocks for the circular concrete wall. The blocks are bonded with cement mortar of grade M10. The finished surface of the structure should be plastered with cement mortar 3 cm thick and finally a layer of cement grout be washed on.

For the top structure of the water cellar with a thick concrete wall, in addition to the dome structure built with cast-in-place concrete or bricks, prefabricated reinforced concrete slabs can be used. Usually, it can be cast using the ground as the mold and then installed by either manually or with a crane.

Fig. 4.9 Excavation and casting of concrete wall of water in unstable subsoil

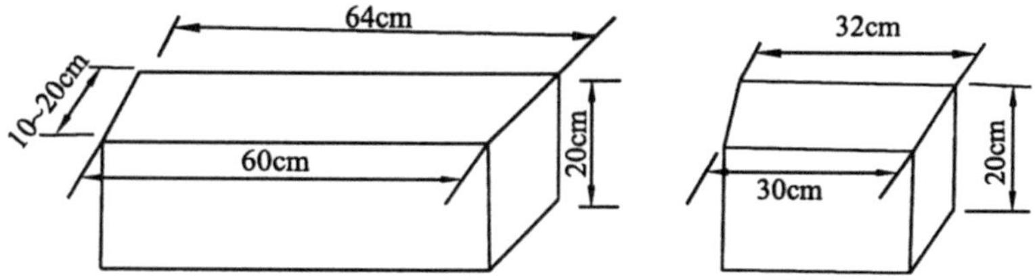

Fig. 4.10 Prefabricated concrete blocks for building circular walls of water cellar

4.3.1.5 How to Build a Rectangular Water Cellar

When building a rectangular water cellar, the "open excavation method" is adopted. To support the earth and water pressure applied on the sidewalls, a gravity-type retaining wall is used. The retaining walls can be built with stones, bricks, or concrete (prefabricated block or cast-in-place).

When the wall is cast in situ, the inner mold for the side wall of rectangular tank is in plain shape, which has much less resistance to the pressure of the fresh concrete compared to a circular shape mold. Besides, thickness of side wall in rectangular water cellar and also the pressure of the fresh concrete when casting is much greater than that in the circular one. Therefore, the bricks can no longer be used for the mold. Instead the conventional timber of steel mold with a supporting framework has to be used. Figure 4.11 is an example of this kind of mold and supporting framework.

When casting the side wall, concreting should be carried out continuously. If any break happens, it should not be longer than 3 h before continuing the concrete casting. If the pause is longer than that, then the surface should be treated to ensure a good connection of the old and new concrete. The concreting has to be stopped for 24 h and the surface of the old concrete is roughened and then cleaned with high-pressure water or with a steel brush. A layer of cement mortar is poured on the treated surface, and the concreting continued. Removal of the mold can be done after 1–2 days in the summer and 2–3 days in the winter after casting.

Fig. 4.11 Mold for casting side wall in rectangular water cellar

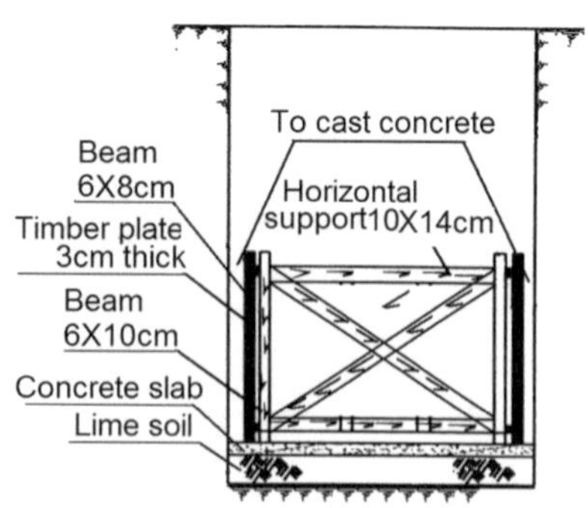

The pressure of the side wall and the bottom slab on the ground is very different. To avoid cracking due to differential subsidence, there is often a joint between the wall and the bottom slab. The wall is usually built prior to the bottom slab. The waterproof sealing materials should be placed before casting the bottom slab. A strip of 3-ply asphalt felt with 4 layers of oil coating and with width equal to the design thickness of the bottom slab is placed along the lowest portion of wall surface. Then the bottom concrete slab is cast.

To avoid using a mold and supporting framework, the wall can also be built using masonry or concrete blocks. In areas with mild climates, bricks can also be used. To ensure a water-tight structure, it is important to lay the masonry in a proper way. Firstly, cement mortar is poured on the ground with a thickness of 2–3 cm and the stones are laid firmly on the mortar with the larger surface downward. Mortar should be poured into all the voids in between the stones. When the first layer of stone is finished, a layer of cement mortar is poured on top of it and then the next layer of stones is laid. It is important to make sure that the joints between stones should be arranged in an alternate way and any through joints should be strictly avoided. When the wall reaches its design height, pointing of the joints should be carried out for the whole surface of the wall. Curing of the joint pointing and the masonry is followed after setting of the mortar. Concrete blocks of regular size and shape therefore make it easier to build.

In some projects, people plaster a layer of cement mortar on the surface of masonry intending to improve the anti-seepage performance. However, since the stone itself is impervious, if the stones are bonded with cement mortar tightly and pointing of the joints between stones is done properly, this layer of cement mortar is not necessary.

For the top structure either a reinforced concrete slab or arch structure can be used. Using the arch structure can save reinforcement but the excavation depth will increase. Usually a reinforced concrete slab is used. These are prefabricated using the ground as a mold and installed manually or by a crane. When the span of the cellar is large, the top structure of slab-beam-column system should be used. To reduce costs and simplify the construction procedure it is suggested to limit the span of the tank to 3 m so the need for a beam can be avoided and the slab directly installed on the wall.

4.3.1.6 Method for Building a Water Cellar in Rock

In the case of a soft rock base, excavation can be carried out manually using a drill and pickaxe. When the base rock is hard, a drilling and blasting method has to be applied. To prevent damage to the bedrock and the adjacent structures, it is better to dig a shallow hole and to use small dose of blasting material. When the excavation approaches the design size of the pit, digging manually instead of blasting is suggested. Blasting with explosives requires qualified and experienced workers and is a relatively dangerous operation. It is strongly advised to engage professionals to do the job to ensure the quality of work and avoid any accident.

If the bedrock is solid and unjointed, plastering cement mortar directly on the rock surface is good enough to obtain a water tight and stable structure. Before plastering the mortar, the rock surface has to be leveled and wetted. Plastering should be undertaken several times. The thickness of the mortar for each layer is around 1 cm and the total thickness should be enough to envelop all the fissures and cracks. M10 is the suggested grade of cement mortar.

If the rock is fractured and unstable, a retaining wall built with concrete or masonry is necessary for supporting the structure. The construction method is similar to that used for construction using a retaining wall in unstable soil.

4.3.2 Method for Building a Water Cave

4.3.2.1 How to Build a Water Cave in Firm Subsoil

Water caves are usually located in areas with relatively firm subsoil. Before building starts, a platform 3–4 m wide is firstly prepared outside of the water cave site for temporary piling, access, and removing the excavated earth. Excavation of the water cave starts from the top arch part and then the lower part. After finishing the excavation of the arch (part I in Fig. 4.12), plastering with cement on the arch and building the supporting arch and entrance structure should be carried out. Then the lower part (part II in Fig. 4.12) can be excavated and again the plastering of the wall and floor undertaken. Afterwards, the arch over the water tank should be excavated (part III in Fig. 4.12) and plastered. Finally the remaining part (part IV) is excavated and the wall and floor completed. Plastering with the cement mortar usually requires three 1 cm layers. After plastering, a layer of cement grout is washed on the surface to improve its waterproofing.

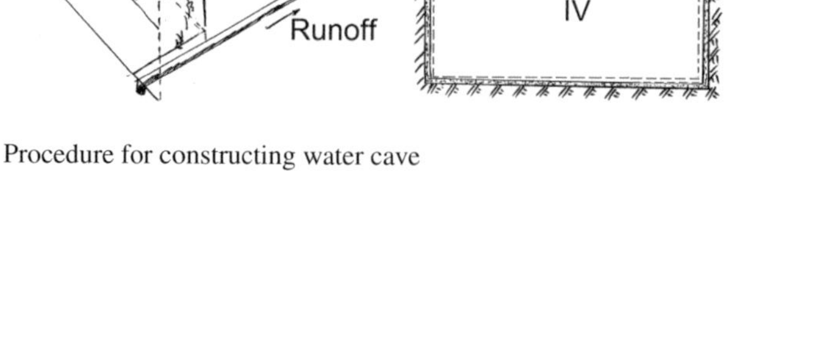

Fig. 4.12 Procedure for constructing water cave

4.3.2.2 How to Build a Water Cave in Bedrock

Water caves in bedrock often take the shape of a tunnel. The most common method of tunneling involves drilling and explosives. The first step is to prepare the tunnel entrance, which is usually lined with concrete or masonry before workers enter to drill further. The height of rock overburden above the arch should not be less than 5–8 m, depending on the firmness of the rock. If the rock is not thick enough then the whole overburden rock should be excavated until the requirement is met. It is very important to check if all shot holes have exploded and there is no blow-out hole remaining. Excessive excavation should be avoided as far as possible. For the rock close to the design dimensions, manual excavation using a drill and pickaxe, etc. is recommended. If the rock is unconsolidated and can easily collapse, supports to the excavated tunnel need to be erected. Tunneling is a highly skilled and professional work. It is strongly advised to carry out construction under the guidance of a qualified specialist team.

4.3.3 Construction of a Surface Water Tank

A surface water tank has its top above or at the same level with ground. Its top can be sealed or open. In China these tanks usually have diameters larger than 5 m. Construction of the surface tank is carried out in a conventional way. The open excavation method is adopted when cutting the pit. Attention should be paid to keeping the slope of the excavated pit stable. If necessary, a slope of 1:0.2–0.5 (horizontal to vertical) should be adopted when excavating the pit. The wall of the surface tank is usually built with masonry or cast-in-situ concrete. In the latter case, both the outer and inner molds are made up of timber or steel plate and should be supported firmly. To avoid using a costly mold and supporting framework for the concrete wall, prefabricated concrete blocks can be used. Building of surface tanks requires a large amount of concreting or masonry work. Normally it is done in several cycles. Therefore, proper treatment for the joints is important.

The top structure of the surface tank can be a simple reinforced concrete slab when the span of the slab is not bigger than 3 m. Otherwise the slab-beam or even the slab-beam-column system should be used. In the rectangular surface tank, a masonry-built arch structure can also be used. In this case, a framework to supporting the arch structure temporarily and a platform for the workers to stand on are necessary. If a concrete arch structure is used it can be prefabricated on the ground and then installed. The reinforced concrete slabs used for the top structure of a rectangular tank can also be prefabricated and installed. To handle and install the slab manually, the tank is designed with a limited width to have the slab shorter and lighter in weight. To build the cover of a circular surface tank, a cast-in situ reinforced concrete slab is usual. Sometimes the components of the slab-beam or slab-beam-column system can also be prefabricated on the ground and then installed manually or with a crane.

4.4 Operation and Maintenance of Rainwater Harvesting System

4.4.1 Operation and Maintenance of Rainwater Catchment

Proper operation and maintenance (O&M) of the rainwater catchment is essential to keep the catchment in a good condition and to prolong its service life and efficiency. O&M of the rainwater catchment includes the following:

(1) Cleaning the courtyard catchment surface before the rain by removing as much dirt and debris as possible.
(2) Close inspection for any cracks in the courtyard catchment surface: Any cracks found should be cleaned and resealed. The sealing material in the joints between the concrete blocks is easy to be damaged and should be checked and repaired carefully.
(3) Regular inspection for any damage to the collection surface built with concrete, bricks, or masonry. Any minor cracks found should be filled with water sealing materials like cement mortar, asphalt, or plastic tar. Before sealing the crack, any dirt inside should be cleaned out. If the crack is too narrow then it has to be enlarged to enable washing out dirts and filling with the sealing materials. When cement mortar is used as filler, the crack should be wetted before filling. If asphalt or oil-based material is used, the crack should be dried by blowing and heating. If crack is wider than 1 cm, or significant displacement of the pavement has happened, the causes of cracking or displacement should be studied carefully. Usually a major crack or displacement is caused by the following:

    - An underground cavity may exist causing subsidence. A cavity in the subsoil may be caused by erosion of fine particles by seepage flow or by soil piping or flowing. In this case, the displaced pavement should be removed to expose the cavity, and then earth is filled in the hole with careful compaction layer by layer. To avoid soil erosion happening again, plastic film can be placed on the surface of the filled part.
    - In cold regions, water may infiltrate into the soil through poorly made or damaged joints, causing frost heaving in the winter and subsidence when thawing during the following spring. For a major displacement of the catchment caused by frost action, repair work should be done in the summer season. The pavement has to be removed and the subsoil leveled and covered with plastic film or asphalt felt. The pavement should be repaired fully to ensure it is watertight.

(4) For highway catchments, a small temporary dam should be built in the drainage ditch for diverting the runoff into the tanks. However, in case of a heavy storm the dam should be removed to ensure the normal drainage function for the safety of the highway.

(5) For catchments using buried plastic sheeting, inspection should be undertaken to see if the covering material is in good condition and maintenance done, if necessary.
(6) For the large-scale purpose-built concrete catchment, fencing around the catchment may be built to protect it from damage by animals or other reasons.
(7) For the collection and conveyance ditch/channel, silt and weeds should be cleared regularly. Maintenance of the scoured portion of the ditch should be done on time. If damage happens in the linings due to subsidence of the ground then repair work for the subsoil should be dealt with first. After the ground is firm then rehabilitation of the lining should be carried out.

4.4.2 Operation and Maintenance of the Storage Subsystem

4.4.2.1 Measures for the Safe and Efficient Storage of Rainwater

(1) Before the rain, the state of the inflow channel/pipe, sedimentation basin, and the screen in front of the tank inlet should be carefully inspected. Any dirt, debris, and branches should be cleaned out.
(2) When storing runoff, the water level in the tank should be monitored. If the water level exceeds the safe level, inflow of the runoff should be immediately diverted. For the underground tank, whose water level is not visible, a detection rope or electrical signaling system may be used.
(3) The facility for diverting the runoff away from the tank such as the diversion ditch and the gate should always be ready for use.
(4) In cold regions, it is not possible to store water in a surface tank in winter unless a special design is used and treatment made to avoid damage caused by frost action. Normally, water is drained before the first frost occurs. The water level in the underground tank should be lower than the maximum frost penetration.
(5) It is important to keep at least 20–30 cm deep of water in the tank to prevent the bottom from drying out and cracking. Do not empty the tank thoroughly except when removing the silt.
(6) A surrounding fence around surface tanks should always be maintained in good condition to avoid accidents and possible drowning.

4.4.2.2 Maintenance of the Tank

(1) Routine inspection of the tank structure should be made to identify any cracks in the wall or base. This should always be carried out before the rainy season starts. The water level should be checked regularly. If the water level has lowered abnormally, it probably means there is some major leakage. In this case,

it is important to locate the leak and take measures to stop it. The following ways can help locate the problem.

- When the water level lowers fast and stops at a certain depth, then the seepage is probably occurring through the wall at this depth.
- If water empties fully and rapidly, then the leakage is probably occurring through the wall and tank floor.
- If necessary, water in the tank has to be drained to identify the place of leakage.

(2) Method to repair crack in the structure:

- In the case of a minor leakage due to small cracks on the surface of the concrete or masonry, the crack can be filled with sealing materials such as the cement mortar, asphalt, or plastic tar.
- If there are honeycomb like voids and pits occur on the surface of the cement mortar or concrete due to poor construction quality, a new layer of cement mortar 3 cm thick may be plastered on the surface or a layer of concrete in 6–8 cm thick added after the original surface has been roughened and thoroughly cleaned.
- If major leakage has occurred, the ground behind the concrete or masonry structure should be carefully inspected to see if any cavity exists in it. A cavity may be caused by soil erosion under seepage flow or soil piping. In a case where a cavity exists in the subsoil, it is necessary to first remove the concrete or masonry structure. The reason for seepage flow or soil piping should be found out and measures should be taken to prevent the problem happening again. The cavities in the soil should be carefully filled up with compacted soil.

4.4.2.3 Sedimentation Management

Sediment in the tank should be cleaned out at least once a year. There are a number of methods to clean out the silt, for instance, to use a sewage pump, to install a drainage pipe at the lower part of the tank, and to dig out the silt manually. The sediment depth in the tank should not exceed 1 m. After each heavy rain, silt in the settling basin should be cleaned immediately. Water remaining in the basin should be used. Any debris on the screen should also be cleaned regularly.

4.4.3 Maintenance of Water Supply and Irrigation Facilities

4.4.3.1 Maintenance of the Hand Pump

Before operation of a high-pressure hand pump, apply lubricating oil to moving parts of the pump, and pour 6–8 drops of edible oil into the cylinder to help smooth its operation.

- When the temperature is lower than 0 °C, empty the pump completely after operation to prevent frost damage.
- During operation of the pump, regularly inspect all connecting bolts and tighten when necessary.
- After use, wash and clean the pump, add lubricating oil to all running parts. Store the unit in a dry and ventilated room.

4.4.3.2 Maintenance of Electric Pumps

- Before operation, check the power cables and connection. Make sure that all the bolts are fastened properly and no oil leaks out.
- The insulation resistance of the motor should be higher than 5 MΩ and should be inspected regularly. An automatic circuit breaker is required to prevent damage in the case of a short circuit.
- After connecting to the power, run the pump without loading for some seconds (not more than 60 s.) and check if the motor starts and the running direction are correct.
- Never use an electric cable for hanging the submerged pump. A cord is tied at the handle of the pump for moving the pump. The pump should not be submerged deeper than 10 m and should be kept at least 50 cm above the tank bottom. The water intake of the pump should be equipped with a mesh to prevent blockage by debris.
- During operation, the water level should be carefully observed especially when the water level in the tank is dropping. Make sure that the pump remains at least 50 cm below the water surface. Running the pump without water would lead to over-heating and damage of the pump.
- When the pump stops operations, it should be taken out and the silt inside the pump should be cleaned out. Then the pump should be stored in a dry and ventilated room.
- Regular service and maintenance of the pump to replace any worn parts is important. When re-assembling the pump, inspection on all seals is required.

4.4.3.3 Maintenance of Pipeline

- If any seepage is detected along a buried pipeline, the operation should be stopped. The seepage point found and repaired as soon as possible.
- Do not bend or fold the pipeline when moving it in the field. Surface pipes should be protected from potential damage by vehicles etc.
- When the operation is finished, the pipes should be inspected for any damage from aging or solar radiation. The pipe should be cleaned and stored in a dry, cool, and ventilated room.

- After irrigation, open the end-cap of pipeline for flushing. Regular flushing is required after each irrigation cycle.
- To prevent creation of a water hammer effect, open and shut off the valves of the pipeline in a slow manner.

4.4.3.4 Maintenance of Drip Systems

- To prevent leakage of drip lines caused by improper installation and moving. Roll up and unroll the drip line properly during installing or storing.
- Gently switch on or off the drip line according to irrigation design when starting operations. Silt inside the tube should be flushed out regularly to prevent clogging.
- When liquid fertilizer is applied, flushing of the laterals should be taken within 10–20 min after irrigation is finished to minimize the precipitation of chemicals on the inner wall of the pipe.
- Inspection and cleaning of the filter is to be carried out regularly. When the pressure difference between inlet and outlet of the filter is beyond 3–5 m, it indicates clogging has occurred inside the filter. The drainage valve should be opened and the silt flushed out or the strainer is taken out for cleaning. After finishing irrigation, uninstall strainer of the filter and store it after cleaning and drying.

4.4.4 Water Quality Management

Although the quality of water supply from RWH systems in China has been greatly improved since the start of the RWH project, from a strict technical point of view, the quality of the stored still does not always meet the national standard for drinking water. According to test results, the chemical and toxicological indexes generally meet the standard but the biological indexes including the total bacteria count and the coli bacillus counts seriously exceed the standard. The main reasons of the quality problem of the stored rainwater are as follows:

- In northern China, the roof and courtyard catchment do not get frequent washing as heavy rainfall is rare. Dirt, bird drops, and vegetation growing in between the tiles breed biological matter, thus polluting the runoff.
- In many of the RWH projects gutters and downpipes have not been installed. Runoff from the roof is mixed with the ground runoff, reducing the quality.
- Measures such as first flush devices and adding chemicals for improving water quality have not yet been adopted by most of the RWH owners.
- The effective and affordable water purification equipment is not yet widely available for cooking and drinking water.

To improve the water quality, the following measures need to be taken:

- For the newly built RWH project, gutters and downpipes should be installed to store the runoff from the roof and ground in separate tanks. The gutters and down pipes that are not available in the old systems should be installed as soon as possible.
- Health education among the rainwater users has to be strengthened. Measures that have been proved effective like the first flush, adding flocculation, and sterilization chemicals have to be promoted.
- The solar cooker for boiling water for drinking and cooking need to be further promoted and the design improved for more convenient use.
- Effective and affordable water purification for treatment of stored rainwater for drinking and cooking purposes needs to be developed.
- Quality monitoring of the stored rainwater has to be carried out on a regular base, with an institution responsible for this mission appointed and relevant regulations drawn up.

References

Irrigation and Drainage Department, Ministry of Water Resources (IDMWR) and Rural Department, Ministry of Finance (RDMOF), ed. Concise textbook for rural rainwater harvesting project. Beijing: China Water Resources and Hydropower Press; 2001 (in Chinese).

Zhu Q, Li Y, John G. Every last drop—rainwater harvesting and sustainable technologies in rural China. Rugby, UK: Practical Action Publishing Ltd; 2012

Chapter 5
Rainwater Harvesting Techniques for Irrigation

Qiang Zhu

Keywords Rainwater harvesting irrigation · Low rate irrigation · Deficit irrigation · Water-saving irrigation

5.1 Principle of Rainwater Harvesting Irrigation

Experience in China over recent decades has shown that RWH cannot only provide water for domestic use but also for supplemental irrigation. As already noted, RWH systems are mini-sized water resource schemes. The amount of rainwater stored in a RWH system is generally very limited requiring very efficient use. RWH-based irrigation is therefore a special kind of water-saving irrigation. It uses much less water than conventional water-saving irrigation and is more appropriately called "low rate irrigation" (LORI). Investigations have shown that LORI water supplied to the crop only consists of 10–15 % of the total crop water consumption over the whole growing season. Most of the water used by the crop is from natural rainfall. Some experts doubt that such a small amount of water can have any effect on the crop yield. Research by Gansu Research Institute for Water Conservancy (GRIWAC) has shown that in fact the crop performance with or without this small amount of supplementary water supply (irrigation) is very different. Table 5.1 shows the results from demonstration projects in the late 1990s and early 2000s, which showed the difference on the crop yield after applying small amount of irrigation (GRIWAC et al. 2002).

Q. Zhu (✉)
Gansu Research Institute for Water Conservancy, Lanzhou, China
e-mail: zhuq70@163.com

Q. Zhu et al. (eds.), *Rainwater Harvesting for Agriculture and Water Supply*,
DOI 10.1007/978-981-287-964-6_5

Table 5.1 Test and demonstration results of RWH-based irrigation on crop yield and WSE. *Source* Zhu et al. (2012)

Crop name	Irrigation water amount (m^3/ha)	Yield (kg/ha)	Yield increase percentage (%)	WSE (kg/m^3)
Wheat	225–300	1990–6843	10.5–88.3	1.65–3.9
Corn	375–405	2940–9050	19.6–88.4	3.11–5.7
Potato	405	27,696	30.6	10.95
Millet	300	2583–2750	20.5	1–1.62
Broom corn millet	300	4011–4258	6.8–13.4	1.5–1.55
Oil sunflower	450	2626–3000	19.8–65	1.65–3.45
Linseed	225	1590–2505	44.7–120.6	3.03–6.08

In the above table, the RWH irrigation amount ranges between 225 and 405 m^3/ha, while in the conventional irrigation in the loess area of Gansu, irrigation amount for field crops is around 1500–2000 m^3/ha (about 6–9 times more). But with this small amount of irrigation water, crop yield can be increased in the range of 10–90 % or about 40 % on average. The reason for the effectiveness of using such a small amount of irrigation water on crop yield can be explained as follows. Only with the water supply from the RWH system, can the crop avoid fatal damage due to serious water stress during critical periods. The crop thus survives to be able to effectively absorb natural rain in the rainy season. If there is no water provided to the crop during the critical dry period, the crop will be seriously damaged or wither completely and any rain falling later in the season would be meaningless. Therefore, the role of LORI is to raise the overall water use efficiency (WUE) of the natural rain, the only water source for the rain-fed agriculture. Here, WUE refers to the crop yield per unit of water consumed, which is composed of the natural rainfall in the growing season, soil moisture attracted by the crop and the RWH-based irrigation water supply. The last two sources are also derived of course from the natural rain stored in the soil and in the rainwater tank, respectively. Research and demonstrations have shown that with LORI the WUE can be higher than for purely rain-fed agriculture by 23–59 % and 15–35 % for wheat and corn, respectively (GRIWAC et al. 2002). This provides strong evidence for the above assumption, which can also be taken as the theoretical basis of RWH irrigation practice.

5.1.1 Why LORI Is so Efficient

The principles of RWH irrigation can be summarized as follows:

(1) RWH irrigation aims at getting the maximum profit (in the form of production or cash) from each unit of rainwater. In other words, the WUE and the water supply efficiency (WSE) should be high.
(2) RWH irrigation is based on the principle of deficit irrigation (limited irrigation). This means the crops water demand is not fully met at all growing

stages, but only partly met at some critical periods of growth. These so-called critical periods are those periods during which crop is subjected to serious water stress when damage to the crop would be unrecoverable even with ample water supply or natural rain at a later stage.

(3) RWH irrigation uses highly efficient irrigation methods which use a very limited amount of water. These include innovative indigenous methods which are simple, affordable and highly efficient as well as modern micro-irrigation techniques including drip, mini spray and bubble irrigation.

(4) The water application is targeted at the root zone of the crop to reduce soil evaporation.

5.1.2 Basic Concepts

The RWH irrigation follows the principle of deficit irrigation. In this section, some basic concepts will be introduced and a case study from Gansu was provided.

It is well known that irrigation has helped to greatly promote agricultural development. In the past time, irrigation management was aiming at getting the highest crop yield by fully meeting the crop water demand throughout the whole crop growing period. Crops use water mainly from the soil moisture to meet its demand. The field capacity (FC) of the soil is the maximum water content that the soil can hold without seepage loss into the groundwater. The wilting point (WP) is the soil water content, at which permanent wilting of the crop happens. Differences between FC and WP are taken as the total available water (TAW) content. It means that crops can survive when the soil moisture is between these two thresholds. When the crop abstracts the soil moisture and makes it down to a certain level, the crop begins to suffer from water deficit, which would eventually affect the yield. This critical soil moisture is called the "readily available water content (RAW)". In Chinese literatures, it is also called the "lower limit of appropriate water content (LLAW)". Crops would not be subjected to any water stress when the soil moisture equals or is larger than LLAW. LLAW varies with crop and soil type and is determined through testing. To create optimum moisture conditions for a highest crop yield, the soil moisture should be kept in the range between FC and LLAW. When soil moisture drops down to LLAW due to water attraction by the crop root system, irrigation with maximum water amount of FC–LLAW should take place. Irrigation management of this kind is called the sufficient irrigation, under which Crop water demand is fully met for the whole crop growing season. Figure 5.1 illustrates the soil moisture curve and the times of water application under condition of sufficient irrigation. The figure also illustrates the method of irrigation scheduling under sufficient irrigation.

Although the sufficient irrigation can produce the highest yield, the irrigation water productivity (IWP) per unit used will not be so high. When water resources become a critical factor for social and economic development, we have to think about how to use water in the most economical way. That is to get the optimum

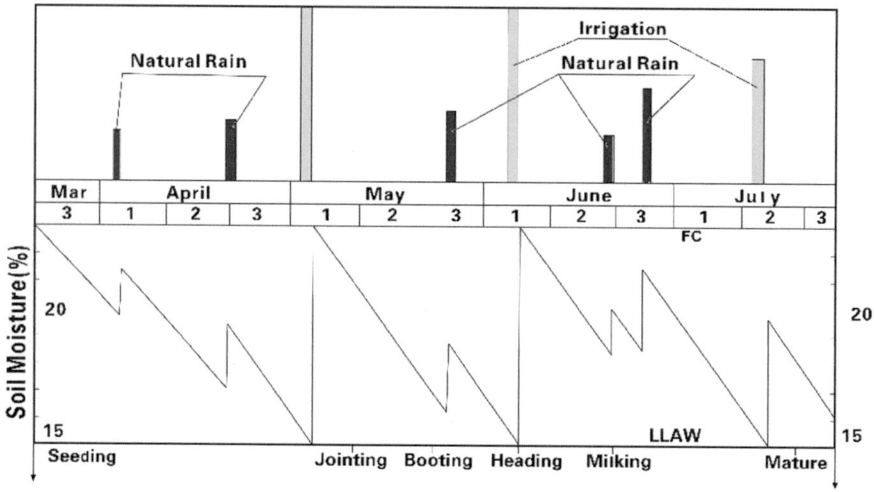

Fig. 5.1 Soil moisture curve and irrigation scheduling

Fig. 5.2 Wheat yield versus crop water consumption in Wuwei Gansu

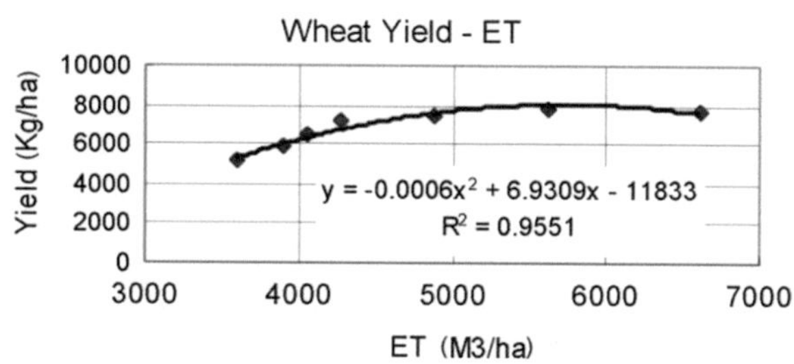

use from the water input. The irrigation water productivity function (WPF) is understood as the response of crop yield to water or the relationship between the output and the water input.

In the past century, people have carried out many studies on WPF. A simple form of WPF is the relationship between the crop yield and irrigation amount over the whole growing season. Figure 5.2 shows a common curve for this relationship, derived from experimental results using wheat yield versus irrigation amount in the Wuwei Municipality, an arid area in the Gansu province (Liu 1987). The curve displays a parabolic function.

The irrigation quota versus yield shown in Fig. 5.2 was for an area of wheat irrigated from a conventional reservoir-canal system source. Figure 5.3 shows the yield of intercropped wheat–corn versus irrigation quota using RWH irrigation (low-rate irrigation) (GRIWAC et al. 2002).

In Fig. 5.3, when the irrigation water amount increases, the yield increases, but the water productivity expressed as yield per irrigation water amount decreases. Figure 5.4 shows the WSE (equals to yield increase divided by irrigation water amount) versus irrigation water amount.

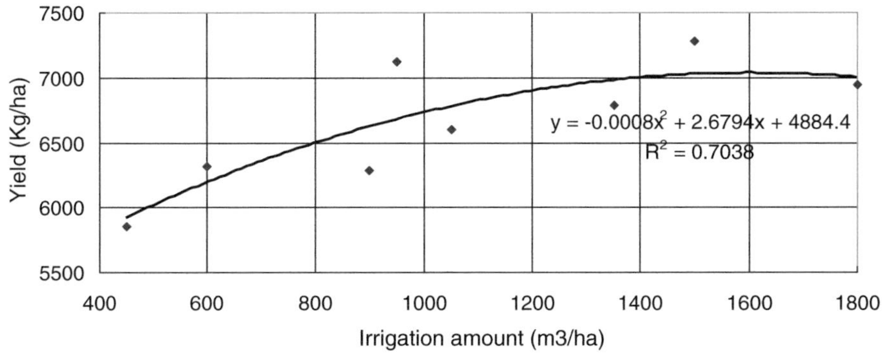

Fig. 5.3 Yield of intercropped wheat–corn versus irrigation amount under RWH irrigation in Dingxi County, Gansu

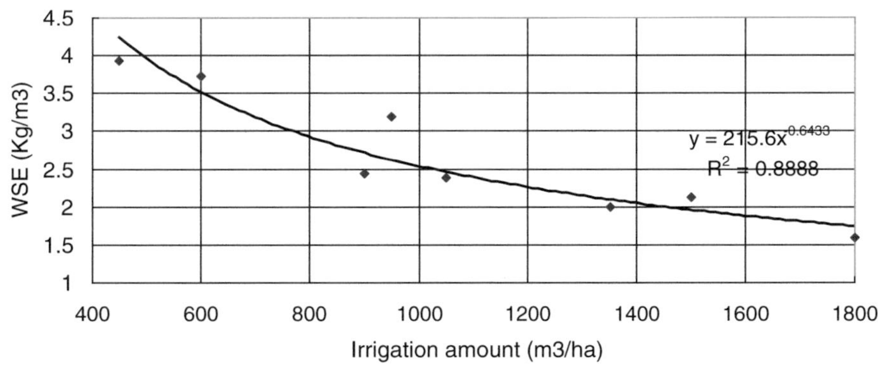

Fig. 5.4 WSE versus irrigation amount

WSE is not only affected by the irrigation water amount but also affected by the time of irrigation taking place. To make clear the effect of irrigation timing on the yield, the simplest form of IWP as shown in Figs. 5.2 and 5.3 is not enough. We should study the relationship between the crop yield and the water consumption in different stages of crop growth. There were many models to describe the timing effects on the yield. Here, we introduce two kinds of model, the multiplying model and the adding model. As representative of these two models, the Jensen (1968) model and Blank (1975) model were shown in Eqs. 5.1 and 5.2.

Jensen model:

$$\frac{Y_{\mathrm{a}}}{Y_{\mathrm{m}}} = \prod_{i=1}^{n} \left(\frac{ET_{\mathrm{a}i}}{ET_{\mathrm{m}i}} \right)^{\lambda i} \tag{5.1}$$

Blank model:

$$\frac{Y_a}{Y_m} = \prod_{i=1}^{n} K_i' \left(\frac{ET_{ai}}{ET_{mi}} \right) \tag{5.2}$$

In the above equations, Y_a and Y_m are the yield under deficit irrigation and sufficient irrigation, respectively; ET_{ai} and ET_{mi} are the water consumption in the stage i of crop growth under deficit irrigation and sufficient irrigation, respectively; n is the number of crop growing stages; i denotes the ith stage of crop growth; λ_i and K_i' are the sensitivity coefficient of yield to water deficit in the ith growing stage in the Jensen and Blank model, respectively. The sensitive coefficient indicates the extent of impact of water deficit in different crop growing stages on the yield. The higher the sensitivity coefficient, the larger the impact of water stress on the yield.

With these two models, the optimum timing of water application can be worked out. If the available water in the tank (or from other water source) is less than the crop water demand, then how to distribute the available water in the different crop growing stages to get maximum profit is one of the topics for the deficit irrigation approach. This can be solved with the dynamic programming or linear programming.

A study on the optimum irrigation scheduling using the deficit irrigation approach and on the economic irrigation water use was carried out in the Minqin County of western Gansu from the year of 1991 to 1993 (Zhu et al. 1994). For this purpose, an experiment on the response of wheat yield to water deficit in various growing stages was conducted. Minqin is an arid area with annual precipitation of about 110 mm, and the agriculture production is completely relied on irrigation. The tested crop was spring wheat. The irrigation amount for fully meeting the crop water requirement is 480 mm (4800 m^3/ha), and the related crop yield is 7280 kg/ha. In the experiment, the crop growing period was divided into six stages, namely, tilling, jointing, heading, booting, milking and milky mature. In the growing period, one to three stages were subjected to water deficit by deleting water application in the stages. The sensitivity coefficients for the 6 stages were analysed based on the test results, which are shown in Table 5.2.

The available irrigation amount was taken as 450, 375, 300, 225, 150 and 75 mm, and each application rate was assumed to be 75 mm (50 m^3/Mu). Using the dynamic programming, the frequency of water application in various growing stages to get a maximum yield for each level of water availability is shown in Table 5.3 (Jensen model).

Table 5.3 shows that the maximum yield per irrigation amount or the IWP was the highest when the irrigation amount was the lowest. However, the economic

Table 5.2 Sensitivity coefficient of water deficit in different crop growing stages. *Source* Zhu et al. (1994)

Model	Stage 1 tilling	Stage 2 jointing	Stage 3 heading	Stage 4 booting	Stage 5 milking	Stage 6 milky mature
Blank	0.11	0.0759	0.221	0.449	0.052	0.109
Jensen	0.0811	0.0613	0.115	0.392	0.067	0.073

Table 5.3 Water application frequency in the growing stages for different water availability. *Source* Zhu et al. (1994)

Irrigation water amount (mm)	Yield (kg/ha)	Yield per irrigation amount (kg/m^3)	Stage 1	Stage 2	Stage 3	Stage 4	Stage 5	Stage 6
450	6487	1.44	1	1	1	1	1	1
375	6480	1.73	1	1	1	1	1	0
300	6197	2.07	0	1	1	1	1	0
225	5292	2.35	0	1	1	1	0	0
150	3765	2.51	0	0	1	1	0	0
75	3038	4.05	0	0	0	1	0	0

irrigation water amount is not necessary related to the maximum IWP. The irrigation water amount that produces a maximum net benefit per water use was regarded as the economic irrigation water use. The net benefit is defined as the production value minus the production cost. The latter was grouped into two parts: the fixed costs and the variable costs. The fixed costs mainly included seeds, fertilizers, chemicals, manures, etc., and the variable costs included mainly the labour cost that was related to the yield as well as the water cost related to the irrigation water use. The total cost was expressed in Eq. 5.3.

$$C = C_o + k_1 Y + k_2 M, \tag{5.3}$$

where C is the total cost, C_o is the fixed cost and k_1 and k_2 are two coefficients related to the cost for labour and machine and the water tariff. The net benefit equals to value of wheat minus the production cost. Based on the price and cost investigated at that time, Table 5.4 shows the net benefit and the net benefit per irrigation water use.

From the table, we can see that irrigation water use of 3000 m^3/ha gave the most economic irrigation water use. The related net benefit per water use is 0.87 CNY/m^3, higher than that with the sufficient irrigation by 36 %. With the increase of grain price and the water tariff at present, it can be assumed that the profit obtained by adopting deficit irrigation would be larger.

The above example showed how to quantitatively determine the economic irrigation quota and the related irrigation scheduling by carrying out irrigation experiment and analysis based on the deficit irrigation approach. In China, RWH-based irrigation has proved to be a highly successful practice for deficit irrigation, in which the water productivity is much higher than that of the conventional irrigation. However, quantitative determination of the deficit irrigation scheduling has

Table 5.4 Net benefit and net benefit/irrigation water amount versus irrigation water amount

Irrigation water amount (m^3/ha)	4800	4500	3750	3000	2250	1500	750
Net benefit (CNY)	3067	3039	3044	2599	1869	689	−183
Net benefit/irrigation water amount (CNY/m^3)	0.64	0.68	0.81	0.87	0.83	0.46	−0.24

not yet been feasible owing to the experimental data for RWH-based irrigation being inadequate. So far the irrigation schedule under RWH condition has been determined mainly by testing and empirical methods.

5.2 RWH Irrigation Scheduling

As previously noted, irrigation water from the RWH system typically provides only 10–15 % of the total crop water requirement for the growing season. While in the above Minqin example, the economic irrigation quota comprises about two-thirds of that the total crop requirement for sufficient irrigation. So the extent of water deficit in RWH-based irrigation is much higher than that in the conventional irrigation. The experimental data of deficit irrigation for the conventional irrigation cannot therefore be directly used for RWH-based irrigation. The irrigation scheduling with RWH system is determined mainly using qualitative methods, namely, to conduct field tests using different irrigation quotas and application timings for different crops. The irrigation schedule is then worked out by comparing yields under different conditions and taking water availability in the tank into consideration. RWH-based irrigation scheduling can be carried out using two methods: the testing and analysis method and empirical method.

5.2.1 Testing and Analysis Method

The steps for determining the irrigation schedule using RWH by this method are listed as follows:

(1) The first step is to find out the critical periods of crop growth by carrying out studies on the local conditions (temporal distribution of rainfall and crop water demand, etc.) and consulting experienced farmers. The crop growing periods are divided into several stages, and impacts of water deficit on the yield at different stages are analysed. The stages during which the crop yield is most sensitive to water stress are taken as the critical period for crop growth. The critical period is different for different crops not only because of the nature of crops and soil, but also because of climatic and particularly rainfall availability in the growing season. For example, in semi-arid and sub-humid areas, usually seeding is the most critical stage for applying water to crops because the germination rate and thus the number of surviving plants with or without water application are very different.
(2) With the knowledge of critical growing periods, a series of test for determining better irrigation scheduling will be carried out. Different sets of irrigation timing, irrigation frequency and the quota will be tested. In China, each water

application using RWH systems typically uses about 5 m^3/Mu (75 m^3/ha) during seeding and 10–15 m^3/Mu (150–225 m^3/ha) for the other growing stages. For example, three tests with irrigation quotas of 15 m^3/Mu (with one application during seeding and one application in another growing stage), 25 m^3/Mu (one application during seeding and two applications during other two growing stages) and 35 m^3/Mu (one application during seeding and two or three applications in other growing stages) can be carried out.

For the first 15 m^3/Mu quota, there are two options for the timing of water application. For example, in the test using spring wheat, the second application may be taken in the jointing stage or heading stage. For the irrigation quota of 35 m^3/Mu, the water can be applied at two stages: jointing and heading stage or at three stages: jointing, heading and milking stages. There can be many combinations of options for this test. To reduce work and the amount for testing needed, relevant literature on irrigation testing should be consulted as should experienced farmers to select the most feasible options for testing.

(3) Testing is carried out to compare the different options. In China, there is a national standard for irrigation testing to ensure the experimental results are reliable. The most important points are the following. For each option, the tests should be repeated three times to ensure a reliable result. All the test plots have to have the same soil, groundwater and fertilizer condition, and even the previous crop should be the same to avoid other multiple effects on the results. Each test plot should have area of about 100 m^2 with a rectangular shape. The plots for different tests should be randomly arranged in the field. Between the neighbouring plots, there should be a protection strip with a width of at least one metre to avoid plots influencing each other.

(4) When harvesting, the yield from each plot should be carefully measured. Comparison of yields for the different options can then be made. Irrigation scheduling should be based on the test results, and meanwhile, the economic factors and the capacity of RWH system should be considered. For example, if two water applications with 15 m^3/Mu can increase yield by 75 kg/Mu of yield, while three applications with 25 m^3/Mu can increase yield by 100 kg/Mu, compared to the non-irrigated yield, respectively, then from an economic point of view, the former solution may be better because it has WSE of 5 kg/m^3, while the latter has only 4 kg/m^3. If we have 25 m^3 of rainwater stored in the tank then it would be better apply it as supplementary irrigation water for 1.67 Mu of land instead of 1 Mu. The benefit would be an increased yield of 25 kg of crop produced.

The results obtained from the testing and analysis methods can be replicated in areas having the same natural and agriculture conditions as the place of testing. In the following section, we will introduce some results on the irrigation scheduling from a research project supported by the State Ministry of Science and Technology (MOST).

5.2.2 Brief Introduction to the Result of Previous Researches

The GRIWAC, the Gansu Academy of Agriculture Sciences (GAAS) and Gansu Agriculture University (GAU) previously carried out the project entitled "Technical integration and innovation research on the efficient rainwater harvesting and utilization in semi-arid mountainous areas" from 1999 to 2002. In this project, there was a unit on the irrigation scheduling for the main grain crops in Gansu, including spring wheat, corn, millet, etc. (GRIWAC, GAAS and GAU 2002).

5.2.2.1 Spring Wheat

In the loess plateau of Gansu province, spring wheat is seeded on around March 15 and harvested at the end of July. The total growing season lasts for about 135 days. Test results revealed the crops water demand was about 3470 m^3/ha. Natural rainfall in the growing season is about 194 mm (1940 m^3/ha) so the water deficit accounts for about 1530 m^3/ha which occurs mainly in May and in first half of June. Water supplied during any part of the growing stage will help enhance the yield. However, the capacity of RWH system is too limited to fully meet the crop water demand and so LORI needs to be adopted. Table 5.5 shows the effect of limited irrigation water in the growing season.

From Table 5.5, we can see that irrigation with a smaller amount of water resulted in a lower yield but higher WSE and WUE. So where the land resource is limited it is preferable to use a larger irrigation water up to 400–500 m^3/ha; while if water is the limiting factor it is better to use a smaller amount e.g. 200 m^3/ha of irrigation water to maximize production per cubic metre of irrigation water applied. Table 5.6 shows the test results of the effect of different irrigation schedules on the yield of spring wheat.

We can draw several conclusions from the results of the irrigation strategy shown in Table 5.6.

1. When the irrigation is applied only once, then irrigation in the booting stage is better than in the jointing stage. WSE and WUE of the former are higher than that of the latter by 10–28 % and 15–24 %, respectively.

Table 5.5 Spring wheat yield versus irrigation amount in the growing season

Irrigation water amount m^3/ha	0	200	400	500
Yield t/ha	1.44	1.92	2.06	2.10
WSE kg/m^3	–	2.38	1.55	1.32
Water consumption m^3/ha	2826	2980	3654	3873
WUE kg/m^3	0.51	0.64	0.56	0.54

Table 5.6 Spring wheat yield versus irrigation schedules. *Source* Zhu et al. (2012)

Test treatment	Yield (m^3/ha)	Yield increase	Soil moisture (%)		Consumed water	WSE (kg/m^3)	WUE (kg/m^3)
			Seeding	Harvest			
WCK	1334	–	14.3	12.5	3410	–	0.39
WJ450	1803	469	–	13.7	3577	1.04	0.50
WJ900	2018	684	–	13.6	3912	0.77	0.52
WB450	1934	600	–	15.7	3111	1.33	0.62
WB900	2096	762	–	15.9	3517	0.85	0.6
WJB450	2270	936	–	12.6	3839	2.08	0.59
WJB900	2375	1041	–	13.8	4010	1.16	0.59

Note The symbols used in the Line test treatment are: "W" means wheat test, WCK—reference for non-irrigated wheat; "J" and "B" means water applied in the jointing and booting stage, respectively, "JB" means water applied in both the jointing and booting stages, "450" and "900" means irrigation water amount of 450 and 900 m^3/ha, respectively

2. When the irrigation is applied twice (in the jointing and booting stages), the yield is higher than that with same irrigation quota but concentrated in one application (either in jointing or booting stage) by 13–26 %. So it is better to divide the irrigation into two applications rather than to concentrate into just one.
3. The yield with irrigation of 900 m^3/ha is higher than that with irrigation of 450 m^3/ha by 8–12 %. But the WSE of irrigation water of the latter is higher than that of the former by 35–56 %. So again, if land is not a limiting factor then irrigation with less water results in higher overall production.
4. Irrigation with 450 m^3/ha separated between the jointing and booting stages provides better irrigation results. The WSE is as high as 2.08 kg/m^3.

5.2.2.2 Corn

Corn is one of the main grain crops in Gansu. It has higher yield than wheat because the available rainfall in its growing season is larger, and the growing time is longer than that of spring wheat. Corn is seeded in mid-April and harvested in mid-September. The growing season lasts for about 150 days. The crop water demand over the whole growing season is about 436 mm, while the mean natural rainfall over this period is around 303 mm. Water deficit amounts to about 133 mm, mainly in late May and June.

Farming practice has shown that corn needs less water in the seedling stage. The young plant has a higher tolerance to water stress, and a certain level of water deficit is even beneficial for deepening the plants root system thus enhancing resistance to drought. The most critical period for water demand comes at the flowering stage (big bell-mouthed stage) when the crop is very sensitive to water shortage. In this period, the WSE can be as high as 5.6 kg/m^3. The effect of different irrigation scheduling on yield is shown in Table 5.7.

Table 5.7 Corn yield versus different irrigation schedules. *Source* Zhu et al. (2012)

Test treatment	Yield (m^3/ha)	Yield increase	Soil moisture (%)		Consumed water	WSE (kg/m^3)	WUE (kg/m^3)
			Seeding	Harvest			
CCK	7131	–	14.4	12.3	4012	–	1.78
CJ600	9270	2139	–	12.3	4605	3.57	2.01
CJ1200	8342	1211	–	13.2	4997	1.01	1.67
CF600	9440	2309	–	11.4	4820	3.85	1.96
CF1200	10,000	2869	–	12.7	5109	2.39	1.96
CJF600	9066	1935	–	12.3	4603	3.23	1.97
CJF1200	9507	2376	–	13	5051	1.98	1.88

Note The symbols used in the Line treatment are: "C" means corn test, CCK—reference for corn without irrigation; "J" and "F" means water applied in the jointing stage and big bell-mouthed stage, respectively, "JF" means water applied in both the jointing and big bell-mouthed stages, "600" and "1200" means irrigation water amount of 600 and 1200 m^3/ha, respectively

From Table 5.7, we can draw some useful points on the irrigation scheduling for corn.

(1) When the irrigation water is applied only once, then irrigation at the big bell-mouthed stage is better than that at the jointing stage. WSE of the former is higher than that of the latter by 8–137 %. WUE for irrigation water of 1200 m^3/ha when irrigating in the big bell-mouthed stage is higher than that in the jointing stage by 17.3 %. However, the result is the opposite for irrigation water of 600 m^3/ha. WUE for irrigation in the big bell-mouthed stage is lower by 2.7 %.
(2) When an irrigation quota of 1200 m^3/ha is applied twice (in the jointing and big bell-mouthed stages), this results in higher yield by 14 % compared with applying the same irrigation quota but concentrated only in the jointing stage. However, the yield is 5 % lower than the same irrigation amount all concentrated at the big bell-mouthed stage. This verifies that the big bell-mouthed stage is most critical for corn irrigation.
(3) When an irrigation quota of 1200 m^3/ha is all concentrated in the big bell-mouthed stage, this results in the highest yield but the WSE is lower than that with a quota of 600 m^3/ha, which gives highest WSE. So we can conclude that using a lower irrigation quota but on more irrigated land results in overall higher production.
(4) Irrigation using 600 m^3/ha and concentrated in the big bell-mouthed stage provides best irrigation schedule. The WSE is as high as 3.85 kg/m^3.

From the results, it can be seen that corn performs better than wheat to deficit irrigation from RWH systems. Both the WSE and WUE are much higher for corn than for wheat. This is because corn has longer growing season than wheat, and the available natural rainfall in the growing season is greater for corn.

Besides, it can be seen in Tables 5.6 and 5.7 that by adopting LORI, the WUE can be higher than for purely rain-fed farming by 28–59 % for wheat and

Table 5.8 Millet yield versus irrigation timing

Irrigation timing	Non-irrigated	Jointing	Heading	Flowering
Yield t/ha	3.76	4.01	4.26	4.11
WSE kg/m^3	–	0.83	1.67	1.17
WUE kg/m^3	1.51	1.54	1.55	1.5

Table 5.9 Yield of millet versus irrigation quota

Irrigation amount m^3/ha	0	100	200	300	400
Yield t/ha	3.76	3.86	3.96	4.12	4.12
WSE kg/m^3	–	1.0	1.0	1.2	0.9
WUE kg/m^3	1.51	1.53	1.53	1.54	1.51

6–13 % for corn. These results prove that RWH can raise the overall rainwater use efficiency compared to rain-fed crops.

5.2.2.3 Millet

(1) Tests on irrigation timing

The results of tests on the impact of different irrigation timings are shown in Table 5.8.

From Table 5.8, it can be seen the best period for irrigation is at the heading stage to get highest WSE and WUE.

(2) The results of tests on the effect of irrigation quota on yield of millet are shown in Table 5.9.

From the tests, the most suitable irrigation quota is 300 m^3/ha.

From the above introduction, we can understand how the RWH irrigation schedule can be determined. Using this method, the RWH irrigation scheduling can be formulated in a scientific way and adapted to local conditions. However, in case the irrigation tests are difficult to carry out due to lack of financial and/or technical capacity, the following empirical way for RWH irrigation scheduling can be undertaken.

5.2.3 Empirical Method to Determine Irrigation Schedules

First, the critical periods during which crop is most sensitive to water stress need to be investigated. This can be done by collecting data on rainfall and crop water demand in the growing season, consulting the experts and experienced farmers and learning the experiences from other areas similar to the local condition. From these investigations, the priorities regarding irrigation timings and amounts for different crop growing stages can be worked out.

Second, the local experiences of RWH irrigation have to be surveyed and studied to find the appropriate irrigation quota and frequency for different crops suitable to the natural conditions. As introduced before, the RWH system in China has a limited capacity. The role of RWH-based irrigation is mainly to help crops tide over serious water stress and avoid serious crop damage. The main water source for crop growth is, nevertheless, still the natural rainfall. In Table 2.3 of Chap. 2, the irrigation quota and frequency for different crops using different irrigation methods recommended by the China National Code of Practice for Rainwater Collection, Storage and Utilization are very small compared to the conventional irrigation. These figures were obtained by investigation of the RWH-based irrigation practices throughout the country and are only suitable for the specific conditions in China. In the other areas, the capacity of the RWH systems and the climatic conditions (rainfall and evapotranspiration) should be evaluated to determine the irrigation quotas and frequency for different crops. The following empirical method explains how to use the practical experiences of RWH irrigation for determining irrigation scheduling.

For example, after investigation, the order of priority for irrigation at the different growing stages is as follows: seeding—heading—jointing—milking. For a tank with water storage capacity of 40 m^3 and an irrigation area of 0.2 ha of wheat, the method of irrigation through holes in plastic film is adopted. Referring to Table 2.3 in Chap. 2, the water amount of 60 m^3/ha $\times$ 0.2 ha $=$ 12 m^3 will be the amount of irrigation water during seeding. The remaining 28 m^3 is for irrigation twice more during the growing season. This should be done in the period of heading and jointing, according to the above priority. This gives the following irrigation schedule: (1) to apply 12 m^3 during seeding in mid-March; (2) apply 14 m^3 in the jointing stage in mid-May; (3) apply 14 m^3 in the heading stage in early June.

5.3 RWH Irrigation Methods

RWH irrigation methods in China are divided into two groups: locally innovated water-saving methods and modern micro-irrigation. These methods all have two features. The first is highly efficient, with evaporation and seepage losses being reduced to a minimum. The second feature is that water is only applied at the crop root zone to avoid soil water evaporation loss as possible.

5.3.1 Locally Innovated Water-Saving Methods

In China, rural RWH projects are carried out mostly in poorer areas where most farmers cannot afford to buy modern irrigation equipment like drip, micro-spray and seepage irrigation systems. Consequently, they have invented many irrigation methods that have high efficiency but at low cost. These methods are still in use on most of the RWH irrigated land in China (Irrigation and Drainage Department of Ministry of Water Resources [IDMWR] and Rural Department of Ministry of Finance [RDMOF] ed. 2001).

5.3.1.1 Irrigation During Seeding

This method has been used in north and northwest China for many years to ensure seed germination. In this method, a small amount of water is applied in the holes where seeds are sown during seeding. The operation procedure involves digging holes or ditches—pouring water into the holes where seeds will drop in—sowing—applying fertilizer—refilling the soil—pressing—laying plastic sheeting—covering the edges of plastic film with soil. The operation can be done manually or by machine. In China, there are many kinds of sowing machines with multiple integrated functions covering all the above operations.

The water amount needed to ensure the healthy emergence of seedlings depends on the soil moisture conditions. According to work carried out in Heilongjiang, a northeast province in China, the irrigation requirements during seeding are shown in Table 5.10.

Irrigation during seeding can ensure the young plants survival for 30–40 days without rainfall. Table 5.11 shows length of time during which the young plant can resist to drought without rainfall after the seeding irrigation has taken place.

Table 5.10 Irrigation requirement during seeding. *Source* Cheng et al. (2009)

Climate status	Soil moisture (percentage of field capacity %)	Irrigation amount during seeding (m^3/ha)
Light drought	62–70	25.5
Medium drought	56–62	52.5
Heavy drought	51–56	75
Extreme drought	<51	105

Table 5.11 Duration (days) of resisting drought following irrigation. *Source* Cheng et al. (2009)

Data source	Irrigation amount during seeding (m^3/ha)	Duration (days) of resisting drought
Inner Mongolia Institute for Water Resources	45	30
	45–60	35
	75–120	40
Jilin Institute for Water Resources	45–60	40

Table 5.12 Effect of irrigation during seeding on yield of corn. *Source* Cheng et al. (2009)

Location	Year	Moisture status	Yield with irrigation (kg/ha)	Yield without irrigation	Yield increase (kg/ha)	Increase (%)
Zhaodong Municipality	1988	Moist	5655	4455	1200	27
	1989	Medium	5100	3915	1050	30.3
Zalandun Municipality	1992	Dry	7560	3105	4455	143
Tuoquan County	1990	Medium	5940	4320	1620	37.5
Keyouqian County	1994	Dry	5400	3090	2310	74.7
	–	Dry	5205	3090	2115	68.4
	1995	Dry	8115	5355	2760	51.5
	–	Dry	9285	5355	3930	73.4

Irrigation during seeding can raise the crop yield significantly. This is because seeding is the most critical period for irrigation. If the soil is dry during seeding, a significant portion of seedlings will not emerge, causing great loss in the yield. It is often found, however, that irrigation with very small amount of water during seeding can even double the yield. Table 5.12 shows the results of yield increases with irrigation during seeding. It can be seen that in a medium dry year, the yield can be increased by 30–40 %, while in the dry year yield can increase by 50–70 % or more.

In China, machines that can fulfil the multiple operations, including ditch digging, seeding, water and fertilizer application, plastic film unrolling and sheeting, soil pressing, and soil covering on the edges of plastic film, etc., have been manufactured in many places. Figure 5.5 illustrates cross section of such a machine.

In some designs, the depth where water and fertilizer are applied can be regulated. Usually water is applied at a point a little deeper than the seeds to link with the soil moisture in the lower layer. The fertilizer is placed several centimetres apart from the seeds to avoid from scorching.

5.3.1.2 Plastic Sheeting for Rainwater Conservation and Concentration

In the north and northwest China, plastic sheeting for avoiding evaporation loss and increasing soil temperature is widely adopted. Tests have shown that plastic sheeting in spring and autumn can increase soil moisture by 57 and 80 mm in the top 200 cm of the soil profile compared to that of bare ground. The ground temperature can also be raised by 2–3 °C in the springtime (GRIWAC et al. 2002). Corn has a higher yield than wheat because it has a longer growing period, and hence, more rainfall is available for it during the growing season. However, in the mountainous area with altitude higher than 2000 m in Gansu, the early frost often

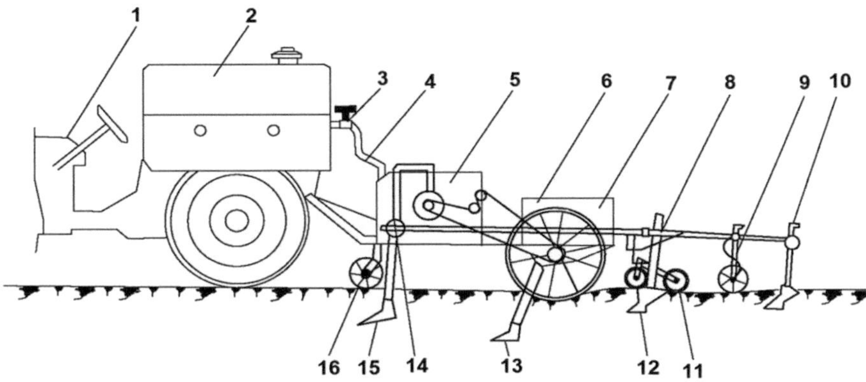

Fig. 5.5 Cross section of integrated sowing and watering machine. *Source* Zhu et al. (2012). *Note*: *1* Tractor, *2* Water tank, *3* Valve, *4* Water supply hose, *5* Second water tank, *6* Fertilizer container, *7* Seed container, *8* Frame for plastic film sheeting, *9* Wheel to press film, *10* Shovel for covering soil on the film, *11* Plastic film unrolling wheel, *12* Plough for ditch to lay the film, *13* Plough to dig ditch for seeding and applying fertilizer, *14* Plug in water outlet, *15* Plough to dig ditch for water application, *16* Wheel to control the depth of seeding

happens in the late September before corn gets mature. By adopting plastic sheeting, corn can be sown earlier by half month and mature earlier. In the mountainous areas, corn has been widely adopted as it can thrive when using plastic sheeting. Another advantage of plastic sheeting is most of the weeds under the film die due to lacking of air so the work of weeding can be greatly reduced.

Plastic sheeting also allows for rainwater concentration. For this purpose, the land is prepared by constructing alternate ridges and ditches. The ridges and the banks of the ditches are covered with plastic sheet, and the crops are planted in the ditches by the side of plastic sheet. Rain falling on the ridge will concentrate into the ditch so the crops can have rainfall increased even doubled. The ratio between the width of the ridge and the ditch is taken as 0.6–1.0. Tests have shown that with this kind of rain concentration, the crop yield can be increased by 18–30 % for potato and corn crops. Figure 5.6 shows a wheat field with the rainwater concentration.

5.3.1.3 Irrigation with Plastic Sheeting

Irrigation with plastic sheeting in Gansu China is undertaken in two ways, namely, irrigation through holes on the plastic sheeting and furrow irrigation with plastic sheeting.

In the first case, the plastic film is placed after sowing and water and fertilizer are applied into the seed holes. When the seeds germinate, the film is punctured in a cross shape by manual to let the young plants emerge. Later the holes are enlarged to let the natural rain and irrigation water enter. The land should be well

Fig. 5.6 Wheat field with rainwater concentration. *Source* Zhu et al. (2012)

Fig. 5.7 Irrigation through holes on the plastic sheeting

prepared with flat and smooth surface to avoid water being retained in depressions. Water application is carried out manually, using a bucket or hose from the tank. The irrigation efficiency of this method is very high. Irrigation water amount is about 1–3 kg for each planting hole and 45–90 m^3/ha.

Figure 5.7 shows this method. On the left, plastic film was being unrolled to cover the land, and on the right, water is applied through the holes on the plastic sheets in the corn field.

Furrow irrigation under plastic film is another kind of RWH irrigation practice with plastic sheeting. The crops are planted on the ridges which are covered with plastic films. The banks of the furrow are also sheeted but the bottom of the ditch remained uncovered to let irrigation water and rain seep into the soil. Evaporation from the ridge and the bank can be avoided. Figure 5.8 illustrates this kind of irrigation. Furrow irrigation with plastic sheeting saves water loss from evaporation. Testing has shown that the irrigation water amount can be reduced by more than 50 % compared to conventional surface irrigation.

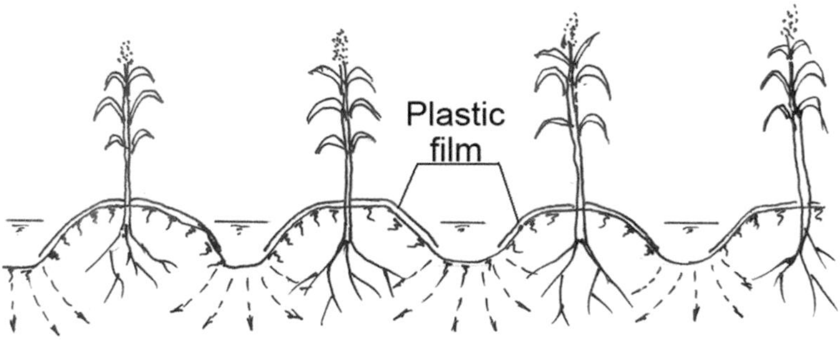

Fig. 5.8 Illustration of furrow irrigation under plastic film

5.3.1.4 Injection Irrigation

Injection irrigation involves the direct injection of water into the root zone. Originally a standard fertilizer/pesticide injector was used for the injection irrigation process. Water was injected into the crop root zone using a manual driven compressor. Later, a specially designed water injector was developed. This injector was linked to the tank with a hose and a small submerged pump was used to provide the water pressure. If the tank is located at a slightly higher elevation above the irrigated field, the water flow can be driven by gravity and no pump is needed. Sometimes, one tank can provide enough water for several injectors.

The advantages of injection irrigation include (1) Water, fertilizer and pesticide can all be applied in one operation; (2) The ground surface is kept dry during irrigation reducing evaporation from the soil; (3) The amount of water applied is very low about 1–1.5 l for each application. For example, for one hectare of corn with a density of 45,000 plants/ha, only about 45–67 m^3 of water is needed. The disadvantage of this method is that it is very labour intensive.

Injection irrigation is most suitable for low-density planting, for crops like melon, grapes, other fruits and corn. Farmers in the Ningxia Hui Autonomous Region have used this method for melons with 6–8 irrigation applications over the growing season each of about 30 m^3/ha. The yields reached over 30 t/ha. For corn, they applied in total 300 m^3/ha over five applications, and the yield amounted to 5.3 t/ha.

5.3.1.5 Seepage Irrigation with Vessel

The seepage irrigation using porcelain vessel with small holes on the wall has been adopted in Luoma Village in Gansu for fruit trees and corn irrigation. Figure 5.9 illustrates the seepage irrigation for trees and corn.

The porous vessel is made of unglazed porcelain. Water in the vessel is absorbed gradually by the soil through capillary action, and water loss due to evaporation and seepage is avoided. Labour for this method of water application

Fig. 5.9 Seepage irrigation for fruit trees and corn. **a** Four vessels around a tree, **b** One vessel for three plants, **c** Percolated porcelain vessel. *Source* Zhu et al. (2012)

is much less than for injection irrigation. However, this kind of seepage irrigation is more suitable for trees than annual crops. Installing the porcelain vessels after seeding and collecting them after harvest, however, requires a lot of labour, and the vessels are easily broken during this process.

5.3.1.6 Irrigation by Hand

The most popular irrigation method using RWH systems in China is manual application as this is low cost and simple, see Fig. 5.10. Water is poured to the crops root zone, either using a bucket or a hose connected to the tank. Where a hose is used, the flow is driven by a pump installed inside the tank. If the tank is located at a higher elevation than the field, water firstly passes through a siphon pipe from the tank and then flows by gravity to the field. The field is often covered with the plastic film and water is applied at each hole in the sheet. The water amount for manual irrigation ranges from 75 to 90 m^3/ha, depending on the crop type and experience of the water user.

Fig. 5.10 Manual irrigation using a hose

5.3.2 *Micro-Irrigation*

Apart from the locally innovated water-saving irrigation methods, modern micro-irrigation techniques have also been widely adopted for the RWH projects in China. Among these, drip systems are the most commonly used, although micro-spraying is also used, but mainly for vegetables and in the greenhouse. Bubble irrigation is also used in fruit orchards. The advantages of micro-irrigation techniques include (1) high irrigation efficiency; (2) water saving; (3) uniform water application; (4) low energy consumption; (5) water and chemical application in one operation and (6) labour saving. The disadvantages of micro-irrigation are (1) higher inputs; (2) periodic clogging of systems; (3) complicated technique for operation and maintenance.

The design of the mini-irrigations is typically a rather complicated procedure. However, for micro-irrigation using RWH, the size of the irrigation system is so small that it is not necessary to carry out complex calculations, but just a case of selecting the suitable equipment and making an appropriate layout. In this Chapter, we are not going to introduce the design procedure for micro-irrigation in general, but instead give some examples of the practical applications after a general introduction to micro-irrigation.

5.3.2.1 General Introduction to Micro-Irrigation

The micro-irrigation is composed of the following components as illustrated in Fig. 5.11

(1) Water source: The water source for micro-irrigation from a RWH system is the water cellar, water cave, rainwater tank, pond or channel.
(2) The pivot: Pivot is for integrating the equipment installed between water source and pipe system. Its function is to take water from the water source

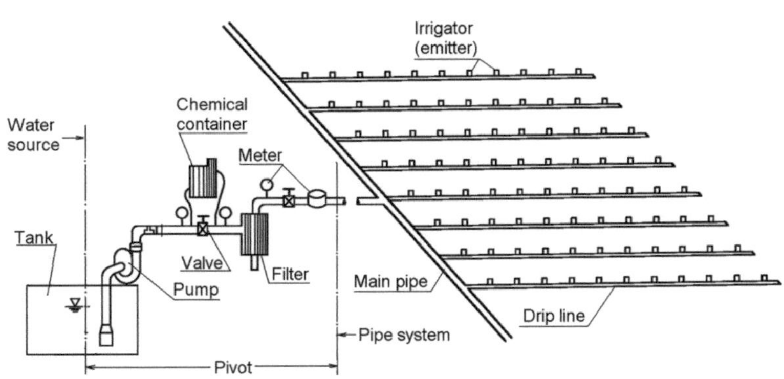

Fig. 5.11 Illustration of micro-irrigation system

and to meet the irrigation demand on the flow rate, water quality and pressure. It is mainly composed of pump-motor/gasoline or diesel engine set, chemical (fertilizer, pesticide) containers, filter and controlling equipment (valve, meter).

(3) Pipe system: The pipe system is for distributing the water (and chemicals) to the field in a uniform and timely manner. It includes the main pipe, branch pipe and laterals as well as the flow regulator, valves and fittings. The RWH irrigation system is small, and usually only a branch pipe and lateral are included.
(4) Irrigator: The irrigator is used to get water from the pipe system and apply to the soil. The pressurized flow in the pipe is changed into water drops or a non-pressurized thin flow at the outlet of irrigator. The irrigator is the key component to ensure uniform irrigation. It includes the emitter in drip system, micro-sprinkler, micro-sprayer and outlet tube in the bubble system.

According to the type of irrigator, the micro-irrigation system is classified into drip, micro-spray and bubble irrigation. The drip is used in orchards, greenhouses and vegetable fields. Sometimes it can also be used for grain crops. Micro-spraying is often used for irrigating orchards and cash crops in the greenhouses, while bubble irrigation is mostly used for fruit trees.

The micro-irrigation systems can also be classified as fixed (stationary), semi-fixed and movable systems.

(1) The equipment in the stationary system is kept in fixed location for the whole irrigation period. The main and branch pipes are usually buried underground. This system has the advantages of being easy to operate and to realize automation but the cost is high. It is adopted in the orchard, greenhouse and especially for crops planted at larger intervals.
(2) In the movable system, all the equipments including the pivot are moved from place to place. One set of equipment can be used for several plots so the input per unit area can be reduced. But moving the equipment requires labour. The movable system is suitable for micro-irrigation in the RWH system. It is suggested to use one set of micro-irrigation system with two tanks or with four tank*s*, see below.
(3) In the semi-fixed system, the pivot and the main and branch pipe are fixed in place and the laterals are moved from field to field. The semi-fixed system is suitable to larger irrigated areas.

5.3.2.2 Equipment Used for the Micro-Irrigation System Under RWH Condition

(1) Drip emitter

In China, drip emitters suitable for RWH system include:

(a) Drip line with built-in emitter

The emitter is installed in the inner wall of the pipe when being manufactured. This is the most commonly used dripper system in China. It has the advantages of high performance with no-clogging, uniform outflow, easy-to-install and affordable. It is widely used both in greenhouses and for field crops. The service life is typically more than 5 years for pipes with a wall thickness of 0.6–0.8 mm and 1–3 years for those with a wall thickness of 0.2–0.4 mm. Examples of drip lines produced in China is listed in Table 5.13. Various kinds of emitters are shown in Fig. 5.12.

(b) Emitter with pressure compensation

Inside the emitter with pressure compensation, a rubble slice is installed that can regulate the opening of the emitter according to the water pressure to keep the flow rate constant. The emitter has working pressure ranged from 60 to 350 kPa. The emitter is installed on the PE pipe with outer diameter of 12–16 mm at an interval of 0.5, 0.75, 1.0, 1.25 and 1.5 m, depending on the design. This kind of emitter is mainly used on slopes or in undulating fields. Due to its higher cost, its use is not so common in RWH irrigation system.

Table 5.13 Examples of drip line produced in China

Outer diameter (mm)	Wall thickness (mm)	Working pressure (kPa)	Flow rate (l/h)	Interval of emitter (m)
16	0.2, 0.4, 0.6	100	2.8	0.3, 0.4, 0.5
12	0.4	100	2.7	0.3, 0.4

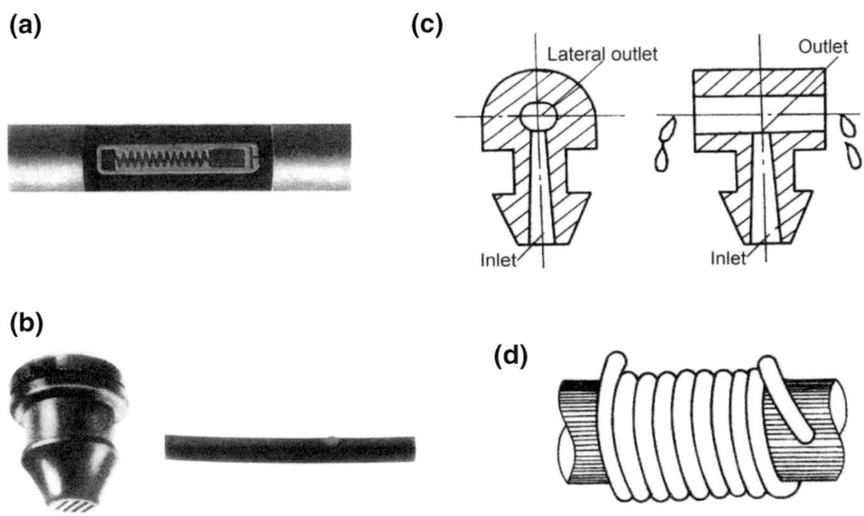

Fig. 5.12 Drip emitter commonly used in China. **a** Built-in emitter, **b** emitter with pressure compensation, **c** orifice-type emitter, **d** mini-pipe emitter

(c) Orifice-type emitter

In this kind of emitter, the cross-sectional area changes rapidly from the inlet to the outlet. The pressurized flow is then changed into water drops. The emitter is installed on the pipe on site by puncturing the pipe using certain tools. Since its simple structure the price is low. The KD-type orifice emitter made in China has working pressure of 100 kPa, and each emitter has flow rate of 10, 15, 20 and 25 l/h.

(d) Micro-pipe emitter

The micro-pipe emitter is made using a PE micro-pipe with diameter of about 1 mm. One end of the micro-pipe is wound on the outside of lateral and acts as the outlet of drip system, while the other end is inserted into the pipe at certain intervals. Since the pressure in the lateral is different along its length, the outflow along the lateral would be unevenly distributed. To get a uniform outflow along the whole drip system, the length of the micro-pipe, which provides resistance to the outflow from the lateral pipe, is changed along the lateral. At the top end of the lateral where the pressure is higher, a longer length of the micro-pipe is used, while at the bottom end, a shorter length is used. By testing the hydraulic resistance of the micro-pipe to the flow and conducting hydraulic calculations, the length of mini-pipe for each emitter is determined, and the outflow of the drip system can be evenly distributed. Compared to the pressure compensation emitter, the micro-emitter is much cheaper. However, the design procedure and the installation of the emitter are much more complicated. Another shortcoming of the micro-pipe emitter is that it quickly ages and is easily broken.

(2) Micro sprayer

The commonly used micro-sprayer has two types: the rotating sprayer and the sprayer with baffle. Normally, the flow rate of the sprayer ranges between 40 and 90 l/h. The maximum rate can be 250 l/h. The maximum spraying radius is 7 m. Figure 5.13 shows these two kinds of sprayers.

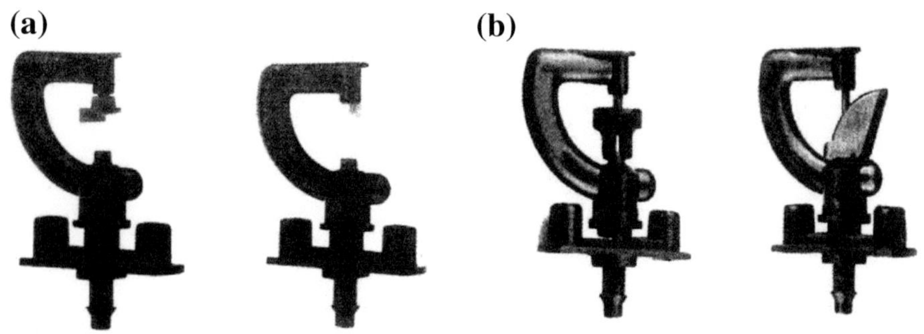

Fig. 5.13 Rotating micro-sprayer and sprayer with baffle. **a** Sprayer with baffle, **b** rotating sprayer

(3) Pipe and fittings

Pipe in the drip system is made of low-density polyethylene with a black colour. The PE pipe is semi-flexible so it is adaptable to the undulating topographic conditions. It has good resistance to ageing and performs well at low temperatures. However, it cannot stand high temperatures. The PE pipe used in micro-irrigation can work under pressures of 0.25 and 0.45 MPa. The fittings include adapters, bends, T-bends, by-pass, plug, etc.

(4) Filter

To prevent the micro-irrigation system from clogging, filters are an essential component. Filters for micro-irrigation systems include screen filters, sand-gravel filters and rotating silt separators. For RWH systems, the micro-irrigation systems are very small, so a screen-type filter made of plastic is used. Commonly, 120# mesh screen is used for the filter in the drip system and 80# mesh in micro-spraying. The screen type filter is shown in Fig. 5.14.

(5) Chemical container

The micro-irrigation system has the advantage of applying water and fertilizer to the root zone in one operation. The fertilizer application equipment suitable for the

Fig. 5.14 Screen-type filter

RWH system is the Venturi type fertilizer sucker and pressure differential fertilizer applier. When the water flows through the Venturi pipe, a vacuum occurs, and the fertilizer solution is sucked into the pipe. In the pressure differential applier, fertilizer solution is driven by the pressure difference built up in the pipe.

5.3.2.3 Layout and Design of the Drip System Under RWH Conditions

(1) Drip system for field crop

The drip irrigation can moisten a strip with a width of about 0.6–0.8 m for sandy soil and 0.8–1.0 m for clay soil. The drip line is placed parallel to the crop line with an interval of the width of moistened strip. To reduce costs when using RWH, the movable drip line system can be adopted, either to move one single drip line or a group of lines at one time. Since the water source of RWH system is very small, the maximum area that one tank with volume of 30–60 m^3 can provide is 2–4 Mu (1334–2667 m^2 or 0.13–0.27 ha). In this case, only the branch pipes that divert water from the tank and the laterals that distribute water to the field are used. To reduce costs, the lateral is used to irrigate the soil on both sides of pipe in flat areas, while on slopes the lateral is arranged along the contour line.

Although the micro-irrigation systems used with RWH are very small, the pivot including a screen-type filter, a valve and sometimes a fertilizer container is indispensable.

In the Ningxia Autonomous Region, the so-called "1-2-4" or "1-2-8" layout pattern for drip systems was adopted. The "1-2-4" pattern denotes to use **one** set of movable drip systems with **two** tanks (each volume capacity of about 30 m^3) for irrigating **four** Mu (0.27 ha) of field crops. The "1-2-8" pattern denotes to use **one** set of drip systems with **two** tanks (each capacity of 50 m^3) for irrigating **eight** Mu (0.53 ha) of field crops. A hand pump or electric pump is used for both of these two systems.

(i) Movable drip system with hand pump for field crop of 2 Mu

The water source of this system is two tanks each with volume capacity of 30 m^3. The RB1.5 type hand pump that can supply flow rate of 1.5 m^3/h with a water head of 15 m is used in this case. The filter is a 120# mesh screen filter with diameter of 1 in. The lateral pipe is a PE black pipe with thickness of 0.4 mm and outer diameter of 16 mm. An emitter with a flow rate of 2.8 m^3/h is installed on the drip line at intervals of 0.4 m. The system includes four laterals each with length of 30–40 m, depending on the land area. The length of the branch pipe is about 40 m. Figure 5.15 shows the layout of this drip system.

(ii) Movable drip system with electric pump for 4 Mu of field crop

The water source of this system is four rainwater tanks each with volume capacity of 30 m^3 or two tanks each with capacity of 60 m^3. The layout and the specification of the filter and drip line are the same as for the above example except that

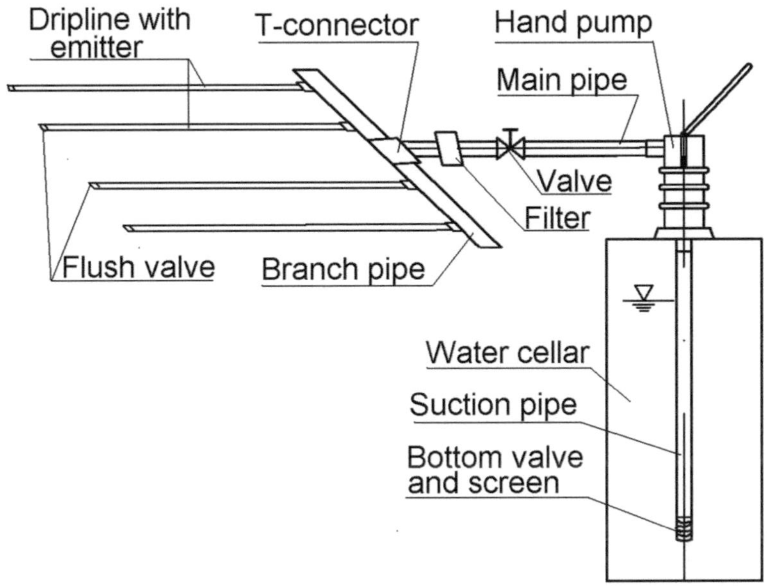

Fig. 5.15 Layout of hand pump drip system

the electric pump is used instead of hand pump, and the number of drip lines is doubled.

(iii) Fixed drip system for greenhouse

In the northern China, the simplified greenhouse is usually in rectangular shape with longer side arranged on east–west direction. The width of the greenhouse is commonly 6–8 m. Following is an example for a 0.6-Mu greenhouse with length of 60 m and width of 7 m. The branch pipe is placed along the longer side and the laterals are arranged on the north–south direction. The laterals are put under the plastic film at interval of 1 m for every two rows of vegetables. Emitters are put every 0.3 m along the lateral. The branch pipe is a PE black pipe with outer diameter of 32 mm and length of 55 m.

According to hydraulic calculations, the total water head of the system is 18 m and working head of the emitter is 10 m. The flow discharge of the whole system is 3.6 m^3/h. The submerged pump QDX4-20-0.55 is used. The equipment of the system is illustrated in Fig. 5.16.

5.3.3 Paddy Irrigation with RWH System

In the humid areas of Southwest China with annual precipitation more than 1500 mm, RWH system has been used for paddy irrigation. The storage facility

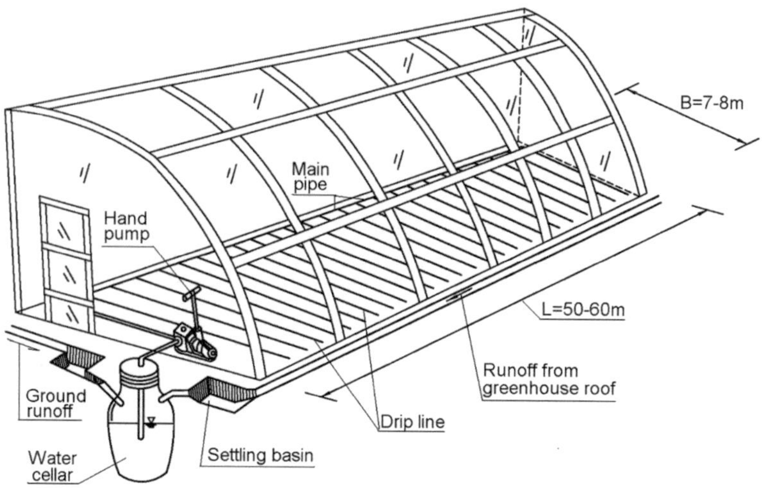

Fig. 5.16 Layout of fixed drip system for a 400-m^2 greenhouse

is usually the surface tank with a round or rectangular section and can be filled up for 3–4 times in one year. According to experiences of Guizhou Province and Guangxi Zhuang Autonomous Region, the tank with a volume of 50 m^3 can irrigate 1 Mu (1/15 ha) of paddy field. In the whole growing seasons, 6–9 applications each with about 15–20 m^3/Mu (10–13.3 mm) are supplied.

Because of the limited capacity of RWH systems and also because paddy irrigation needs more water than the dry land irrigation, it is important to follow water-saving principles for the paddy irrigation using RWH as water source. Since the 1980s, many institutions in China have successfully developed the water-saving techniques for paddy irrigation (Peng et al. 1998). The principle of water conservation in paddy irrigation is to reduce the time and depth of water submersion of the rice field and let the land be exposed more frequently to the air. People have summed up the essential points in four words: "Shallow, thin, moist and expose". "Shallow" means that rice seedlings should be transplanted under shallow water. "Thin" means that submersion of the field in the growing season should be under thin layer of water. "Moist" means that in many stages of rice growing, the soil is only to be kept saturated or partially saturated but not submerged. "Expose" means that the field should often be exposed to the air with natural or artificial drainage of the water. During transplanting and the returning green, the field is submerged under a thin water layer. But after that the field is to be exposed and soil is to be kept under saturation. The following is instruction for water management in various stages of rice growing.

Transplanting of seedlings has to be done with submersion in water but the water layer should not be deeper than 20 mm. While in the follow-up stage of the 'returning green', if the depth of submerged water drops to less than 5 mm, irrigation is necessary but the water depth after irrigation or after rainfall should not be deeper than 40 mm. After the above period, submersion of the field is no longer needed. On the contrary, the land should be controlled in case of submersion. If the water layer in the field caused by natural rain or irrigation is kept longer than 5 days, then drainage measures should be taken to expose the field. The only need is to keep a certain amount of moisture in the soil.

In the tiller stage, the soil moisture needs to be kept in the range of 50–100 % of saturation. If the soil saturation is lower than 50 %, then irrigation with a depth of no more than 20 mm should be carried out. In the later tiller stage, the plant starts transitions from vegetative growth to reproductive growth. At this stage, the demand of plant for nutrients and water increases. The lower limit of soil moisture is 60 % of saturation. During the rainy season, it is important to have drainage if waterlogging happens after heavy rain.

In the period of jointing to booting, the rice grows most rapidly and also needs the most water. The best moisture range for this period is 70–100 % saturation. During irrigation, it is necessary to maintain suitable soil moisture levels, and drainage should be undertaken to expose land to air within 3–5 days after irrigation or after heavy rain.

In the heading and flowering period, photosynthesis and metabolic processes are proceeding intensely, and the rice plant needs a lot of water. The lower moisture limit is 70–80 % of saturation and the upper limit is 100 % saturation.

In the milking stage, light irrigation by flooding the surface for every 3–5 days is suitable. It is better to keep the surface of the field dry, while keeping the soil moist underneath. Good ventilation in the soil is essential for promoting milking, increasing grain weight and enhancing the yield.

The water conservation in paddy irrigation can not only save water but can also increase the yield and improve the grain quality. Tests carried out between 1982 and 1989 showed that with water saving irrigation methods, the water consumption of rice reduced by 40 %, in which the evapotranspiration reduced by 29.4 and seepage loss reduced by 48.6 %. The irrigation water amount reduced by 52 % while the yield increased by 9.5 %. Furthermore, the rice quality in protein, fat and amino acid content were also improved (Peng et al. 1998).

From the above introduction, it can be seen that the water-saving techniques for paddy irrigation need to be flexible and include drainage systems. The response of the system to the changes in conditions of climate, soil and crop growth should be prompt. Usually, the field ditch network for both irrigation and drainage purposes is built in the most fertile high-yielding rice paddies. In China, these are the fundamental agriculture fields and are protected from land requisition for industrial use under the law.

References

Blank H. Optimal irrigation decisions with limited water (D). Ph.D. thesis, Department of Civil Engineering, Colorado State University, Fort Collins, CO; 1975.

Cheng M, et al. Rainwater harvesting water saving irrigation technology in the semi-arid loess hilly area of north China. Zhengzhou, China: Yellow River Water Resources Publications; 2009.

GRIWAC (Gansu Research Institute for Water Conservancy, Dryland Agriculture Instituteof Gansu Academy of Agriculture Sciences, Gansu Agriculture University and DinxiCounty Government, Technology integration and innovation study on highly efficientconcentration of natural rainfall in the semi-arid mountainous area, Lanzhou: GRIWAC, 2002 (unpublished report, in Chinese)

Irrigation and Drainage Department of Ministry of Water Resources (IDMWR) and Rural Department of Ministry of Finance (RDMOF), ed. Concise textbook of rural rainwater harvesting project. Beijing: China Water Resources and Power Publications; 2001 (in Chinese).

Jensen ME. Water consumption by agriculture plants. In: Kozlowski TT, editor. Water deficits and plant growth, vol. 2. New York: Academy Press; 1968. pp. 1–22.

Liu Z. Study on the water saving irrigation for spring wheat. Xinjiang Auton. Reg. Water Res Sci Technol 1987;4–5 (in Chinese).

Peng S, et al. Water-saving techniques for paddy irrigation. Beijing: China Water Resources and Hydro Power Publications; 1998 (in Chinese).

Zhu Q, Liao L. Study on the optimum irrigation scheduling and economic irrigation quota. Gansu Water Res. Hydropower Tech 1994;58:21–7 (in Chinese).

Zhu Q, Li Y, Gould J. Every last drop—rainwater harvesting and sustainable technologies in rural China. Rugby, Uk: Practical Action Publishing Ltd; 2012.

Chapter 6
Rainwater Harvesting and Agriculture

Shiming Gao and Fengke Yang

Keywords Tillage techniques · Contour farming · Stubble mulch · Plastic sheeting · Soil fertility

In the long struggle against drought and water scarcity, the Chinese people have created some very practical cultivation techniques for conserving water in the soil. These consist of selecting the best suited crops, and improving the soil structure and fertility by using farm manures. It also involves making optimal use of rainwater through intensive cultivation and soil moisture preservation. Key techniques for doing this include changing the topography of the land by constructing terraces, furrows, ridges, as well as using advanced techniques for storing and increasing efficiency of scarce water resources.

6.1 Tillage Techniques

Tillage helps create top soil with the best proportion of solid, liquid, and gaseous (air) components for regulating the soil moisture, nutrients and temperature in order to optimize crop growth. Tillage is an important water conservation technique, especially in low rainfall areas, as it helps maintain most of the rainwater within the field by increasing the capacity of soil to store moisture.

S. Gao (✉) · F. Yang
Gansu Academy of Agricultural Sciences, Lanzhou, China
e-mail: gao-shm@sohu.com

F. Yang
e-mail: yang_fk@163.com

Q. Zhu et al. (eds.), *Rainwater Harvesting for Agriculture and Water Supply*,
DOI 10.1007/978-981-287-964-6_6

6.1.1 Deep Ploughing

Deep ploughing is intended to create a suitable cultivation profile for crops, through increasing soil aeration, promoting microorganism activity and the availability of soil nutrients.

Experimental data from Shandong Agricultural Academy of Science has shown that deep ploughing can make soil bulk density decrease by 0.1–0.2 g/cm^3, increasing the non-capillary porous proportion by 3–5 %, therefore greatly improving the root growth environment. In a field with deep ploughing to 34 cm, the roots of wheat can reach to the depth of 150 cm, but in the field of shallow ploughing (17 cm deep), the root depth is only 90 cm.

Research from the Gansu Academy of Agricultural Sciences (GAAS) demonstrated that the hydrolyzed nitrogen content of deep ploughed soils was 6.8 mg/kg higher than that at the end of autumn ploughing. Deep ploughing can boost crop root growth by making use of phosphorus located deeper in the soil.

Deep ploughing can also make better use of rainfall by minimizing the runoff, evaporation and transpiration losses and increasing the water storage capacity of the soil. Research results support this and have shown that deep ploughing can increase soil water holding capacity by 2–5 %. Tests in the Li village, Henan Province showed that increase of water content in the top 0–60 cm of the soil amounted to 5 %, equivalent to 375 m^3 per hectare.

Deep ploughing can bury crop residues providing organic fertilizer, clean the soil surface and create a suitable seed bed through harrowing and levelling practices.

Generally, deep ploughing is conducted using the mouldboard plough in the following ways:

- Semi-turning ploughing. Soil layers are turned by 135° and connected with each other, leaving the soil surface with a tile-like shape.
- Fully turning ploughing. Turing soil particles by 180°.
- Turning soils in different layers, using the complex type plough with small front plough to plough soils in different layers.

The timing of deep ploughing should coincide with the onset of the local rainy season, to intercept more rainwater. The deep ploughing is classified into summer, autumn and spring ploughing.

6.1.1.1 Summer Deep Ploughing

Early summer ploughing can catch and store more rainwater in the rainy season. Experimental results from GAAS show that the soil water content is different with the date of deep ploughing. When the date of ploughing is at the first, second and third ten day periods of July, the water content in the top 0–100 cm of soil is 20.63, 22.75 and 11.75 mm, respectively. This is much more than when ploughing in the first ten days of August. The best time of deep ploughing in summer is in the first ten days of July. The research results of Shanxi show that wheat yields increase up to 62.4 % when ploughing just before the summer, when compared to ploughing at the end of summer. This can result in a net yield increase of up to 1500 kg/ha.

6.1.1.2 Autumn Ploughing

Generally, deep autumn ploughing should be conducted immediately after the autumn harvest in order to reduce soil surface evaporation and to intercept more rainwater. An experimental result in Qinghai Province shows that the water content in 0–100 cm soils was 293 mm when ploughing is done immediately after the beginning of autumn, while it was 29.3, 56 and 65.6 mm less, respectively, when the date of ploughing was 4th, 7th and 10th day after beginning of autumn.

In cases with no soil erosion, it is better to till the farmland immediately after autumn, leaving the stubble mulch on the soil surface throughout winter. According to experiments by GAAS, the moisture in 0–30 cm soil in the next spring when autumn ploughing was conducted is 3.75 % (for millet crop) more than farmland with no tillage.

6.1.1.3 Spring Ploughing

Earlier spring ploughing is good in preventing the soil moisture from evaporation loss. The depth of deep ploughing is about 13–16 cm for animal power and 20–25 cm for mechanized ploughing.

6.1.2 Sub-soiling

Sub-soiling (deep loose ploughing) is a tillage method for loosening the soil without turning over of the soil.

The function of sub-soiling can be summed up as follows.

- to kill weeds and turn the surface layer.
- to reduce crop diseases and insect hazards.
- to prevent soil moisture evaporation loss during the process of tillage and to benefit sowing.
- to create good soil structure that is favorable to regulating the process of mineralization and humus formation.

In the "loose" part of the soil, the aerobic micro-organisms are active, which in turn benefits the mineralization process and effective nutrients release. In the "solid" part of the soil, the anaerobic micro-organisms are active, which in turn is helpful to humus formation so that the potential fertility of the soil is raised and the release and storage of nutrients is balanced. Meanwhile, it can reduce the constraints to soil water supply and storage. In the "loose" part of the soil layer, a greater amount of rainwater can be intercepted and stored in the deep soil layer. In the "solid" part of the soil layer, the soil water content can remain high for a long time. After ploughing (sub-soiling), the crop stubble remains in the soil surface and significantly reduces water loss and soil erosion.

6.1.3 Inter-tillage for Preserving Soil Moisture

Inter-tillage is a ploughing operation which can eliminate weeds by breaking the soil crust which may form due to precipitation, irrigation and other field operations. The best time for inter-tillage is between the crop seedling stage and the full ground cover stage. Inter-tillage has many positive effects on the crop growth environment: making better use of rainfall by minimizing evaporation and runoff, increasing water holding capacity of soil, improving nutrient availability and eliminating weeds. Research results from Henan Province showed that inter-tillage can increase the wheat yield by 4.3 %.

6.1.4 Harrowing and Levelling

Harrowing is a common cultivation measure conducted before sowing, after tillage and crop harvest as well as during the seedling stage. Levelling to form a smooth soil surface is an effective way to prevent soil water loss through evaporation. These two tillage practices are always conducted together and have the function of breaking up soil lumps, levelling the field and lightly compacting the soil. It can also reduce soil surface pore density, which helps preserve the soil moisture of root zone.

6.1.5 Contour Farming

Contour farming involves undertaking tillage, planting and other farm operations along the contour. Contour tillage will impede the down slope flow of water, allow water to infiltrate into the soil and reduce soil erosion. Compared to conventional hill slope cultivation, experimental result shows that contour tillage on slopes of 2° can reduce water loss (runoff) by about 51–57 % and increase water content in 0–70 cm soil by about 2.8–9.6 % (Figs. 6.1 and 6.2).

Fig. 6.1 Contour farming in mountainous areas

Fig. 6.2 Landscape of contour farming before harvesting

6.1.6 Contour Ditch and Ridge Cultivation

6.1.6.1 Contour Cultivation on Levelled Hill Land

This method is suitable on hill slopes above 25° for growing wheat, millet and potato. The cultivation is done along the contour (Fig. 6.3) with ridges and ditches.

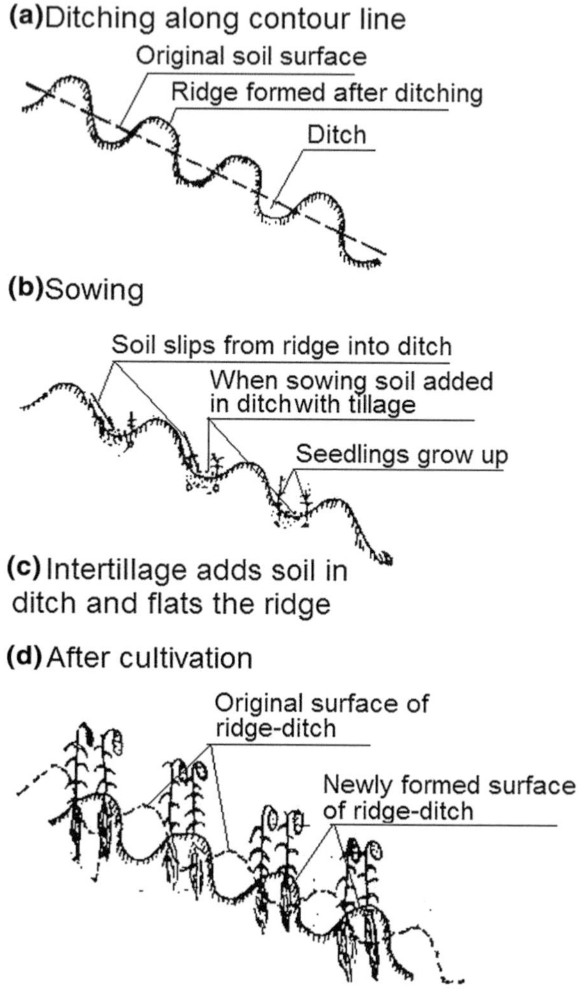

Fig. 6.3 Sketch of contour cropping on the slope land

6.1.7 Ridge Cropping Along the Contour

The ridge cropping is a method that can increase rain infiltration, reduce runoff and prevent soil erosion. Experimental results show that ridge cropping can reduce runoff by 77 % compared to traditional tillage. Potato yield increases by 8–21 % and millet yield by 77 % (Fig. 6.4).

During sowing, ridges and ditches can be formed along the contour lines of the slope by ploughing and seedlings planted in the ditches. Rainfall concentrates in the ditches and helps the crop grow. Later inter-tilling and earthing up the plants together with slippage of the earth from the ridges make the ridges flatter and the

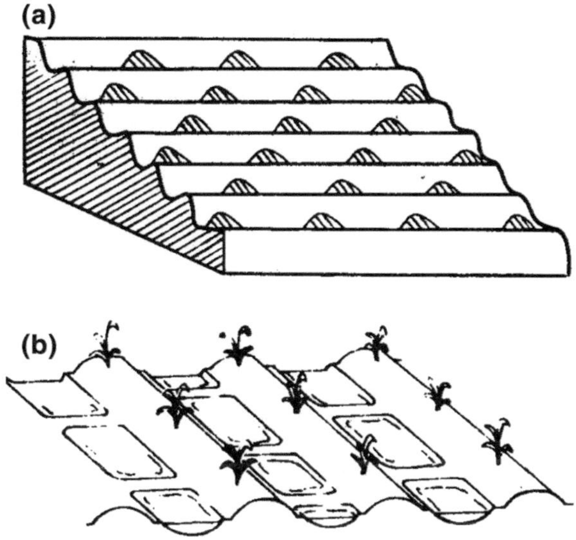

Fig. 6.4 Sketch of ridge plot contour tillage. **a** Ridge-sector contour tillage field laying; **b** Planting patterns of ridge-sector contour tillage field

ditches shallower and finally the ridges and ditches transform mutually. Annual repetition of this will gradually turn an even slope into terraces, see Fig. 6.3.

Cultivation with Ridge Plot Contour Tillage
On flatter slopes or tableland when sub-soiling along the contour line, the furrows are deepened and lateral earth dikes built to form ridge plots. This kind of cultivation can increase the rainfall retaining capacity of the land. The ridges and lateral dikes should be compacted and planted with grasses to prevent collapse. After the plots are levelled, manure should be applied when sub-soiling to fertilize the soil. Crops are planted on the ridges, which also help concentrate rainwater runoff for meeting the crops demand, see Fig. 6.4.

6.1.8 Furrow Cultivation with Soil Fertility Improvement

This method involves sowing furrows 30–50 cm in width along contour. The fertile surface soil and deeper earth in the furrows are mixed. Fertility is then improved by applying organic manure and chemical fertilizers. The procedure of this kind of cultivation is as follows and shown in Fig. 6.5.

Step 1: Build a dike on the down slope boundary of the field

Remove the fertile topsoil to a depth of 15 cm along a 46-cm-wide strip along the contour and spread on the field by the side of the furrow. In the furrow, remove a further 20 cm of soil and place this on the downside towards the field boundary to form a dike with a height of 20 cm after compaction

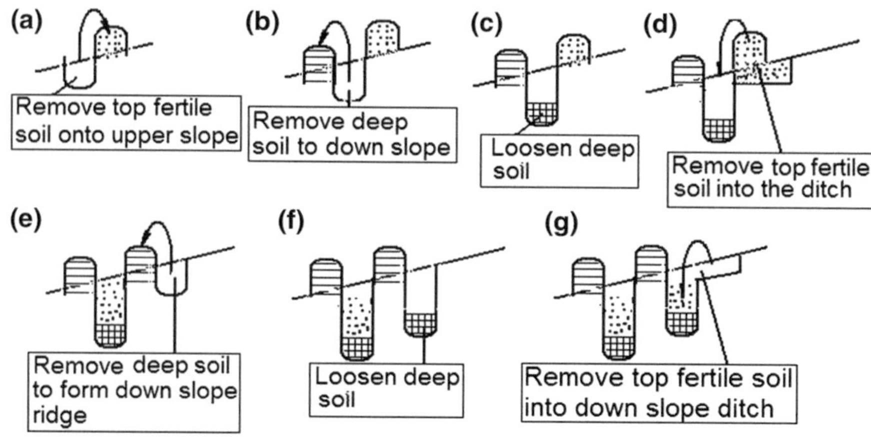

Fig. 6.5 Tillage system for fertilizer concentrating and soil improving

Step 2: Loosen the furrow and fill with fertile soil for planting

A 15-cm layer of earth in the bottom of the furrow with width of 46 cm is further loosened. Then the previously excavated fertile topsoil with a depth of 15 cm and width of 80-cm by the side of the furrow is cut and filled into the furrow for planting crops

Step 3: Build the second ridge and furrow

By the side of the furrow, loosen the soil to a depth of 20 cm along a 34-cm-wide strip. The earth on the neighbouring 46-cm-wide strip is then shovelled onto the outside of the 34 cm strip to form second ridge with height of 15 cm

Step 4: By the side of the second ridge, repeat Step 2 to form the second furrow built same way as the first. The same procedure is repeated until the ridge–furrow system is finished for the whole field

6.2 Mulch Cultivating and Soil Moisture Preservation

6.2.1 Role of Farmland Mulch

Organic mulch has many roles in improving soil conditions, by significantly reducing evaporation, reducing runoff and increasing the moisture availability of the soil.

Organic mulch has the following functions.

(1) Mulching reduces the exchange between air and soil and thus decreases evaporation from the soil surface.

(2) Particles from the rotted mulch mix into the soil increasing its organic content and water holding capacity. More humus also means higher nutrient holding capacity.
(3) Stubble mulch reduces the solar radiation received by the surface and decreases soil temperature.
(4) Decomposing materials from the mulch release nutrients increasing the quantity of living organisms in the soil, and increasing humus and promoting formation of soil aggregates.

6.2.2 Types and Application Methods for Farmland Mulch

Mulching can be divided into stubble mulch and plastic sheeting

6.2.2.1 Stubble Mulch

Stubble materials are cheap, locally available, and easy to transport and apply. Cereal straw and stalks from maize plants are the most common materials used for farmland mulching (Fig. 6.6).

Data from the China Academy of Meteorology and many institutions revealed that stubble mulch has a great effect on raising soil temperature and improving soil moisture conditions.

Fig. 6.6 Stubble mulch

Autumn stubble mulching has a distinct function on regulating soil temperature. It increases soil temperature by 0.5–1.9 °C, reduces the frost penetration depth by 5 cm and advances the spring thaw by 10 days. It also advances the return greening time of winter wheat by 4–5 days. The soil temperature during spring and summer in mulched fields was a little bit lower by 0.8–2.1 °C. This has an active role on increasing soil moisture conservation and on crop disease prevention and control, as well as on pest elimination.

Research by the China Academy of Agriculture Sciences also showed a reduction of soil bulk density and pore density and increase in nutrients after two-year mulching.

Mulching can significantly improve the water and fertility conditions in the field. The yield of winter wheat and summer corn can be increased by 5.5–20 % and water use efficiency can be increased by 14–24 % (Zhang et al. 1999).

6.2.2.2 Plastic sheeting

Plastic sheeting (plastic mulch, see Fig. 6.7) has several advantages, and these include:

- Raising soil temperature; promoting crop establishment and growth, particularly in the initial growth stage.
- Speeding crop maturing rate, ear formation and increase the weight of the grain seeds;
- Improving soil moisture conservation and prevention of dense weed growth;

Fig. 6.7 Crop field covered with plastic film

- Boosting crop root development, increasing crop nutrient up-take, and improving light distribution;
- Reducing soil erosion and soil susceptibility to erosion.

Before laying plastic sheeting, the fields should be prepared by breaking soil clods and levelling. Plastic sheets can be laid on the levelled ridges (Fig. 6.8). The best time for laying the plastic sheeting is either before or after sowing. The soil water content in 0–80 cm soil layer in fields with plastic sheeting in autumn is higher than that in spring (Table 6.1).

Fig. 6.8 Crop field with plastic sheeting on the ridges

Table 6.1 Effects of different mulch treatments on soil water content (%) in spring wheat field at depth of 0–80 cm

Treatment	Before mulch	Mulch after 30 days	Before sowing	Seedling	Jointing	Before harvest	Average soil water content (%)
Autumn cover	16.4	15.9	15.3	18.6	21.9	20.9	18.2
Spring cover	16.4	14.2	13.2	17.4	21.0	20.5	17.1
Non covered	16.4	14.2	13.2	16.5	20.2	19.7	16.8

Laying plastic sheets at the end of the rainy season in autumn has several advantages:

- It helps catch and store rainwater in the autumn and prevents soil moisture from evaporation both in the autumn and the spring.
- Plastic mulch on the ridges along with ditch can withhold rainwater and enhance water and soil conservation within the field.
- Combined with supplementary irrigation, this ensures a good harvest even in the severe drought years (see Fig. 6.8).

6.3 Soil Fertility and Field Improvements

Soil fertility and field improvements are the most critical factors that determine the crop yield and contribute significantly to raising the water use efficiency. Tests in Shangqiu, Henan Province showed that with same water consumption, yield in the wheat field with high fertility can be higher than that with moderate fertility by 23–27 % (Zhang et al. 1999).

6.3.1 Using Biotechnology and Crop Rotation to Raise Soil Fertility

Planting certain nitrogen fixing crops can help raise soil fertility, and these include:

- Planting leguminous crops
- Planting green manure crops
- Planting Chinese herbs

6.3.2 Use of Organic Fertilizers

The experimental results from the Chinese Academy of Agricultural Sciences (CAAS) have shown the following benefits of using organic fertilizers:

- Humus content in the topsoil of irrigated fields increases by 0.04 % per year after three year applications each of 37,500 kg/ha organic fertilizers. The organic content in non-applied field decreased by 0.06 % per year during the experimental time period.
- Organic fertilizers easily become instant fertilizers, especially phosphorus. Quick available phosphorus content in high grade barn manure can reach up to 0.1–0.5 %, which is thousand times greater than the background content in the field.

Experimental data from Northwest University of Agriculture and Forestry showed that organic fertilizer can raise soil water holding capacity during the summer fallow period. The water content in a 2-m soil layer in dry plateau of Shanxi Province is 139.2 and 56.6 mm with and without applying organic fertilizer, respectively. Organic fertilizer application can also raise crop productivity.

6.3.3 Rational Application of Chemical Fertilizers

According to data from soil surveys and field experiments, farmland can be divided into different groups with different soil productivity. A reasonable target yield can be designed and then the amount of nitrogen, phosphorus, potassium and micronutrients fertilizers will be applied following a pre-determined timetable.

After determining the target yield, the application of fertilizers and corresponding nutrients can be estimated as follows:

The fertilizers (kg) = [The total nutrients for the target yield (kg) − nutrients supplied by soil (kg)]/[The content of nutrients in fertilizer (%) × fertilizer efficiency (%)].

The total amount of nutrients for target field (kg) = target yield/100 × nutrients for production of 100 kg grain.

The efficiency of fertilizers (%) = [Nutrient elements absorbed by crop in the field of applied nutrients − Nutrient elements absorbed by crop in the non-applied field]/[Total nutrient elements in fertilizers applied].

The amount of nutrients supplied by soil (kg) = Tested content of nutrients in soils (mg/kg) × 2.25 × rectifying coefficient. Here 2.25 is a coefficient for converting the unit of tested soil nutrient content in mg/kg to kg/ha.

The rectifying coefficient = Yield of crops planted in field without fertilizers application × Nutrients taken by crop/the nutrient content in the soil (mg/kg) × 2.25.

6.4 Techniques for Conserving Soil Water with Chemical Treatment

Use of chemical treatments can hinder or reduce evaporation and transpiration. The advantages of applying chemical compounds are that these can improve plant turgidity by minimizing transpiration loss, conserve soil water, and thereby reduce irrigation requirements. At present, three kinds of chemical treatments are widely used: cover treatments, absorbent treatments or evapo-transpiration resistance treatments.

6.4.1 Evapo-Transpiration Resistance Treatment

Fulvic acid is widely used for evapo-transpiration resistance treatment. Experimental results show that spraying once during the water sensitive stage can reduce wheat transpiration rate within 10 days. The water content in wheat leafs is 3.7 % higher than that of non-sprayed wheat.

Tests showed that spraying wheat with fulvic acid can increase yield by 10.4–18 % (Zhang et al. 1999).

Fulvic acid can be used in dressing seeds or spraying crops in the growth stages sensitive to water. For wheat, it is used in the heading and early milking period.

Seed Dressing with Fulvic Acid

For closely planted crops, such as wheat, the diluting proportion is seeds:fulvic acid:water = 50 kg:0.2 kg:5 kg

For more widely spaced crops, such as melon, the diluting proportion is seeds:fulvic acid:water = 50 kg:0.1 kg:50 kg

Fulvic acid is first dissolved in pure water with 10 % of seed weight, and then mixed with seeds until becoming black. After heaping up for 2–4 h, the seeds can be sowed in field. In case of delay in sowing, the seeds dressed with fulvic acid should be air-dried.

Seeds dressed with fulvic acid can germinate 1.5–2 days earlier and increase plant bio-mass. The proportion of fulvic acid solution for top spraying is 750 g fulvic acid dissolved in 900 kg water for one hectare of wheat and 1125 g fulvic acid dissolved in 900 kg water for one hectare of corn. The best time to spray is before 10 am and in the afternoon. In general, spraying once is enough for a crop, but in serious droughts, 2–3 applications are better, spraying once every 10 days.

6.4.2 Water Absorbing Resin

Absorbing resin is an artificial macromolecule, which can absorb a certain amount of water a hundred times of its own weight. It can form a gel, and improves soil structure and its fertility preserving capacity. Water absorbing resin can increase water content by 7–20 % with a 0.5 % concentration. High water absorbing resins applied to soils with different textures (sandy loam, loam and clay) may produce different water absorbing effects. Field tests have shown that spraying corn with 7500 g resin in a 0.5 % concentration can increase soil water content in the top 0–10 cm soil layer by 4.8–11.2 %.

6.4.3 Chemical Cover Treatments

Chemical cover treatment can preserve soil moisture, increase soil surface temperature and improve growth conditions for seedling stage of crop. Experiments conducted in Daxing District of Beijing in 1972 showed that by adopting chemical cover treatment the field evaporation rate reduced by 33.3–48.9 % (quoted from Zhang et al. 1999).

The average temperature of soils in the top 0–20 cm after spraying chemical treatments increases 2.6–4.5 °C on sunny days, 3.1 °C on cloudy days and 2.1–2.2 °C when the sky is overcast.

Chemical spray treatment is prepared by diluting 1 kg of chemical compound with 5–6 kg water. Therefore, 1200–1500 kg is needed to totally cover one hectare of farmland. The time to spray the chemical solution is 10 days ahead of sowing in spring. The water used for diluting the chemical treatment should be soft. When using water with high content of mineral matter, washing powder should be added.

Before spraying chemical cover treatment, the farmland should be prepared, irrigated and fertilized, so that subsequent irrigation and fertilizer application are not necessary.

References

Zhang Z, Gong, S, Wang X. Rainwater harvesting techniques. Beijing: China Water Resources and Hydropower Publications; 1999. pp. 79–94. (in Chinese).

Part II
Rainwater Harvesting Experiences from Around the World

Chapter 7
Rainwater Harvesting: Global Overview

Andrew Guangfei Lo and John Gould

Keywords Rainwater harvesting case studies · Prospects of rainwater harvesting

7.1 Introduction

Rainwater harvesting has been a common practice by different civilizations for about four thousand years mainly used for drinking purposes or agricultural practices. The Egyptians built storage tanks ranging from 200 to 2000 cubic meters (some of them are still in use today); in Thailand, the practice is about 2000-year old, and in rural areas of the world, pots and different types of methods have been used to irrigate the soil and store rainwater for centuries. Looking at the practice today, it is possible to see that the basic methods of collecting and storing rainwater utilized years ago have not changed much since then but only the characteristics of the systems in terms of technologies and construction materials and means of utilization of the stored water. Unfortunately, in many countries, technological development has resulted in the abandonment of traditional methods and approaches, which in turn has resulted in increasing the pressure on freshwater resources. Cheap and accessible water generally leads to waste and careless management.

Professor Andrew Lo—Past President of the International Rainwater Catchment System Association

---

A.G. Lo (✉)
Chinese Cultural University, Taipei, Taiwan
e-mail: andrewlo@faculty.pccu.edu.tw

J. Gould
Lincoln University, Canterbury, New Zealand
e-mail: johnegould@gmail.com

Q. Zhu et al. (eds.), *Rainwater Harvesting for Agriculture and Water Supply*,
DOI 10.1007/978-981-287-964-6_7

There are many advantages in the practice of rainwater harvesting, for example, water could be used at the same place it is collected and stored, the owner could be the user as well as the manager encouraging better practices; negative environmental impacts are minimal (the opposite to big water impoundments or reservoirs). Rainwater is relatively clean and can be used for different purposes with little or no treatment, and the technology for the systems is cheap and accessible.

Rainwater harvesting also has a range of possibilities as an interim or primary source of water for daily use on many small islands to a water source during periods of drought in semi-arid regions especially in some countries in Africa, Latin America and Asia, Gould and Nissen-Petersen (1999).

There are modern and traditional systems for rainwater harvesting, for example, simple small-scale roof collection systems (Thomas and Martinson 2007), larger systems (usually used for providing water for schools, stadiums, airports; during the second world war even whole airfields were used as catchments to collect rainwater in some countries a practice which continues today), collection systems for high-rise buildings in urbanized areas, land surface catchments systems, and storm-water collection systems to prevent pollution of water sources from roads, industrial sites, and agriculture. The use of catchments comprised of exposed bedrock has been a common method of RWH for thousands of years as well as ground catchments for concentrating runoff agriculture. In the Negev desert in Israel, storage tanks for runoff from hillsides allowed for the development of orchards in areas where rainfall was as little as 100 mm yearly.

7.2 Examples of Rainwater Harvesting and Utilization Around the World

7.2.1 Bangladesh

In Bangladesh, rainwater collection is seen as a viable alternative for providing safe drinking water and is particularly relevant in areas where groundwater suffers from high levels of arsenic or is affected by seawater intrusions. In the late 1990s, about 1000 rainwater harvesting systems were installed primarily in rural areas, by the NGO Forum for Drinking Water Supply & Sanitation. This Forum is the national networking and service delivery agency for NGOs, community-based organizations, and the private sector concerned with the implementation of water and sanitation programs in "neglected" and "underserved" rural and urban communities. Its primary objective is to improve access to safe, sustainable, affordable water and sanitation services and facilities in Bangladesh.

The rainwater harvesting tanks in Bangladesh vary in capacity from 500 to 3200 l, costing from Tk.3000–8000 (US$50–150). The composition and structure of the tanks also vary, and include ferrocement tanks, brick tanks, reinforced concrete ring tanks, and sub-surface tanks.

The harvested rainwater is used increasingly for drinking and cooking and is accepted as a safe, easy-to-use source of water among local users. Water quality testing has shown that water can be preserved for four to five months without bacterial contamination. The NGO Forum has also undertaken some recent initiatives in urban areas to promote rainwater harvesting as an alternative source of water for all household purposes.

One recent initiative which began with a pilot project in 2011 is the Amamizu project which has been undertaken with Japanese support (www.skywaterbd.com). This project aimed to transfer the highly successful ferrocement Thai jar design to Bangladesh. This began with a hand on training course in Bangladesh through which Thai artisans skilled in jar construction trained Bangladeshi counterparts. To date, around 200 jars have been constructed in rural communities where they have been very well received.

7.2.2 *Bermuda*

The island of Bermuda is located 917 km east of the North American coast. The island is 30 km long, with a width ranging from 1.5 to 3 km. The total area is 53.1 km^2. The elevation of most of the land mass is less than 30 m above sea level, rising to a maximum of less than 100 m. The average annual rainfall is 1470 mm. A unique feature of Bermuda roofs is the wedge-shaped limestone "glides" which have been laid to form sloping gutters, diverting rainwater into vertical leaders and then into storage tanks. Most systems use rainwater storage tanks under buildings with electric pumps to supply piped indoor water. Storage tanks have reinforced concrete floors and roofs, and the walls are constructed of mortar-filled concrete blocks with an interior mortar application approximately 1.5-cm thick. Rainwater utilization systems in Bermuda are regulated by a Public Health Act which requires that catchments be whitewashed by white latex paint. The paint must be free from metals that might leach into water supplies. Owners must also keep catchments, tanks, gutters, pipes, vents, and screens in good repair. Roofs are commonly repainted every two to three years and storage tanks must be cleaned at least once every six years.

7.2.3 *Botswana*

Botswana is landlocked, and much of the country is covered by deep Kalahari Desert sand. It has few perennial rivers or natural lakes and much of its groundwater resource is deep and in many areas is saline. Despite the relatively low, highly variable and seasonal rainfall ranging from <250 to 750 mm annually, due to the scarcity of other water resources, there is a significant interest in rainwater harvesting as a supplementary water supply.

In recent decades, thousands of roof catchment and tank systems have been constructed at primary schools, health clinics, and government houses throughout Botswana by the town and district councils under the Ministry of Local Government, Land and Housing (MLGLH). The original tanks were prefabricated galvanized steel tanks and brick tanks. The galvanized steel tanks have not performed well, with a short life of approximately 5 years. The brick tanks are unpopular, due to leakage caused by cracks, and high installation costs. In the early 1980s, the MLGLH replaced these tanks in some areas with 10–20 m^3 ferrocement tanks promoted by the Botswana Technology Centre. The experience with ferrocement tanks in Botswana is mixed, some have performed very well, but some have leaked, possibly due to poor quality control. Since the late 1980s, molded plastic tanks typically ranging from 1 to 20 m^3 have become more popular. Due to the relatively low rainfall and a dry season lasting more than 6 months and prolonged droughts common, rainwater tanks need to be large, and household roof catchment systems can generally only provide a supplementary domestic water source in most cases.

In recent years, the government has become increasingly keen to promote food security and encourage greater agricultural self-sufficiency and has supported construction of small-scale ponds fed from surface runoff, for providing water for livestock and irrigation.

7.2.4 Brazil

Rainwater harvesting and utilization is now an integrated part of educational programs for sustainable living in the semi-arid regions of Brazil. The rainwater utilization concept is also spreading to other parts of Brazil, especially urban areas.

Brazil is a water-rich nation; nevertheless, it has a large area that is semi-arid in the North East of the country where it has only been possible to survive by managing rainwater. This semi-arid region covers 975,000 km^2 over 1133 municipalities over 9 states and has 23 million inhabitants. The region receives less than 800 mm rain per year on average but has great annual variability, high rates of evaporation, and regular droughts. During the last 15 years, the Government of Brazil has worked with civil society to implement a large number of projects to enhance livelihood and increase water security in the area.

Northeast Brazil is an area in which people have been quite poor and survived on marginal farms. Since 2003 ASA (Articulação no Semi-Árido Brasileiro)—[Speaking up for the semi-arid areas in Brazil]—a local NGO network has brought together some 3000 civil society organizations associated with the semi-arid areas and supported construction of over 400,000 rainwater tanks for domestic water supply, Fig. 7.1. This has been part of the one million cistern program (P1MC). Subsequently, to further enhance livelihoods, the P1 + 2—One Piece of Land and Two Types of Water—Program was launched in 2007. So far, it has completed 9000 cement-plate tanks, 420 sub-surface dams, and 302 rock catchments serving some 12,000 families. This program encourages the use of rainwater for small-scale irrigation as well as water-conserving technologies such as drip irrigation,

Fig. 7.1 A typical 16 m^3 Ferrocement Rainwater tank collecting runoff from a clay tile roof in Pernambuco State, Brazil (Photo: Courtesy Andrew Lo 1999)

planting pits, contour tillage, vegetative soil protection, and use of manure. Larger rainwater storage cisterns (52 m^3) for supplemental irrigation of vegetable gardens and for poultry raising have been constructed either with purpose built catchments (210 m^2) or using runoff diverted from roads.

Other agencies which have played important roles in facilitating initial technical development and promulgation of rainwater harvesting technologies are the ABCMAC—the Brazil rainwater harvesting association—and EMBRAPA—a major agricultural research agency. The importance of civil society and its ability to construct an intermediary to negotiate with government on behalf of its constituencies has also been critical to the success of the program.

The project aims to develop the institutional and policy context for mainstreaming RWH, and in particular, boost the role of RWH for enhanced food production and food security. Knowledge management through the development of learning systems, and mobilization and strengthening of existing RWH networks are critical instruments in this respect. Innovative approaches may be needed to raise the profile of RWH techniques and approaches in national development and ensure their mainstreaming in quality-enhancing operations.

In total, more than half a million rainwater tanks have been built in this region so far. Its success has been due to the combination of support from the Government and significant community involvement including advocacy by the large NGO and civil society component. The program has borrowed ideas on technology choice from East Africa (ferrocement tank design) and strategies and practices used in the highly successful 1-2-1 program in Gansu, China as inspiration.

There are many useful lessons from the Brazilian experience which would be applicable in the context of future and on-going projects in Africa, and the possibilities for some technical exchanges are already being explored with organizations such as the RAIN Foundation and SEARNET.

7.2.5 Cambodia

Climatic conditions in Cambodia are well suited to rainwater harvesting with a mean annual rainfall of 1400 mm and only 2 or 3 months of the year with little or no rainfall. Given the relative abundance of rain, small tanks can provide an effective household supply even from a small roof catchment. Although ground and surface water supplies are abundant in many areas, problems with arsenic, iron, and nitrate contamination of ground water and pesticide or bacteriological contamination of surface water, mean that water quality is often problematic and around half the rural population are still lacking access to safe supplies. The Government and NGOs active in providing improved water and sanitation have been working hard to meet Cambodia's own post-Millennium Development Goal targets for 2025, namely, to provide 100 % coverage for improved rural water supply.

Since 2003 it has been widely recognized that RWH has an important role to play in meeting the water supply needs of those still not adequately served, especially in rural areas. This was part of the motivation behind the establishment of the Cambodian NGO Rainwater Cambodia in 2004, which with significant support from international donors has assisted in improving water supplies for over 64,000 people through providing household rainwater supplies and larger tanks at 139 schools and 16 health centers.

A minimum household storage capacity of 3000 l is recommended and this is normally provided through one or more ferrocement water jars. At schools and health clinics, rainwater tank volumes typically ranged from 14,000 to 35,000 l. In order to improve the quality of the stored rainwater, first, flush devices with a self-closing ball valve are fitted to all new systems, as shown in Fig. 7.2.

Traditionally, rainwater has always been collected and stored in smaller jars up to 500 l, which has made the upgrading and extension of this already familiar technology and water source relatively straight forward. In addition to Rainwater Cambodia, many other NGOs and CBOs (Community-Based Organizations) working in Cambodia such as UNICEF, Oxfam, World Vision, Water Aid, etc., are also involved in the construction of household rainwater tanks.

7.2.6 Japan (Tokyo)

In Tokyo, rainwater harvesting and utilization is generally promoted to mitigate water shortages, control floods, and secure water for emergencies, Group Raindrops (1995).

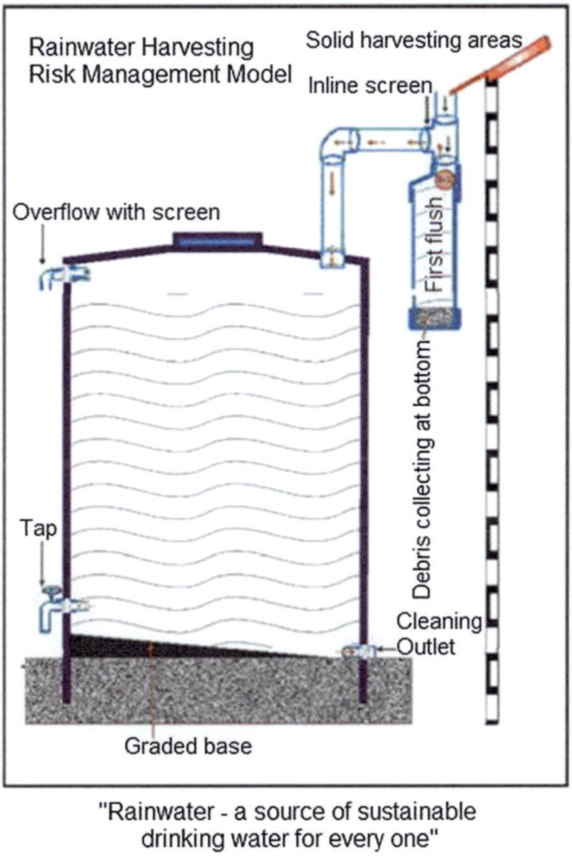

Fig. 7.2 Diagram showing cylindrical tank design being promoted along with ferrocement jars for household rainwater storage in rural Cambodia (*Source* with permission from Phreng and Keo 2014). Note the self-closing first flush device to ensure high water quality

The Ryogoku Kokugikan Sumo-wrestling Arena, built in 1985 in Sumida City, is a well-known facility that utilizes rainwater on a large scale. The 8400 m^2 rooftop of this arena is the catchment surface of the rainwater utilization system. Collected rainwater is drained into a 1000 m^3 underground storage tank and used for toilet flushing and air conditioning. Sumida City Hall uses a similar system. Following the example of Kokugikan, many new public facilities have begun to introduce rainwater utilization systems in Tokyo including the 622 m high Sky Tower which opened in 2012.

At the community level, a simple and unique rainwater utilization facility, "Rojison" has been set up by local residents in the Mukojima district of Tokyo to utilize rainwater collected from the roofs of private houses for garden watering, fire-fighting, and drinking water in emergencies. To date, about 750 private and public buildings in Tokyo have introduced rainwater collection and utilization systems. Rainwater utilization is now flourishing at both the public and private levels.

7.2.7 Kenya

Kenya has a long history of RWH stretching back centuries including traditional terracing systems such as *Fanya juu* and ground catchments and tanks made from lime called *Djabias* used along the coast. The first roof tanks date from the late nineteenth century when the introduction corrugated iron allowed for metal roof catchments and tanks at remote farmsteads. In the mid-twentieth century, the Government of the day also started building rock catchment systems serving communities in the semi-arid Kitui district. Due to the very varied nature of geographic and climatic conditions in the country, there has been scope for a very broad range of RWH technologies to be developed both for domestic supply, agriculture and livestock. Many of the RWH initiatives have been spearheaded by NGOs and community-based organizations as well as the Government and private sector. Several water tank manufacturers are based in Nairobi, and plastic, metal, and other types of rainwater tank are sold throughout East Africa and beyond.

Following the 6th International Rainwater Catchment Systems Conference, in Nairobi in 1993 (the first held in Africa), the Kenya Rainwater Association was established in 1994 the first national RWH association in Africa. Since that time tens of thousands of RWH systems have been constructed in Kenya by a wide range of organizations, resulting in millions of people benefiting from the technology over the intervening years.

Organizations like ASAL Consultants have promoted a wide range of RWH technologies including ferrocement roof and ground tanks, rock catchments, sand and sub-surface dams, and road catchments among others—details of these can be found at www.morewaterforaridlands.org. RWH for both household supplies and agriculture has also been promoted by SEARNET—an NGO with a regional mandate whose headquarters are in Nairobi. With a network covering 18 countries, SEARNET has actively been involved disseminating information and advice on RWH technologies across East and Southern Africa.

7.2.8 Thailand

In the mid-1980s, millions of rural people in NE Thailand faced a recurrent seasonal water shortage during the dry season. The solution to this was found by upgrading a traditional ceramic clay 100–200 l water jar, into an upgraded larger ferrocement jar. Storing rainwater from rooftop runoff in ferrocement jars is an appropriate and inexpensive means of obtaining high-quality drinking water. The relatively high rainfall exceeding 1200 mm in most of the region, large corrugated iron roofs, and the low cost of materials including cement and skilled labor, conditions were ideal for the widespread adoption of household rainwater jars. The jars range from 100 to 3000 l and are equipped with lid, tap, and drain. The most popular size is 2000 liters and these are used to store rainwater for a six-person household (Fig. 7.3).

Fig. 7.3 Ferrocement jars in Thailand (Photo courtesy Andrew Lo 1999)

Two approaches are used for the acquisition of water jars. The first approach involves technical assistance and training villagers on water jar fabrication. This approach is suitable for many villages, and encourages the villagers to work cooperatively. Added benefits are that construction of this environmentally appropriate technology is easy to learn, and villagers can fabricate water jars for sale at local markets. The second approach is applicable to those villages that do not have sufficient time or skills for making water jars involve access to a revolving loan fund to assist villagers in purchasing jars. Villagers are also trained on how to ensure safe supply of water and how to extend the life of the jars.

Initially implemented by the Population and Community Development Association (PCDA) in Thailand, the demonstrated success of the rainwater jar project has encouraged the Thai government to embark on an extensive national program for rainwater harvesting.

Several million of these ferrocement 'Thai jars' have been constructed in N.E.Thailand, since the late 1980s. The design has also spread to several other countries and has been widely adopted in neighboring Cambodia (see above) and continues to spread with its recent introduction in Bangladesh.

7.2.9 China, Gansu Province

Gansu is one of the driest provinces in China. The annual precipitation is about 300 mm, while potential evaporation amounts to 1500–2000 mm. Surface water and groundwater are limited; thus, agriculture in the province relies on rainfall and people generally suffer from inadequate supplies of drinking water.

Since the 1980s, research, demonstration and extension projects on rainwater harvesting have been carried out with very positive results. In 1995/96, the “121” Rainwater Catchment Project implemented by the Gansu Provincial Government supported farmers by building one rainwater collection field, two water storage tanks, and providing one piece of land to grow cash crops. This project has proven successful in supplying drinking water for 1.3 million people and developing irrigated land for a courtyard economy. As of 2000, a total of 2,183,000 rainwater tanks had been built with a total capacity of 73.1 million m^3 in Gansu Province, supplying drinking water for 1.97 million people and supplemental irrigation for 236,400 ha of land.

Rainwater harvesting has become an important option for Gansu Province to supply drinking water, develop rain-fed agriculture, and improve the ecosystem in dry areas. Seventeen provinces in China have since adopted the rainwater utilization technique, building 5.6 million tanks with a total capacity of 1.8 billion m^3, supplying drinking water for approximately 15 million people and supplemental irrigation for 1.2 million ha of land.

Note: A more detailed account of the rainwater harvesting programs in Gansu is given in the book Every Last Drop (Zhu et al. 2012).

7.2.10 Taiwan

The average annual rainfall amounts to 2515 mm in Taiwan, about 2.6 times the world average. The annual per capita rainfall resource, however, is only about 4300 m^3, which is less than 1/6 of the world average. The uncertain hydrologic and special geographic characteristics have created limiting natural and environmental conditions that usually concentrate 78 % of rainfall between May and October every year. The water supply difference between dry and wet periods is very significant. Only during spring seasons, when the surface runoff is less, almost all the rainfall can be used. During large storm events, stream runoff rises and recedes quickly resulting in more than 77 % flowing directly to ocean every year. The effective rainfall use is, therefore, extremely small even with very high rainfall amount.

Population and industry have become concentrated as a result of recent urbanization in Taiwan. A series of water resources problems have emerged along with the social, environmental, and hydrologic changes, as well as the urban and industrial development. The government policy is therefore, geared toward more sustainable development to combat the inadequate and degrading water resources and water environment.

Trend of Rainwater Harvesting in Taiwan: In Taiwan, rainwater harvesting originated in the Yuan Dynasty. Farmers constructed dikes along streams to divert water to ponds and use for irrigation. In the Ming and Ching Dynasties, irrigation ponds spread over the entire island. Toward the late Ming Dynasty, General Cheng Chin Kung pushed the Dutch people away and increased more farmland and constructed more irrigation ponds. This trend continued throughout the Ching Dynasty. During the last half century, large industrial parks have emerged and the government has had to develop large reservoirs to satisfy the increasing demand.

But, it has also resulted in serious detrimental impacts to the environment. With the increasing environmental awareness, the traditional large water resources engineering projects have been replaced by small-scale ones, such as rainwater utilization, desalination, and water recycling along with small-scale engineering works such as river dams and artificial lakes.

Accomplishment of Rainwater Harvesting Promotion in Taiwan: In recent years, the Water Resources Agency, Ministry of Economic Affairs has focused the promotion of rainwater harvesting in agriculture. This effort has been spread to the industrial sector. The "Criteria to Promote Rainwater Harvesting Systems" and the "Incentive Program to Collect and Recycle Rainwater" are two government policies to promote construction of rainwater harvesting systems in government buildings, schools, communities, factories, and farms. The incentive support includes collection devices, storage devices, simple water treatment devices, pipes, water pumps, and water meters. Positive results were obtained from many different sectors:

Agricultural Rainwater Utilization—Hill slope areas comprise about 3/4 of the total land area. Most agricultural land is located on sloping land where irrigation by stream diversion is very difficult. Therefore, the development of small to medium size rainwater harvesting systems is very suitable in hill slope and is one of the main characteristics of rainwater utilization in Taiwan. In the old days, farmers constructed natural and man-made farm ponds to store rainwater which enabled the beginning of agricultural development in the Taoyuan Plateau. With widespread farm pond building, the area was famous for its "hundred-mile farm ponds." Even with the completion of the Shihmen Reservoir to provide irrigation water, the hundred-mile farm ponds still play an important role in supplying irrigation water in this area.

Since 1998, a total of 3800 small to medium size water tanks have been built with a total storage capacity of almost 140,000 m^3 and a total investment of 15.319 million NT$. Most storage tanks are of the 50 cubic meter size and use aluminum alloy as building material. The average rainwater collected exceeds 5 million m^3 and the beneficial irrigation area is about 5868 ha. With an annual interest rate of 10 % and the use-life of 15 years, the investment cost is about 3–7.5 NT$ per m^3. Compared to the investment cost of 22 NT$/$m^3$, the beneficial effect of agricultural rainwater use should be worth promoting.

Industrial/Residential Rainwater Utilization: Promotion of industrial/residential rainwater uses begins in 1996. Up till now, the total investment reaches 50.73 million NT$. The total storage capacity of the water tanks is about 20,538 m^3. The substituted water supply is about 700,000 m^3. The main rainwater use is for toilet flushing (about 84.1 %), followed by irrigation (8.8 %), cleaning (5.5 %), and landscaping (1.3 %). The promotion of industrial/residential rainwater uses has spread to public buildings, schools, and industry all over the island. Several examples are selected here for further illustration:

(1) Taipei Zoo Rainwater Harvesting Project:
Taipei Zoo is one of the most popular sightseeing attractions in Taiwan. The Zoo takes advantage of the high elevation and rainfall characteristics of the

area and constructed almost 900 m^3 storage capacity water tanks of various sizes to provide water supply for toilet flushing, irrigation, and animal washing. The average rainwater use exceeds 400,000 m^3.

(2) Hydraulic Laboratory in Cheng Kung University Rainwater Utilization Project:
The average water use in the Hydraulic Laboratory in Cheng Kung University is about 14,000 m^3. A rainwater harvesting system was constructed to comply with the government water conservation policy. The roof top (about 14,000 m^2) of a large experimental water pond is used to collect rainwater and divert to a nearby round-shaped water tank (3 m deep, 15,000 m^3 capacity) for storage. The collected rainwater satisfies the total water demand of the entire laboratory.

(3) Tzu Chi Hospital Complex Rainwater Harvesting Project:
The concept of rainwater utilization is being introduced to the first construction phase of the Tzu Chi Hospital Complex. The collected rainwater is used for landscape irrigation purpose. The average annual rainwater use reaches more than 1700 m^3
Rainwater Harvesting Incentive Program in Taiwan: In order to encourage the public to construct rainwater harvesting systems, the Water Resources Agency of the Ministry of Economic Affairs developed both the "Criteria to Promote Rainwater Harvesting Systems" and the "Incentive Program to Collect and Recycle Rainwater" policies. Details of both policies are listed as follows:
– Suitable applicants: Agricultural, residential, and industrial water users
– Supports: Stainless steel or aluminum alloy water tank/Ferrocement or brick water tank
– Application rules:

- This criterion is applicable to well-designed not yet constructed rainwater harvesting systems, not already constructed rainwater harvesting systems.
- Each application is meant for one set of water tank. The amount of support is given according to total storage capacity up to 50 m^3. Cost exceeding this limit should be provided by the user.
- The support will be given to the highest priority applicant first until the annual budget has been exhausted. The application priority will be determined according to the scheme set by the Agency.
- This incentive program will not support applicants who have applied other incentive support. Applicants receiving support from this program should not apply other incentive support.

(4) Application priority criteria

- Land-subsidence or severe water shortage areas
- Remote or labor island (farm, school, or factory) applicants
- Willing applicants that accept interview or survey and provide water use information

(5) Support standard for rainwater harvesting system
Future Prospect of Rainwater Harvesting in Taiwan: After many years of great effort from the government and the public, significant rainwater harvesting achievement has been gained. However, many problems still exist in rainwater harvesting promotion that needs to be solved. These include

- Lack of rainwater harvesting awareness
- Insufficient rainwater resource knowledge leads to inability to accept rainwater as an important means to alleviate the water shortage problem.
- Unbalanced development
- At present, rainwater harvesting use in agriculture and public water supply is developing very fast, although its uses in flood mitigation, hydrologic cycle maintenance, and water quality conservation have just started.
- Lack of integrated and systematic research
- In order to achieve highly efficient use, researches in rainwater storage, utilization, and management techniques have to be integrated and further promoted.
- Lack of standardization and guidance
- Local standards need to be determined according to local conditions, being tested, demonstrated, and applied. National standards can then be established to accelerate development of rainwater harvesting systems.

Faced with the high-tech industry that requires extremely high water demands in the twenty first century, the future rainwater harvesting strategies to provide high quality and stable water resources are as follows:

- Establishment of Rainwater Harvesting Association in Taiwan: organizing interested experts to join force to promote and utilize rainwater more efficiently.
- Implementation of rainwater harvesting legislation: legalization of rainwater harvesting techniques will enable actual promotion and implementation.
- Promotion of regional integrated rainwater harvesting studies and related design standards: establishing rainwater utilization standards because different rainwater harvesting systems perform differently.
- Research and development in related rainwater harvesting techniques: establish and integrate rainwater harvesting techniques and water use management.
- Continued effort in technology promotion and extension activities

Rainwater utilization training courses and symposia should be held regularly on top of occasional rainwater utilization exhibitions, rainwater study tours, and mass media education promotion programs. Production of rainwater utilization system design and construction guidelines, successful case studies video tapes should be distributed to the industries, government agencies, and academic institutions to further promote rainwater uses.

7.3 Future Development Prospects of Rainwater Harvesting

7.3.1 Future Needs

Between 1940 and 1990, the world's population has more than doubled, from 2.3 to 5.3 billion. Simultaneously, per capita use of water doubled, from 400 to 800 cubic meters per person per year as reported by Clarke (1993). By 2012, the global population topped 7 billion and the per capita water use reached 1385 cubic meters. The rapidly growing pressure on water resources is alarming. Water shortages are indeed becoming one of the world's most pressing problems. Developing alternative water supplies to cope with water shortages are a pressing challenge for water resources decision makers. Rainwater harvesting has long been identified as the practical means for cooperation between public and private sectors to help solve water supply problems. However, historically, cooperation between public and private sectors was not very successful because few countries have had policies or guidelines designed for the development and operation of their rainwater utilization efforts.

Nevertheless, there are a few significant examples from around the world where severe regional water shortage problems have been addressed through the carefully planned and strategic application of rainwater harvesting systems. Some of these have already been touched on earlier in this chapter. The impressive achievements in the 1980s and 1990s in N.E. Thailand where around 18 million rural residents benefited from the construction of several million 2-cubic meter reinforced thin-wall concrete jars to store rainwater for drinking water supply (Wirojanagud and Chindaprasirt 1987). In Gansu Province, China, the provincial rainwater utilization project initiated in the mid-1990s has helped to solve their water shortage problems and has enhanced rural health and agricultural production, Zhu and Wu (1995). The million tank initiative in N.E. Brazil has also been a good example of what can be achieved when communities, community-based organizations. and government come together to address water shortage issues using rainwater harvesting technologies. While there are many other examples of countries where the application of rainwater harvesting has had a significant impact on addressing water shortages including several already mentioned earlier, in a lot of cases progress has been achieved through a more piece meal approach with numerous relatively small individual community initiatives often backed by non-governmental organizations collectively helping to address the challenges of water shortage in a significant way. Among the countries where this approach is common are Kenya, Uganda, and India. Since the late 1990s, for example, there has been a growing interest and application of rainwater harvesting to address a growing water crisis in India. One organization, the Centre for Science and Environment has been a key in coordinating efforts to raise the profile of rainwater harvesting and promote its use. Numerous workshops, conferences, and publications have been produced covering these developments in India, but two books in particular Dying Wisdom: (Agarwal and Narain 1997) and Making Water Everyone's Business (Agarwal et al. 2001) have been especially influential in promoting the technology.

The spirit of teamwork and cooperation between private and public sectors has been strongly maintained in these rainwater harvesting development programs. Future cooperation of rainwater harvesting development and management in various levels of public and private sectors is very bright.

7.3.1.1 Rainwater Harvesting: Global Networks and Initiatives

There have been a number of initiatives to build a global network, where RWH practitioners and researchers can share experiences and information. The International Rainwater Catchment Systems Association established in 1991 oversaw the biennial International conference series that had actually been initiated by Professor Yu Si Fok, with the first conference in Hawaii in 1981.

RAIN Foundation (www.rainfoundation.org) since it was established in 2003, the Netherlands based RAIN Foundation has been working with its partners to develop, spread, and implement rainwater harvesting systems worldwide with the aim that everyone in the world should have access to safe water. The RAIN Foundation has supported RWH initiatives through its partners in several developing countries including Burkina Faso, Ethiopia, Kenya, Uganda, Mali, and Nepal. Perhaps most significantly, they are effectively using social media to link RWH practitioners and researchers through hosting various online forums 'webinars' and linking interested individuals and projects through their RWH mail group. Since around 2000 RWH has increasingly been considered as a mainstream water supply technology which should be considered alongside other conventional options such as groundwater and reticulated-treated surface water supplies, rather than an alternative option of last resort. More recently, the potential of using RWH for supplementary irrigation to tied crops over critically dry periods is being recognized, an approach that was pioneered and successfully demonstrated in Gansu, China.

At the Stockholm Water Week, the urgent need to encourage the better management of rainwater to meet the growing food and water security needs around the world was strongly articulated in a joint statement by a high profile group of concerned scientists and experts in the field, see Box 7.1.

BOX 7.1: Concerned Scientists and Experts Declaration on Water, Hunger and Sustainable Development Goals

Managing rain: the key to eradicating poverty and hunger

We scientists and experts, joining the 2014 World Water Week in Stockholm, are deeply concerned that sustainable management of rainwater in dry and vulnerable regions is missing in the goals and targets proposed by the UN Open Working Group (OWG) on Sustainable Development Goals (SDGs) on Poverty (Goal 1), Hunger (Goal 2), and Freshwater (Goal 6).

We commend the OWG for setting ambitious and aspirational global development goals of eradicating poverty and hunger and promoting equity, ensuring peace and transparent global governance, within the context of global sustainability for climate, oceans, and ecosystems.

Our concern arises from the failure to recognize the ominous congruence between, on the one hand, poverty, malnutrition, rapid population growth and economic reliance on agriculture, and the water challenges and predicament in semi-arid tropical and subtropical climates on the other. These drylands are the most water vulnerable inhabited regions of the world, hosting the world's poorest countries.

This is a challenge of global importance. Drylands cover 41 % of the world's land surface, host 44 percent of the world cultivated systems and are home to 2.1 billion people in nations with the world's highest population growth rates. Here, food production and human livelihoods rely on limited, highly variable, unreliable, and unpredictable rain. When it rains, it often pours in intense convective storms that generate flash floods with eroding surface runoff, making fruitful rain-fed agriculture and traditional irrigation extremely challenging. However, even in these areas, there is generally enough rainfall and thus potential to drastically improve food production, if only we can guide more of the water to beneficial, productive uses.

By 2050, business as usual will mean 2 billion smallholder farmers, key managers and users of rainwater, eking out a living at the mercy of rainfall that is even less reliable than today due to climate change. Setting out to eradicate global poverty and hunger without addressing the productivity of rain is a serious and unacceptable omission. The proposed SDGs cannot be achieved without a strong focus on sustainable management of rainwater for resilient food production in tropical and subtropical drylands.

Sustainable development for the poorest dryland farmers depends on the ability to build resilience and raise agricultural production within the capacity of local and severely underutilized rainwater. Management practices and techniques, such as rainwater storage, efficient supplementary irrigation, and integrated management of water, land, crops, and nutrients, can provide significant productivity gains and sustainable intensification of smallholder agriculture for livelihood improvements, community development, and food security. This could also open the possibility for investments, stimulating further agricultural development, benefiting from experiences in mid- and high-income countries.

We therefore call upon the United Nations General Assembly to add in any Hunger Goal a target on sustainable and resilient rainwater management for improved food production, through the adoption of sustainable watershed management practices at all scales aiming for an increase of over 50 % in the yield of food per unit of rainwater.

Signed in Stockholm, August 2014

Malin Falkenmark, SIWI, Johan Rockström, SRC, Torgny Holmgren, SIWI, Mohamed Ait Kadi, GWP, Tony Allan, King's College, Naty Barak, NETAFIM, Jeremy Bird, IWMI, Lisa Sennerby Forsse, SLU, Fred Boltz, Rockefeller Foundation, Peter Gleick, Pacific Institute, David Grey, University of Oxford, Jerson Kelman, University of Rio de Janeiro, Roberto Lenton, University of Nebraska, Julia Marton-Lefèvre, IUCN.

7.3.2 National Policy for Rainwater Utilization

Fok (1993) stressed the importance of national policy for rainwater harvesting development and pointed out that the best national policy should be based on grass-root communities' needs with cultural, social, environmental, and economic considerations. With limited financial support from central, provincial, and local government and foreign aid, rainwater harvesting users should contribute their own share of cost and labor for the development and operation of their rainwater harvesting systems. In many developing countries, women are generally charged with household chores including fetching water. Therefore, their involvement in formulating policy for rainwater harvesting development is crucial. With an established national rainwater harvesting policy, both private and public sectors can form a workable partnership in developing and managing their rainwater harvesting systems. In addition, the cooperation among rainwater harvesting users to aim for a better community will proceed naturally. The public sector can assist rainwater harvesting users to manage their own water supply. This water resources management strategy will be a first in the twenty first century. Decision makers in water resources will be in a win–win situation with the development of a national policy for rainwater harvesting. We have to keep in mind that each country has its own social-environmental conditions. Therefore, each nation has to design its own national policy for their rainwater harvesting development.

7.3.3 Utilization of Rainwater in Megacities

Population growth in the future will be concentrated in urban areas and particularly in the megacities in developing countries. The need for potable water is one of the many requirements that have to be met within these megacities that will have to cope with the influx of large populations from rural areas. As almost 90 % of the migration of population to megacities is expected to occur in developing countries, there is the need to look into simple systems for augmenting existing water supplies. Rainwater catchment systems appear to be one of the most appropriate abstraction systems that can be adapted in emerging megacities in both developing and developed countries.

The two major areas where runoff from megacities can be collected successfully are tapping of runoff in urban catchments in major water supply schemes and, secondly, utilizing smaller catchments fruitfully to abstract runoff for non-potable uses. In both these systems, the quality of the raw water bears paramount importance as it has been shown that there can be wide variations depending on surface area, collection methods, location, etc. (Appan 1999).

In the case of the larger urban storm-water collection schemes, there is the need for maintaining large tracts of the urban catchment so as to ensure that pollution levels are under control. Besides, it is advisable to avoid the collection of the dry weather flows and also the first flushes during storms. It is also preferred that such systems should be integrated with conventional collection systems so that quality control of raw water becomes easier.

For the smaller rainwater catchment systems, the same principles of design will apply wherein suitable relationships are established between the rainfall (inputs) and withdrawal rate (output) and the storage (water tank). There are quite a few successful cases where such systems have been adapted for use in public institutions, industries, airports, etc.

In the future, the migration of people to urban areas, in both developing and developed countries, is inevitable. And the challenge of meeting the water requirements can be largely met by an appropriate understanding, study, and application of rainwater catchment systems.

7.3.4 Future Prospect of Rainwater Harvesting Development

Recent economic setbacks offer political leaders a chance to review their regional cooperation opportunities and to examine their natural renewable resources. Definitely, water resources should be among their top considerations. Decisions on how to develop sustainable water resources is a great challenge. Water is the natural resources that have been most neglected and abused in the past century. The World Health Organization (WHO) has documented that contaminated water supplies are costing thousands of lives and causing millions to become ill on a daily basis resulting in great loss in productivity in each developing country. Fok (1992) has listed the advantages of rainwater harvesting, namely: low cost, fast implementation, catching rainwater at place of water use, and users developing their own water supply without relying on public sectors funding. In addition, users exercise water conservation and maintenance of water quality. These are some of the factors that public water decision makers look for in sustainable water resources development. A great deal of technological advancement has been published in many previous IRCSA International Conferences Proceedings. These publications are the most comprehensive collection of rainwater harvesting technological advances in international, regional, national, provincial, and local levels. The prospect of establishing the national rainwater harvesting development policy in every country can be accomplished through continuous and relentless promotion of rainwater harvesting.

7.3.4.1 Short-Term Prospects

The short-term prospects include the following:

(1) The formulation of rainwater harvesting development policies in every nation. In addition, each country will have a set of rainwater harvesting development and operation guidelines for their private and public sectors to follow.
(2) Each country develops online source information for rainwater harvesting development, products, and operation. Rainwater harvesting hardware that meets safe environmental standards will become widely available at prices that are affordable in a range of different sizes, shapes, and construction materials.
(3) Public sectors offer funding for rainwater harvesting research. In addition, there are a great number of non-profit private foundations specialized in community development, family health, natural disaster prevention, fire protection, etc., promoting rainwater harvesting.
(4) Public sector actors in water supply and water resource management adopt rainwater harvesting as a feasible alternative for water supply and water resources management. Financial institutions become willing to approve mortgages for rainwater harvesting development because rainwater harvesting systems are recognized as part of a building and their construction is under a valid building permit.
(5) Rainwater harvesting recognized as a fire protection tool. The stored water can be used for fire-fighting as well as other emergencies.
(6) Finally, from an integrated natural resources management point of view, rainwater harvesting is included in future city planning and rural development plans as standard.

7.3.4.2 Long-Term Prospects

The long-term prospects over the next thirty to fifty years may include

(1) Innovative technology in weather prediction helps users of rainwater harvesting systems plan when to store needed rainwater.
(2) Rainwater harvesting technology used to help restore grasslands in arid areas.
(3) Rainwater harvesting provides scarce water for reforestation in semi-arid lands.
(4) Many international non-profit private foundations contribute funds to support rainwater harvesting activities.
(5) Rainwater harvesting incorporated in new buildings that have collection systems for renewable natural resources such as energy and water. The buildings are constructed with an integrated resources management plan. The portion of the collected rainwater is dissociated into hydrogen and oxygen for energy production. Other portion of the collected rainwater is used for cooling. Wastewater is treated for reuse. The buildings do not use water to flush toilets because new technology has produced dry toilets to dispose wastes. In fact, all the wastes are treated and reused.

7.3.5 Further Development of the Technology

There is a need for the water quality aspects of rainwater harvesting to be better addressed. This might come about through

- Wider adoption of affordable first-flush devices that are installed as standard and greater involvement of the public health departments in the monitoring of water quality
- Monitoring the quality of construction at the time of construction.
- Provision of assistance from governmental sources to ensure that the appropriate-sized cisterns are built
- Promotion of rainwater harvesting as an alternative to both government- and private sector-supplied water, with emphasis on the savings on water bills
- Provision of assistance to the public in sizing, locating, selecting materials and constructing tanks, and development of a standardized plumbing and monitoring code
- Development of new materials to lower the cost of storage

Preparation of guidance materials (including sizing requirements) for inclusion of rainwater harvesting in a multi-sourced water resources management environment.

References

Agarwal A, Narain S, editors. Dying wisdom: rise, fall and potential of india's traditional water harvesting systems. In: State of India's environment, vol. 4. India: Centre for Science and Environment; 1997, 404pp.

Agarwal A, et al, editors. 'Making water everybody'd business: practice and policy of water harvesting. India: Centre for Science and Environment; 2001, 404pp.

Clarke R. Water: the international crisis. Cambridge: MIT Press; 1993.

Fok YS. Rooftops: the under-utilized resource. In: Proceedings of regional conference of International Rainwater Catchment Systems Association, vol. 1. Kyoto, Japan; 1992, pp. 164–174.

Fok YS. Importance of national policy for rainwater catchment systems development. In: Proceedings of the 6th international conference on rainwater catchment systems, Nairobi, Kenya, pp. 27–31, 1–6 Aug 1993; 1993.

Gould J, Nissen-Petersen E. Rainwater catchment systems for domestic supply: design, construction and implementation. UK: IT Publications; 1999 335p.

Group Raindrops. Rainwater & you—100 ways to use rainwater. Organizing committee for Tokyo international rainwater utilization conference. Organizing committee for Tokyo international rainwater utilization conference (1994); 1995.

Phreng K, VK Keo. Rainwater harvesting formalization for rural cambodia. 37th WEDC international conference, Hanoi, Vietnam; 2014.

Proceedings of the Tokyo international rainwater utilization conference—rainwater utilization saves the earth.

Rain Foundation. www.rainfoundation.org.

Thomas T, Martinson B. Roofwater harvesting: a handbook for practitioners. Delft: IRC International Water and Sanitation Centre; 2007, 157p.

UNEP. Global environment outlook 2000. Earthscan Publications Ltd.; 1999.

Wirojanagud P, Chindaprasirt P. Strategies to provide drinking water in the rural areas of Thailand. In: Proceedings of the 3rd international conference on rainwater cistern systems, pp. B5-1–B5-11, Khon Kaen, Thailand; 1987.
Zhu Q, Wu F. Rainwater catchment and utilization in the arid and semi-arid area in Gansu, China. In: Proceedings of the 7th international rainwater catchment systems conference, vol. 1, pp. 1-18–1-28, Beijing, China, 21–25 June 1995; 1995.
Zhu Q, Yuanhong L, Gould J. Every last drop: rainwater harvesting and sustainable technologies in rural China. UK: Practical Action Publishing; 2012, 160p.

Chapter 8
Rainwater Harvesting for Domestic Supply

John Gould

Keywords Rainwater tank design · Rainwater demand · Rainwater supply · First flush devices

For most rainwater catchment systems, the storage tank represents the single greatest cost, especially for roof tanks where an existing roof structure in effect provides a free catchment area. The choice of a suitable tank design to match an existing catchment and local conditions is important, and careful consideration should be given to selecting the right one.

8.1 Types of Water Storage Structure

Rainwater storage reservoirs can be subdivided into three distinct categories:

- Surface or above-ground tanks which are common in the case of roof catchment systems, where the catchment surface is elevated, e.g., For roof catchments (Figs. 8.1 and 8.2);
- Sub-surface or underground tanks which are often associated with purpose-built ground catchment systems (Figs. 8.1 and 8.3); and

J. Gould (✉)
Lincoln University, Canterbury, New Zealand
e-mail: johnegould@gmail.com

Q. Zhu et al. (eds.), *Rainwater Harvesting for Agriculture and Water Supply*,
DOI 10.1007/978-981-287-964-6_8

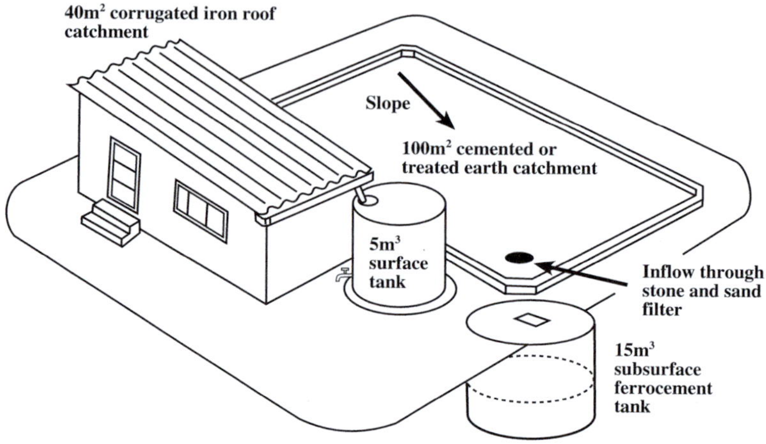

Fig. 8.1 Examples of surface (roof catchment) and sub-surface (ground catchment) rainwater tanks at rural households in Botswana

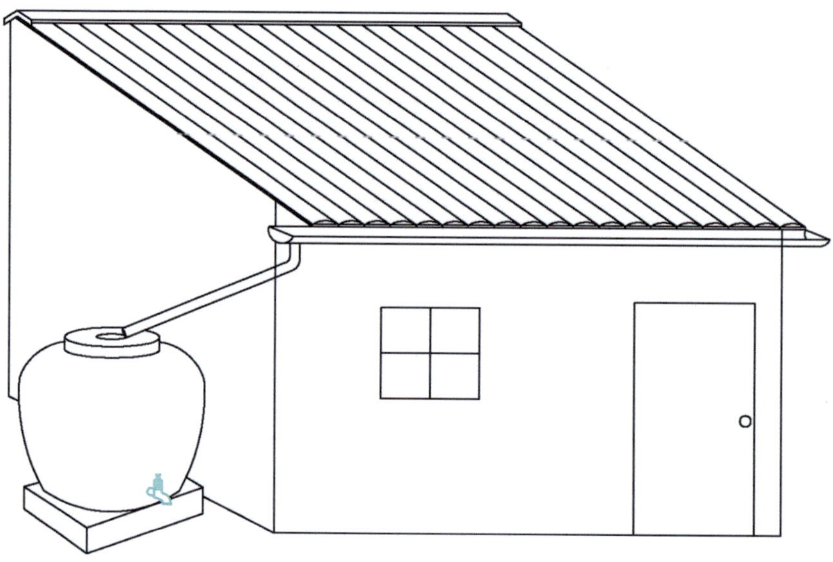

Fig. 8.2 2 m^3 'Thai Jar' ferrocement roof tank common in S.E. Asia (Gould and Nissen-Petersen 1999)

- Dams with reservoirs for larger catchment systems using natural catchments, e.g., Rock catchment dams, earth dams, and sub-surface or sand dams in sand rivers (Fig. 8.4). Examples and further information on these can be found at www.morewaterforaridlands.org

Since the water storage structure (tank or reservoir) is generally the most expensive part of the system, careful selection, design and construction are essential.

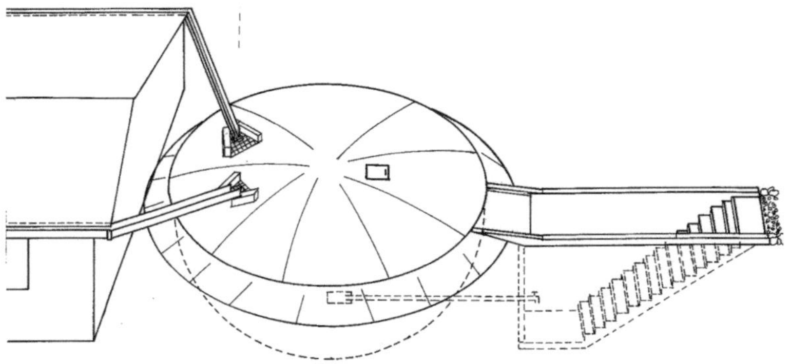

Fig. 8.3 80 m^3 hemispherical sub-surface tank from Kenya (Courtesy and permission from Nissen-Petersen)

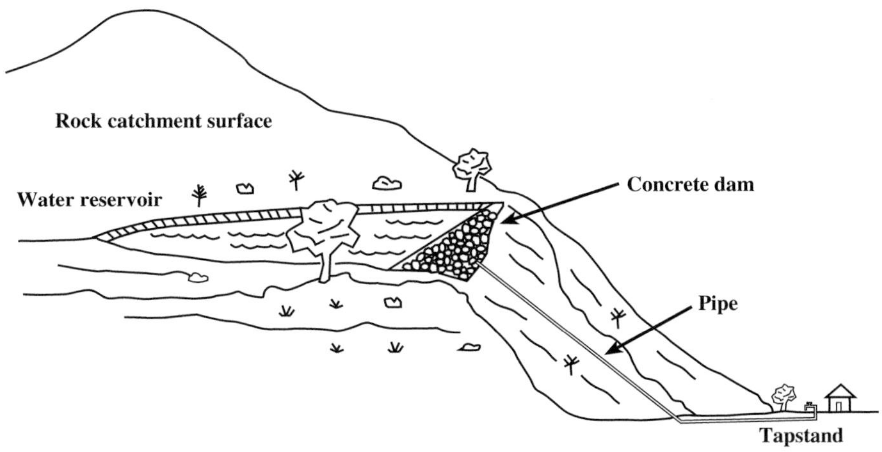

Fig. 8.4 Rock catchment dam common in East Africa

8.1.1 Surface Tanks

These can be constructed from a wide range of materials including

- Metal
- Wood
- Plastic
- Fiberglass
- Bricks
- Interlocking, compressed soil or stone blocks
- Ferrocement

- Concrete
- Rubber
- Others, e.g., ceramic

The key features of any tank are that it should be watertight, durable, affordable, and not contaminate the water in any way.

Surface tanks may vary in size from 1 m^3 to more than 40 m^3 for households and up to 100 m^3 or more for schools, hospitals, etc.

Smaller surface tanks can be made centrally and transported to site. It may be easier to ensure quality control in their construction. This is especially important for tanks which require a high level of workmanship, such as ferrocement. In Thailand, millions of 1–2 m^3 ferrocement jars have been constructed at small rural enterprises and delivered to households (Fig. 8.2).

A key advantage of surface tanks over sub-surface tanks is that water can be extracted simply and easily through a tap. If the tank is elevated, water can even be piped by gravity to where it is required. The main disadvantage of surface tanks is that they are relatively more expensive when compared with sub-surface storage.

Where roof catchment systems are incorporated in the initial plans and designs for houses, it is possible to make substantial cost savings by incorporating the rainwater storage reservoirs into the house structure itself, e.g., in the U.S. Virgin Islands, Bermuda, and parts of China, e.g., Zhejiang Province.

8.1.2 Sub-surface Tanks

Several of the techniques used for building surface tanks can also be used for sub-surface tanks with the soil being backfilled around the outside of the tank on completion. Where the soil is firm, some of the forces of the water against the side of the tank are absorbed by the soil and the walls do not have to be as strong as for an equivalent surface tank.

For ferrocement tanks, it is possible to line a carefully excavated hole with chicken wire and barbed wire reinforcement and plaster directly onto it. This dispenses with the need for a formwork for the walls of the main tank and helps reduce costs significantly.

Soil water and soil pressure will also exert external pressure on the tank walls. While this is counter-balanced by the internal water pressure when the tank is full, when the tank is empty this may produce substantial pressures and the walls must be strong enough to resist these. To overcome this, tanks must be cylindrical or hemi-spherical in shape.

Seasonal rises in groundwater levels may also create a situation where an empty tank can float like a boat. Care, therefore, has to be taken when siting the tanks! Raising the sides of the tanks above ground level and ensuring the tanks are never completely empty can also help counter this problem. Surface water should also be diverted away from the tanks by raising the ground around them.

Where impervious soils exist such as clay, it is often possible to construct unlined sub-surface reservoirs. Invariably, these suffer from problems of seepage, evaporation, and poor water quality.

The main advantage with sub-surface tanks is that they are generally cheaper per unit volume than surface tanks and use less space. The main disadvantage is that to access the water, some form of pump or other extraction method is required unless steps and a tap-stand are constructed from where the water can be extracted by gravity. Although, sub-surface tanks are quite common in some regions such as East Africa where they provide supplies for schools or small scale irrigation, nowhere are they as numerous as in China. In Gansu and neighboring provinces in China, over 2.4 million upgraded *Shuijiao* (sub-surface tanks) have been constructed since 1990. These are based on the traditional bottle shaped water cellars which have been used for centuries and were dug into loess soils and lined with clay. The upgraded *Shuijiao* are lined with cement mortar and have a concrete base and cover.

8.1.3 Dams and Sub-surface Storage

There are several different types of dams (rock catchments, earth dams, sub-surface dams, concrete dams, etc.) used for storing direct surface runoff.

(1) **Rock Catchment Dams**

Rock catchment dams are one of the cheapest and most effective types of rainwater storage system in areas with suitable sites (Fig. 8.4). Where impermeable exposed disjointed bedrock exists, potential dam sites can normally be found in natural valleys or hollows which can easily be converted into storage reservoirs by constructing rubble stone-masonry dams. Granitic inselbergs common in Africa and other parts of the tropics are ideal locations for rock catchments which normally vary from 500 to 10,000 m^3 in volume. Water can be piped by gravity to tap stands or storage tanks at the base of the outcrops or to nearby villages to improve accessibility.

(2) **Earth dams**

Earth dams consist of raised banks of compacted earth and can be constructed to retain water where it regularly flows into small valleys, depressions or on hillsides. The dam wall is normally 2–5 m in height and has a clay core and stone aprons and spillways to discharge excess runoff. Storage volumes can range from hundreds to tens of thousands of cubic meters.

(3) **Hafirs**

Hafirs are excavated reservoirs normally 500–10,000 m^3 in volume. Originating in Sudan, these sources still provide important traditional water supplies in many parts of semi-arid Africa for both people and livestock. Hafirs are located in natural depressions, and the excavated soil is used to form banks around the reservoir to increase its capacity. Bunds and improvements to the catchment apron may help increase runoff into the reservoir, but seepage and evaporation often result in drying out late in the dry season.

(4) **Sand River Storage Systems**

These are found mainly in semi-arid regions where most rivers are ephemeral and surface water flow may only be visible for a few weeks a year. Once the water subsides, the sandy river bed is exposed but while the river may appear dry, water is normally found flowing very slowly under the sand. These sand rivers provide important traditional water sources in many semi-arid parts of the developing world. To increase the upstream storage capacity in the river bed, a sub-surface dam can be built across the river channel using earth or stone-masonry with its top level with the sandy river bed. Sand dams are similar to sub-surface dams except that the top of the dam wall exceeds the level of the sandy river bed. These dams are built in stages with the dam wall height being increased by 0.3 m after floods have deposited sand to the level of the spillway. This allows sand to be trapped upstream of the dam wall, thus increasing the overall storage capacity of the river bed. Coarse river sand provides the best and greatest storage potential which often amounts to several thousand cubic meters for a single dam. The advantage with sand river storage is that it normally represents an upgrading of a traditional and hence socially acceptable water source. Since the water is stored under the sand, it is protected from significant evaporation losses and also less liable to be contaminated. The construction of river-intakes and hand-dug wells with hand pumps on the river bank can further help improve water quality.

(5) **Groundwater Recharge**

Groundwater recharge can be enhanced by rainwater harvesting where local conditions permit. This can sometimes result as a by-product due to downstream seepage from reservoirs following earth dam construction. In other places, runoff is deliberately diverted to recharge or maintain groundwater levels to ensure existing wells do not dry up.

8.2 Selection of Water Storage System

The most appropriate choice of storage system in any situation will depend on local conditions.

Several factors influence the choice of rainwater tank or reservoir, and these include:

- The amount of water storage required
  Is it for main or only supplementary supply?
  For single family, school, or whole community?
- Type and size of catchment
  Small fixed sized roof or expansive ground or rock catchment surface
- Rainfall amount and distribution
  Semi-arid low-rainfall or humid high-rainfall climate
  Seasonal climate with wet/dry season or rainfall all year

- Soil type and permeability
  Impermeable soils in arid areas may favor constructing sub-surface tanks
- Availability and cost of construction materials
  Cheap or freely available river sand, hard core, or rocks may make certain designs more affordable than others
- Availability and cost of off-the-shelf tank designs
  The cost and advantages with constructing storage tanks/reservoirs on site need to be balanced against the cost of off-the-shelf tank designs
- Affordability
  The high cost of rainwater storage tanks relative to incomes is often a key limiting factor
- Local skills and experience
  If local skills and experience are absent, a significant investment in training may be needed
- Availability of other water sources
  The quality, accessibility, and cost of development of other water sources

The comparative costs of locally available alternatives will also be a key factor in choosing the most appropriate tank or reservoir option.

Various environmental factors may preclude certain types of tank. For example, metal tanks are not suitable in areas of saline soils and coastal areas.

8.3 Determining Demand, Runoff Coefficient, and Supply

In this section, we will examine various ways of sizing a household rainwater catchment tank in order to endeavor to meet household water demand while minimizing system cost.

Usually, the main calculation when designing a domestic rainwater catchment system will be to size the tank correctly to give adequate storage capacity. The storage requirement will be determined by a number of interrelated factors. They include:

- Local rainfall data and weather patterns
- Roof (or other) collection area
- Runoff coefficient (this varies depending on roof material and slope)
- User numbers and consumption rates

Whether rainwater harvesting is done to meet occasional, intermittent, partial, or full household water supply requirements will also play an important part in determining the storage size.

8.3.1 Determination of Demand

Estimating household annual water demand may, at first, seem straightforward, i.e., multiplying mean daily water use per person by the number of household members by 365 days. For example, if household water use is 20 L of water per person per day for a family of 5, an annual household water demand of $100 \times 365 = 36{,}500$ L (36.5 m^3) might be expected.

In reality, it is more complex because adults and children use different amounts of water and seasonal water use varies significantly. For example, more water is used in the hottest and/or driest seasons. The number of family members staying at home may also vary at different times of the year. To try to take into account all such variables, household surveys need to be designed very carefully and detailed information sought.

Where demand estimates are being used as the basis for designing rainwater systems, they should therefore be treated with great caution, especially if the rainwater systems are the major or only source of supply. In such situations, adding a 20 % or more "safety margin" is appropriate.

Another factor is that people will tend to use rainwater more sparingly when water levels in household tanks get low. This informal rationing process is very important as it can significantly reduce the likelihood of the tank becoming completely empty and reduce the time period of any such system failures when they occur.

8.3.2 Runoff Coefficient

The runoff coefficient[1] (*C*) for any catchment is the ratio of the volume of water which runs off a surface to the volume of rainfall which falls on the surface.

$$C = \frac{\text{Volume of runoff}}{\text{Volume of rainwater}} \tag{8.1}$$

All calculations relating to the performance of rainwater catchment systems involve the use of a runoff coefficient to account for losses due to spillage, leakage, infiltration, catchment surface wetting, and evaporation which will all contribute to reducing the amount of rainwater which actually enters the storage reservoir.

For a well-constructed roof catchment, especially one made from corrugated iron sheets or tiles, the runoff coefficient for individual rainfall events may often be over 0.9, i.e., >90 % of rainfall collected (Ree 1976).

The long-term runoff coefficient for the system will, however, probably be less due to occasional, but substantial losses, resulting from gutter overflows during torrential storms (or temporary blockages by debris such as leaves), and the collection efficiency of both roof and ground catchment systems may be reduced when precipitation occurs as snow or hail or is affected by very strong winds. For this reason, it is appropriate to use a runoff coefficient of 0.8 as standard when

[1] Runoff coefficient may also referred to as the Rainwater Collection Efficiency (RCE).

Table 8.1 Runoff coefficients for different catchments in Gansu Province, China

Roof catchment		Ground catchment	
Clay tile (hand-made)	0.24–0.31	Concrete lined	0.73–0.76
Clay tile (machine made)	0.30–0.39	Cement soil mix	0.33–0.42
Cement tile	0.62–0.69	Buried plastic sheet	0.28–0.36
		Compacted loess soil	0.13–0.19
Corrugated iron[a]	0.8–0.85		

Source of data from Zhu and Liu (1998)
[a]Estimate for comparison and not included in the study

designing roof catchment systems. This figure is also recommended in a guide on the use of rainwater tanks produced in Australia (Cuncliffe 1998). Runoff coefficients for traditional roofing materials such as grass thatch and local clay tiles are generally lower than this as are those for most ground catchment systems.

Natural land surfaces will normally have runoff coefficients below 0.3 and even as low as zero. Massive rock outcrops used for rock catchment systems are the one exception and may have runoff coefficients of as much as 0.8 according to Lee and Visscher (1992).

A major study in Gansu Province in China (Zhu and Liu 1998) to determine the runoff coefficients of different local roof and ground catchment surfaces in areas with mean annual rainfall varying from 200–500 mm came up with the following results, Table 8.1.

These figures show the runoff coefficient for the particular catchment type which could be expected 95 % of the time (it may be less in occasional extreme drought years). The higher figure in the range relates to catchments in areas with mean annual rainfall of 400–500 mm, while the lower figure is for areas receiving just 200–300 mm. The long-term estimate of the runoff coefficient for corrugated iron roofs is included for comparison and was not part of the study. Due to their higher cost, metal roofs and gutters were not present in the study area at that time.

8.4 Determining the Rainwater Supply

The actual amount of rainwater which can be supplied varies greatly, as it depends on the quantity and distribution of rainfall, the size of existing or affordable catchment surfaces, and the volume of the storage tank. In situations where existing roofs are to be used, the catchment area is fixed, and for a particular location, the amount of rainfall cannot be changed. In these instances, the only variable the designer can use to influence the available rainwater supply is the volume of the storage tank.

Rainfall Data

Rainfall data can normally be obtained from National Meteorological Departments, the Ministry of Agriculture, Universities, and Research stations, and may be obtainable directly through the Internet. Care should be taken when selecting the best rainfall data source station, as in some areas, mean rainfall may vary markedly over short distances, especially in mountainous terrain.

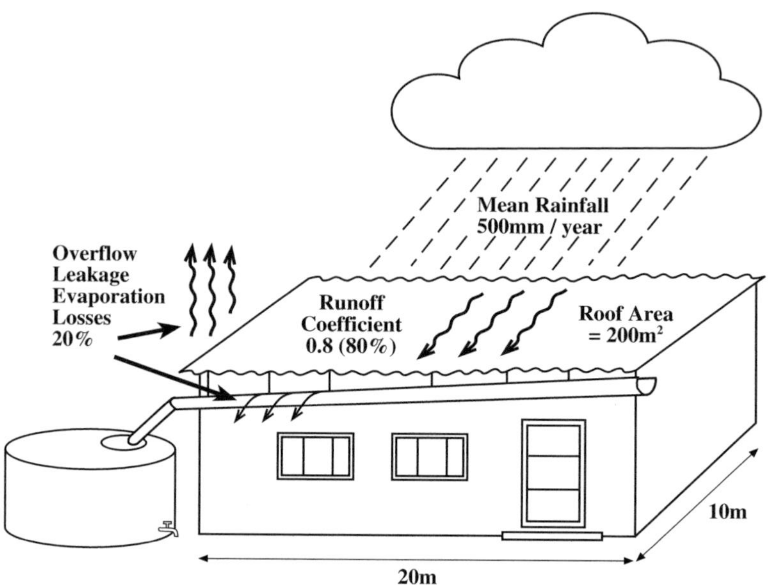

Fig. 8.5 Typical roof catchment system (used in sample calculation)

Rainfall is very variable, especially where annual precipitation is less than 500 mm. It also varies with location, so that data from a weather station 20 km away may be misleading when applied to the site of the rainwater harvesting system (see Fig. 8.5).

To determine the potential rainwater supply for a given catchment, reliable rainfall data for a period of at least 10 years are required. In drought prone climates, a longer historic rainfall record is preferable and a 20-year series is ideal. A longer rainfall data series may give a false picture of current rainfall conditions if regional climatic changes have occurred.

8.5 Calculating Potential Rainwater Supply from a Simple Roof Catchment

The size of the supply of rainwater depends on the amount of rainfall, the area of the catchment, and its runoff coefficient. For a roof or sloping catchment, it is the horizontal plan area which should be taken (see Fig. 8.5). The runoff coefficient takes into account any losses due to leakage, evaporation, and overflow and is normally taken as 0.8 for a well-constructed roof catchment system. Rainfall is the most unpredictable variable in the calculation since in many areas, there is considerable variation from one year to the next. An estimate of the approximate mean annual runoff from a given catchment can be obtained using the following Eq. (8.2).

$$S = R \times A \times C_r, \tag{8.2}$$

where
S = Mean rainwater supply in cubic meters (m^3)
R = Mean annual rainfall in millimeters (mm/a)
A = Catchment area in square meters (m^2)
C_r = Runoff coefficient

For example,

$$\begin{aligned} S &= 500 \text{ mm/a} \times 200 \text{ m}^2 \times 0.8 \\ &= 0.5 \text{ m/a} \times 200 \text{ m}^2 \times 0.8 \\ &= 80 \text{ m}^3\text{/a} = 80{,}000 \text{ L/a} \\ &= 219 \text{ L/day.} \end{aligned}$$

8.6 Sizing the Water Storage Structures

There are a number of different methods for sizing system components. These methods vary in complexity and sophistication. Some are readily carried out by relatively inexperienced first-time practitioners; others require computer software and trained engineers who understand how to use this software.

The choice of method used to design system components will depend largely on the following factors:

- The size and sophistication of the system and its components
- The availability of the tools required (e.g., Computers)
- The skill and education levels of the practitioner/designer

The actual amount of rainwater supplied may vary greatly from year to year and also depends on the volume of the storage tank and the rate of water use.

8.6.1 Methods for Sizing Rainwater Tanks

Below three different methods for determining, the required storage volume is outlined.[2]

[2] The examples shown here are adapted from the Domestic Roofwater Harvesting Research Program part of the Development Research Unit at the University of Warwick UK. The website at: www2.warwick.ac.uk/fac/sci/eng/research/civil/crg/dtu-old/rwh/ is an excellent resource for finding out information on domestic rainwater collection and has many links to other useful websites.

(1) **Demand Side Approach**

A very simple method is to calculate the largest storage requirement based on the consumption rates and occupancy of the building.

As a simple example, we can use the following typical data:

Consumption per capita per day – $C = 20$ L

Number of people per household – $n = 5$

Longest average dry period = 100 days

Annual consumption = $C \times n \times 365 = 36{,}500$ L $\left(36.5\ \text{m}^3\right)$

Storage requirement = $C \times n \times 100$ days $= 20 \times 5 \times 100 = 10{,}000$ L $\left(10\ \text{m}^3\right)$.

This simple method assumes sufficient rainfall and adequate catchment area, and is therefore only applicable in these situations. It is a method for acquiring rough estimates of tank size.

If good quality data are available on household water consumption and the maximum length of any dry periods, this method can be very useful. It has the advantage of not requiring accurate rainfall data or sophisticated modeling tools.

(2) **Supply Side Approach**

In low-rainfall areas or areas where the rainfall is of uneven distribution, more care has to be taken to size the storage properly. During some months of the year, there may be an excess of water, while at other times there will be a deficit (see Fig. 8.6). If there is sufficient water throughout the year to meet the demand, then sufficient storage will be required to bridge the periods of scarcity. As storage is expensive, this should be done carefully to avoid unnecessary expense.

The example given here is a simple spreadsheet calculation for a site in North Western Tanzania. The rainfall statistics were gleaned from a nurse at the local hospital who had been keeping records for the previous 12 years. Average figures for the rainfall data were used to simplify the calculation, and no reliability calculation is done. This is a typical field approach to rainwater harvesting storage sizing.

Demand:

- Number of staff: 7
- Staff consumption: 45 L per day $\times$ 7 = 315 L per day
- Patients: 40
- Patient consumption: 10 L per day $\times$ 40 = 400 L per day

Total demand: Staff consumption + Patient consumption = 715 L/day or 21:75 m^3/month

Mean annual rainfall = 1056 mm

Example Medical dispensary, Ruganzu, Biharamulo District, Kagera, Tanzania 1997 Supply:

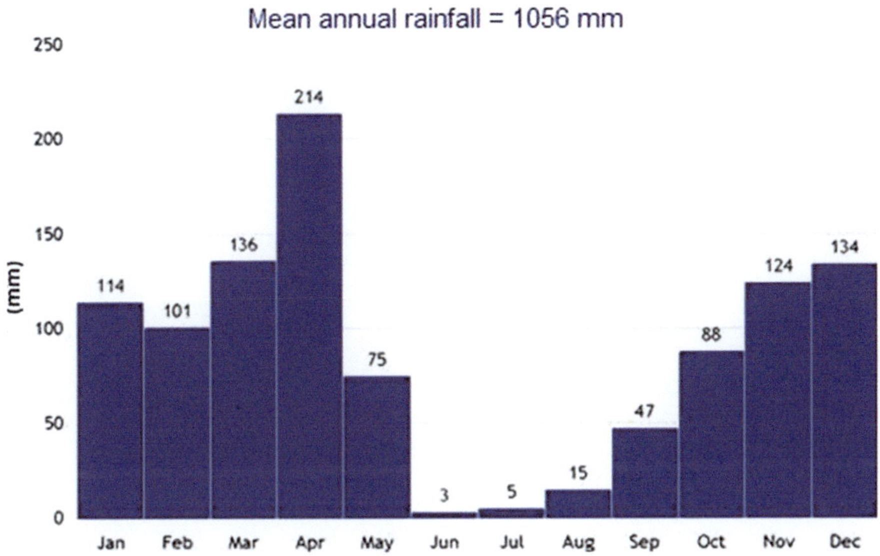

Fig. 8.6 Mean monthly rainfall for Biharamulo District

- Roof area: 190 m^2
- Runoff coefficient (for new corrugated GI roof): 0.9
- Average annual rainfall: 1056 mm per year
- Annual available water (assuming all is collected) $= 190 \times 1.056 \times 0.9 =$ 180.58 m^3

$$\begin{aligned}\text{Daily supply available} &= 180.58/365 \\ &= 0.4947\ \text{m}^3/\text{day} \\ &= 495\ \text{L per day} \\ &\quad \text{or } 15.05\ \text{m}^3\ \text{per mean month}\end{aligned}$$

So, if we want to supply water all the year to meet the needs of the dispensary, the demand cannot exceed 495 L per day. The expected demand cannot be met by the available harvested water. Careful water management will therefore be required.

Figure 8.7 shows the comparison of water harvested and the amount that can be supplied to the dispensary using all the water which is harvested. It can be noted that there is a single rainy season. The first month that the rainfall on the roof meets the demand is October. If we therefore assume that the tank is empty at the end of September, we can form a graph of cumulative harvested water and cumulative demand and from this we can calculate the maximum storage requirement for the dispensary as shown in Fig. 8.8.

Table 8.2 shows the spreadsheet calculation for sizing the storage tank. It takes into consideration the accumulated inflow and outflow from the tank, and the

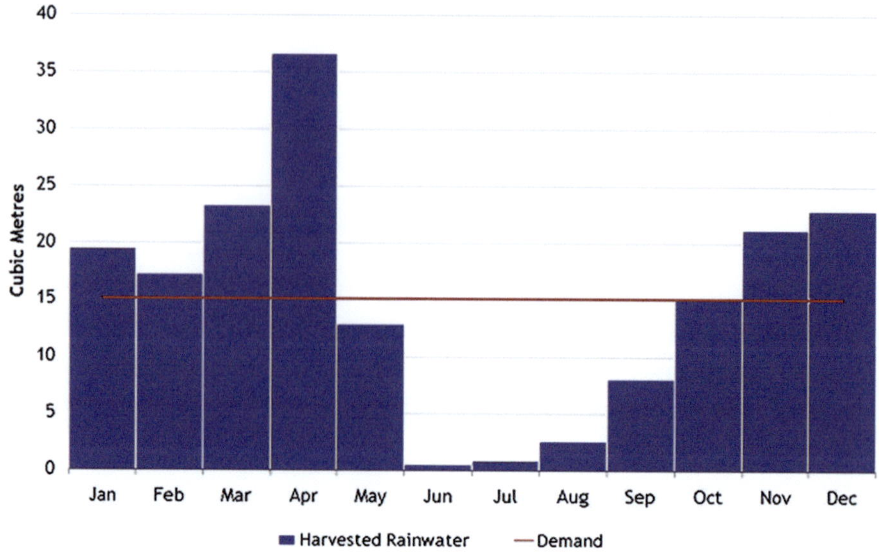

Fig. 8.7 Comparison of the harvestable water and demand each month

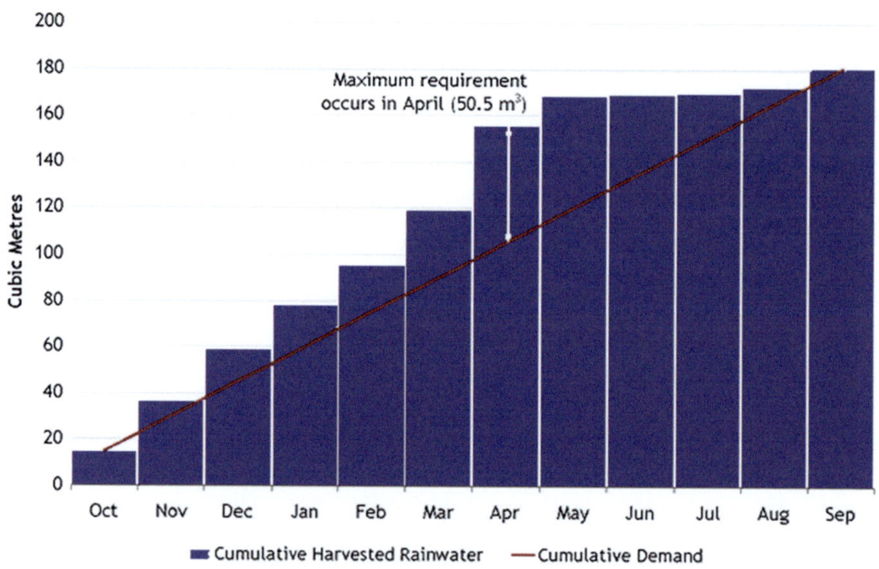

Fig. 8.8 Predicted cumulative inflow and outflow from the tank (*Note* The maximum storage requirement occurs in April)

capacity of the tank is calculated as the greatest excess of water over and above consumption. This occurs in April with a storage requirement of 50.45 m^3. All this water will have to be stored to cover the shortfall during the dry period.

Table 8.2 Spreadsheet calculation for sizing the storage tank

Private month	Rainfall (mm)	Rainfall harvested (m^3)	Cumulative rainfall harvested (m^3)	Demand (based on total utilization)	Cumulative demand (m^3)	Difference between column 4 and 6
Oct	88	15.05	15.05	15.05	15.05	0.00
Nov	124	21.20	36.25	15.05	30.10	6.16
Dec	134	22.91	59.17	15.05	45.14	14.02
Jan	114	19.49	78.66	15.05	60.19	18.47
Feb	101	17.27	95.93	15.05	75.24	20.69
Mar	136	23.26	119.19	15.05	90.29	28.90
Apr	214	36.59	155.78	15.05	105.34	*50.45*
May	75	12.83	168.61	15.05	120.38	48.22
Jun	3	0.51	169.12	15.05	135.43	33.69
Jul	5	0.86	169.97	15.05	150.48	19.49
Aug	15	2.57	172.54	15.05	165.53	7.01
Sep	47	8.04	180.58	15.03	180.58	0.00
Totals		180.58		180.58		

8.6.2 Tank Efficiency and the Case for Diminishing Returns

On days when rainfall is heavy, the flow into a tank is higher than the outflow drawn by water users. A small tank will soon become full and then start to overflow. An inefficient system is one where, taken over say a year, that overflow constitutes a significant fraction of the water flowing into the tank. Insufficient storage volume is, however, not the only cause of inefficiency: poor guttering will fail to catch water during intense rain, leaking tanks will lose water, and an 'oversize' roof will intercept more rainfall than is needed.

$$\text{Storage efficiency } (\%) = 100 \times \left(1 - \text{overflow}/\text{inflow}\right)$$

$$\text{Provided that inflow} < \text{demand}$$

$$\text{System efficiency } (\%) = 100 \times \text{water used}/\text{water falling on the roof}$$

In the dry season, a small tank may run dry, forcing users to seek water from alternative sources. Unreliability might be expressed as either the fraction of time (e.g., of days) when the tank is dry or the fraction of annual water use that has to be drawn from elsewhere. A rainwater harvesting system may show unreliability not only because storage is small, but because the roof area is insufficient. Figure 8.9 shows how reliability, expressed as a fraction of year, varies with storage volume (expressed as a multiple of daily consumption) for two locations close to the Equator and therefore both with double rainy seasons.

From this graph, one can see that increasing storage size, and therefore cost, gives diminishing returns. For example, look at the left-hand column of each triplet (Kyenjojo with roof sized such that average annual water demand is only

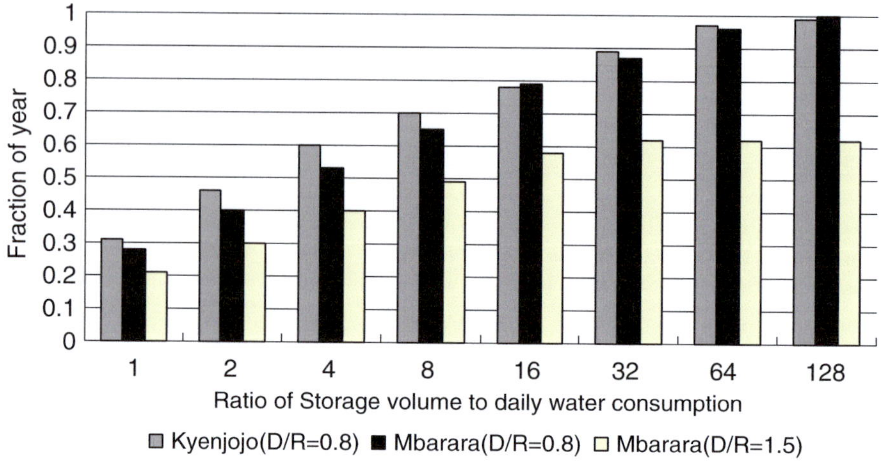

Fig. 8.9 Availability of rainwater supply as a fraction of the year

80 % of average annual roof runoff). Assuming a say 100 L per day demand shows that increasing storage from 1 day (100 L) to 16 days (1600 L) raises the reliability from 31 to 78 %, but storage has to be increased as high as 128 days (12,800 L) to achieve 99 % reliability. Such high reliability is so expensive that it is an unrealistic design objective for a rainwater harvesting system in a poor country. In any case, as we shall see below, users may change behavior so as to reduce the effective unreliability of their systems.

8.6.3 System Features that Affect Tank Sizing

- An oversized roof slightly compensates for an undersized tank.
- 'Partial' rainwater harvesting systems, either where it is accepted that rainwater will not meet needs throughout the year or where rainwater is only used to meet specific water needs like cooking/drinking, can be built with surprisingly small tanks.
- The reliability level appropriate to the design of a rainwater harvesting system rises with the cost (in money, effort, or even ill-health) of the alternative source that is used when the tank runs dry.
- If users are able and willing to adjust their consumption downwards during the dry season, or when they find water levels in their tank lower than average, tanks can be sized smaller. A simple but effective approach to such rationing is shown in Box 8.1 below.

8.6.4 Rationing

Whether carefully controlled and executed or simply carried out as a natural reaction to the tank becoming empty and a looming water shortage some form of water rationing is almost inevitable in situations where a given water demand cannot always be met by any particular rainwater collection system.

In areas where seasonal water shortages are common, it therefore makes sense to incorporate a systematic and carefully planned rationing system. A rationing schedule can be very simple and need not be hard to execute as the example in the box below illustrates where simply reducing a standard abstraction rate of 3 large buckets (60 L) to 2 (40 L) when the level of water in the tank fall below a third full can greatly increase the time period that the system can continue supplying water. When the tank is more than two-thirds full, the daily water allowance can be increased to 4 buckets (as this will ensure less water is lost due to tank overflow in periods of higher rainfall). In periods of severe drought or at the end of the dry season, an emergency rationing measure of taking just 1 bucket could be introduced when the tank is less than 1/10 full to further extend the period during which water can be abstracted.

Box 8.1: Simple and effective Approach to Rationing Water Supply

Rainwater Tank showing suggested daily abstraction rates to extend the period which the tank can supply water.
Daily water supply

When tank >2/3 full	4 buckets
	3 buckets
When tank <1/3 full	2 buckets

Rationing schedules of this type may only be needed in situations where no other water sources are available, or during the dry season or drought periods. In situations where a household is entirely dependent on a rainwater tank as its sole water supply, it would be prudent to introduce even more stringent ‘emergency rationing’ when the tank becomes less than 10 % full. Some computer models available free online such as the rainwater tank performance calculator discussed below include options for incorporating rationing into the analysis.

8.7 Computer Models

There are several computer-based programs for calculating tank size quite accurately.

8.7.1 SimTanka

One such program, known as SimTanka, has been written by an Indian organization and is available free of charge on the World Wide Web. The Ajit Foundation is a registered non-profit voluntary organization with its main office in Jaipur and its community resource center in Bikaner, India.

SimTanka is a software program for simulating performance of rainwater harvesting systems with covered water storage tank. Such systems are called Tanka in western parts of the state of Rajasthan in India.

The idea of a computer simulation is to predict the performance of a rainwater harvesting system based on the mathematical model of the actual system. In particular, SimTanka simulates the fluctuating rainfall on which the rainwater harvesting system is dependent.

Rainwater harvesting systems are often designed using some statistical indicator of the rainfall for a given place, like the average rainfall. When the rainfall is meager and shows large fluctuations, then a design based on any single statistical indicator can be misleading. SimTanka takes into account the fluctuations in the rainfall, giving each fluctuation its right importance for determining the size of the rainwater harvesting system. The result of the simulation allows you to design a rainwater harvesting system that will meet demands reliably, that is, it allows you to find the minimum catchment area and the smallest possible storage tank that will meet your demand with probability of up to 95 % in spite of the fluctuations in the rainfall. Alternatively, you can use SimTanka to find out what fraction of your total demand can be met reliably.

SimTanka requires at least 15 years of monthly rainfall records for the place at which the rainwater harvesting system is located. If you do not have the rainfall record for the place, then the rainfall record from the nearest place which has the same pattern of rainfall can be used.

The included utility, Rain Recorder, is used for entering the rainfall data. Daily consumption per person is also entered and then the software will calculate optimum storage size or catchment size depending on the requirements of the user. SimTanka also calculates the reliability of the system based on the rainfall data of the previous 15 years.

SimTanka is free and was developed by the Ajit Foundation in the spirit that it might be useful for meeting the water needs of small communities in a sustainable and reliable manner. But no guaranties of any kind are implied. For more information or to download the software, see their website at www.indev.nic.in/ajit/Water.htm (Source: the information given here is taken from this website).

In reality, the cost of the tank materials will often govern the choice of tank size. In other cases, such as large rainwater harvesting programs, standard sizes of tank are used regardless of consumption patterns, roof size, or number of individual users.

8.7.2 Rainwater Tank Performance Calculator

Another excellent such program which is available free online can be found at www2.warwick.ac.uk/fac/sci/eng/research/civil/crg/dtu-old/rwh/model and provides a simple yet powerful design tool. This tank performance calculator provides a simple, easy to follow, step by step approach to assess tank performance.

As the cost of a domestic rainwater harvesting system depends mainly on the size of the tank, it is important to design the tank to ensure optimum performance at tolerable cost.

This program calculates the approximate system reliability and efficiency for a selection of tank sizes including one that you can define, using your monthly rainfall data and roof area.

The user also defines how the rainwater will be used by giving a nominal daily demand and choosing between three water management strategies:

- Constant demand.
- Varies with tank volume.
- Varies with season.

The program requires the following data: nominal daily demand, roof area, and tank volume. Ten years of monthly rainfall data (in mm) need to be entered in the following format. Each number has to be followed by a comma (e.g., 74, 58, 106, 195, 164, 103, 104, 104, 128, 127, 120, 67).

8.7.3 Points to Bear in Mind Regarding Computer Models

The use of computer based models allows great flexibility when producing output for system design since the model can be tailored for any particular system under given rainfall conditions. The format of the output can also be customized to requirements and the performance of specific designs simulated under various demand scenarios. For example, the implications of introducing a dry season rationing schedule on storage requirements could easily be tested using a long rainfall record.

Due to their speed and flexibility, computers provide a powerful tool when simulating future rainwater supplies and anticipated storage requirements on the basis of past rainfall data. While computer models are becoming increasingly capable of mimicking the performance of real systems, they are only as good as the data used. Obtaining good quality rainfall data in a format that can readily be used in any model will greatly reduce the time, effort, and cost of any computer-based modeling exercise. As with other methods, an accurate and lengthy rainfall record is essential. A minimum of 20 years of rainfall data is preferable especially in drought prone areas. While mean monthly data can be used, weekly or daily records will give a more accurate prediction of system performance.

Where feasible, these should be used, especially if they are available in digital format that can be readily fed into a computer program. Normally, the data will need to be reformatted to meet the specifications of any particular program. Care should be taken to check the data carefully, so that account can be taken of any missing or spurious values.

Where rainfall records are short, it is possible to use computers for data simulation to extend the record. This may be convenient but the resulting data will be no more accurate than the historical data on which it is based. For those who really like to do their own thing, there is even advice on how to use common computer packages or write your own program to determine the performance of any roof catchment system.

It should be borne in mind, however, that the tidy and uncomplicated world of the computer simulation can turn out to be very different from reality, where leaks or the pranks of small children can play havoc with the neat predictions of the most sophisticated computer model.

8.8 Structural Design of Water Storage

8.8.1 Selecting an Appropriate Tank Design

Ideally, a tank as well as having the appropriate volume with respect to the catchment area, rainfall conditions and demand, should have a functional, durable, and cost-effective design. Field experience has shown that a universally ideal tank design does not exist. Local materials, skills and costs, personal preference, and other external factors may favor one design over another.

(i) Key requirements common to all effective tank designs[3]

- A functional and water tight design.
- A solid secure cover to keep out insects, dirt, and sunshine.
- A screened inlet filter.
- A screened overflow pipe.
- A manhole (and ideally a ladder) to allow access for cleaning.
- An extraction system that does not contaminate the water, e.g., tap/pump.
- A soak-away to prevent spilt water forming puddles near the tank.
- A maximum height of 2 m to prevent high water pressures (unless additional reinforcement is used in the walls and foundations).

(ii) Other features might include:

- A device to indicate the amount of water in the tank.
- A first flush mechanism.
- A lock on the tap.
- A second sub-surface overflow tank to provide water for livestock.

[3]List adapted from Latham and Gould (1986).

If rainwater catchment systems already exist in an area, it is important to take time to visit and inspect a few systems. The owners or users of the systems should be questioned regarding their assessment of their own tank designs. This process allows the advantages and disadvantages of different locally produced designs to be compared and weighed against the sales pitch given by manufacturers of commercial systems. The exercise will also provide some useful lessons, which may help avoid potential future problems that might result from the selection of an inappropriate design.

8.8.2 Deciding on Most Suitable Type of Tank

Having established the required volume for the storage reservoir, it is also necessary to decide whether to opt for buying an "off the shelf" commercially available tank or to construct the system on site.

A number of factors are likely to influence this decision. These include the cost, durability, acceptability, and appropriateness.

Transport costs can also be high, especially where distances are large and roads are poor. Practical considerations are also important. For example, in many circumstances, such as in remote rural locations, 10 m^3 usually represents the upper limit for tanks which can be constructed at central locations and transported in one piece to the location required.

Plastic and fiber glass tanks although expensive in some countries are light and relatively easy to transport. Metal tanks are also commonly delivered by road, where delivery to remote locations is required, such as in Australia. Costs can sometimes be reduced by nesting several progressively smaller tanks together, if covers can be detached. Delivery of small ferrocement tanks is possible if care is taken and road conditions permit.

There are some advantages to buying ready-made commercial tanks. They can be quickly erected at the site and be operational within days of the decision to install them, compared to weeks or months for tanks requiring construction on site. Because commercial tanks are normally built in large numbers, they are subject to economies of scale and quality control. This can, in some cases, make them more durable and cost-effective than tanks built on site especially if poor levels of workmanship are suspected.

Commercially available tanks do not require the availability of skilled labor, construction materials, and an appropriate design. Since commercially made tanks, especially those requiring special equipment, such as for molding plastic, are often produced only at major centers, transport costs to remote rural locations are often high. The other main disadvantage with commercial ready-made tanks is that they are normally more expensive than tanks constructed on site. This is especially the case in situations where free or low-cost labor and local building materials are available. The free collection of river sand and aggregate, for example, is common practice in many community self-help projects in developing countries.

Systems designed and constructed by an individual householder or community can also be tailor-made to meet the specific local requirements. In the context of a community project, there are several other benefits to be derived from constructing systems on site. While unskilled voluntary labor may help bring down system costs, the utilization of skilled paid labor from the community will provide employment. The involvement of the community in the construction of any water systems helps develop skills and self-reliance. This also ensures the community is more likely to be able to properly operate, maintain, and repair the systems in the future.

8.8.3 Tank Shape, Dimensions, and Type

As a general rule, water tanks should ideally be cylindrical or spherical in shape. This is because cylindrical or spherical shapes optimize the use of materials and increase the wall strength. Spherical tanks are difficult to construct and require some sort of stand for support. A good compromise shape for strength and cost-effectiveness is the Thai jar (Fig. 8.10). The jar shape gives maximum strength since the walls are curved in both vertical and horizontal directions yet it requires no special stand. Although, "jumbo jars" up to 6 m^3 have been built, to construct larger surface tanks in this way would be very difficult so for practical purposes, a cylindrical shape is the best compromise. This does lead to comparatively large stresses along the joint between the wall and the base which must be strong enough to withstand this (Watt 1978).

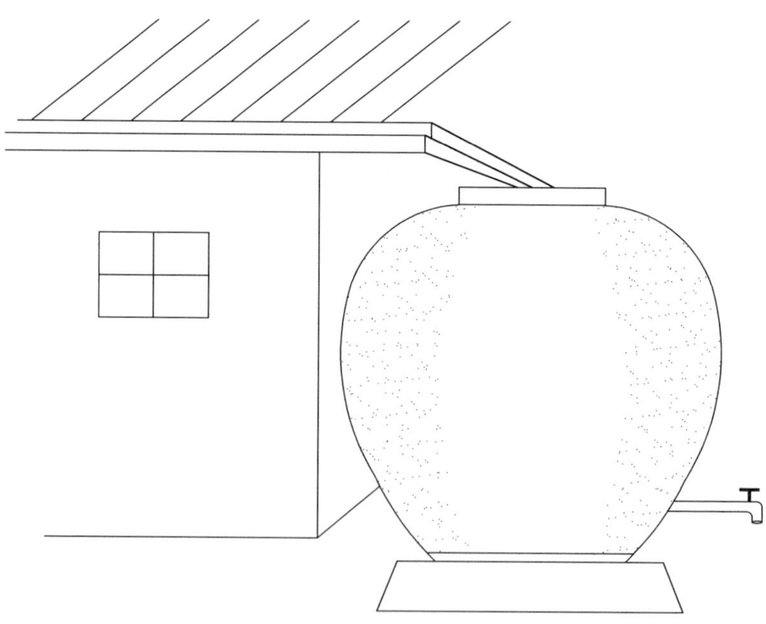

Fig. 8.10 Thai Jar—the *curved shape* gives it strength and cost efficiency

For surface tanks, the cylindrical shape is by far the most common. To maximize the storage volume while minimizing the cost, the tank should be reasonably evenly proportioned. Tall tanks with narrow widths and very low tanks with large diameters require more materials and cost more per unit volume. There are, nevertheless, a number of other factors which need to be considered. Unless additional reinforcement is added, the height of water tanks for roof catchment systems should not exceed 2 m.

This is because the internal force of water against the tank walls increases with the depth of the water. If the height of a tank is increased, then the reinforcement must be increased accordingly to prevent collapse of the tank. Where tanks exceed 2 m in height, special attention should be paid to reinforcing the lower sections adequately and if in doubt, expert advice should be sought (Fig. 8.11).

When deciding on whether a surface or sub-surface tank is more appropriate, the following points should be considered. Although underground storage reservoirs are generally cheaper, some form of pump or gravity-flow connection to an excavated or lower level tap stand is generally required to extract water. While substantial cost savings may be possible particularly in the case of larger excavated tanks, other factors such as local soil conditions need to be considered. Where the sub-soil is rocky, excavation may not be feasible and where it becomes waterlogged, there is a risk that a nearly empty sub-surface tank could start to float and rise out of the ground.

The decision regarding the final choice of storage tank will depend on a wide range of factors including the availability of materials and locally available labor skilled in tank construction.

8.8.4 Cost-Effectiveness

The cost-effectiveness is often the determining factor when deciding to choose a particular technology or design. The high initial capital costs required for rainwater catchment systems, particularly in more arid and seasonal climates where large

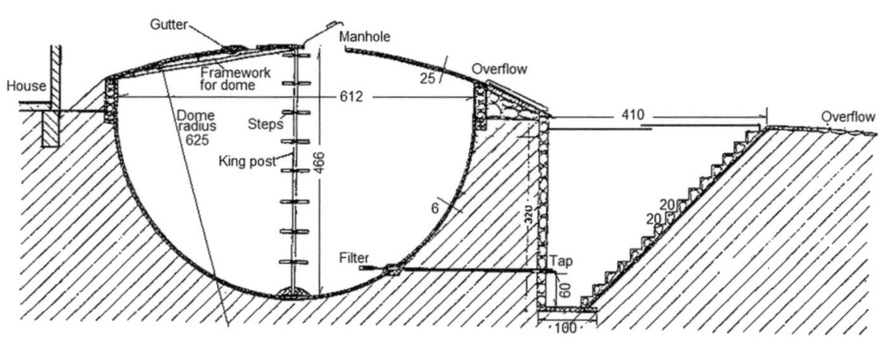

Fig. 8.11 Hemispherical 90 m^3 sub-surface ferrocement tank design from Kenya (Permission and courtesy from Erik Nissen-Petersen)

storage reservoirs are required, further increase the need to ensure cost efficiency. Tank sizing and the selection of a tank with suitable shape and dimensions are crucial in optimizing cost-effectiveness. The choice of a durable design with a long life expectancy and low maintenance costs is also critical.

While a $1000 tank may initially appear a much cheaper alternative than a $2000 tank, if the more expensive one has a life expectancy of 25 years compared with 10 years for the cheaper one, the costs should be reassessed in terms of $2000/25 = $80/year versus $1000/10 = $100/year, respectively. In this case, the more expensive tank would seem to be cheaper over the long term.

The use of discount rates which involve making assumptions regarding the declining future value of current funds upsets such simplistic analysis, as do uncertainties surrounding the actual life expectancies of different designs. Nevertheless, it is clear that two different designs cannot simply be compared at face value without taking other factors into account.

8.8.5 Availability and Suitability of Materials and Skilled Labor

The availability of different raw materials is crucial to the decision regarding the eventual choice of tank design. If key materials are not locally and cheaply available, it may be worth considering a commercially available design, rather than substitute a critical building material with something less suitable.

The availability of suitably experienced and qualified labor for tank construction is vital if a project involves tank construction on site. Certain designs, such as ferrocement require particular care and attention to detail, which, if ignored, could eventually jeopardize a project. If the necessary skills are not available in the project area, it is sometimes possible to develop the necessary skills through training courses.

Unless large numbers of tanks are going to be built, it may be difficult to justify any such major investment. It should nevertheless be recognized that investment in training and skills development may have many positive spin-offs for a community in a variety of areas unrelated to the project itself.

8.8.6 Siting of Tanks

Key issues to be considered when siting a rainwater tank or reservoir:

- Avoidance of any potential health hazards, e.g., never locate tanks near toilets/pit latrines, waste disposal facility, or other source of pollution.
- Avoid sites where surface runoff is evident due to the risk from soil erosion, which could cause damage to poorly sited tanks (if such sites have to be used, bunds and/or cut-off drains should be constructed to divert flood waters away from the base of the tank).

- Tanks should generally be located so they can collect water from as large a roof/catchment area as possible. This will often be the determining factor regarding the siting of the tank, e.g., between two buildings.

8.8.7 Design Flaws, Implementation, and Operation

To ensure the success of any rainwater harvesting project, it is essential to take time to design systems carefully and in a way appropriate to the local conditions. This process should involve the community in all aspects of planning, project design, implementation, operation, and maintenance.

Experience has shown that simply providing rural communities with the necessary "hardware," however, technically sound it may be, is not enough to ensure projects will succeed. In the case of jointly owned and operated communal projects, particular care and attention is needed to make sure procedures are in place to guarantee systems will be operated and maintained properly and projects will be sustainable. This will ensure that clear lines of responsibility and accountability are established and that resources both in terms of trained personnel and finance are available to guarantee regular maintenance and timely minor repair work is done before major system failures occur.

Encouraging the proper operation and maintenance of individual household systems such as roof catchments is generally much more straightforward than for communal systems since it is in the householders' own personal interest to maintain their own systems properly and since all the benefits of the system accrue to the household, most are highly motivated with respect to proper system upkeep.

The best way to minimize potential problems is to ensure that any design used has been thoroughly field tested. If a new design is being implemented, it should undergo field trials through a pilot project phase and any necessary modifications or improvements to the design made before wider replication take place. Many projects in the past have also failed because of unrealistic expectations of the willingness of communities and individuals to provide free or cheap labor.

8.8.8 Importance of Field Testing New Designs

The adoption and widespread replication of new designs, however, promising they may seem at the development and demonstration phase, is extremely risky if they are not first subjected to thorough field testing through carefully monitored pilot projects.

The failure of various low-cost cement tank designs using organic reinforcement (bamboo, sisal, and basketwork) in both east Africa and southeast Asia during the early 1980s provides an important lesson in this respect. The development and hasty promotion of low-cost bamboo and basketwork reinforced tanks designs, in Thailand and Kenya, respectively, and their widespread adoption resulted in one of the most serious failures of rainwater tank technology to date (Latham and Gould 1986).

In the case of the bamboo reinforced tank, insufficient field testing and premature promotion of the design resulted in extensive replication in Indonesia and Thailand (where as many as 50,000 were constructed). Unfortunately, after a couple of years, many instances of tank failure were reported due to damage of the bamboo reinforcement, resulting from termite, bacterial, or fungal attack. Apart from the problem of cracking and leakage of the 5–12 m^3 tanks, the risk and related danger of a tank bursting and causing injury, or even death, had to be taken seriously.

As for the basketwork reinforced 'Ghala basket' tanks widely promoted in Kenya in the early 1980s, several thousand of these were constructed, but within a couple of years, most tanks suffered failure due to rotting or termite attack of the organic "reinforcement." The lessons in both these instances are clear. It is vital that new designs are thoroughly field tested before widespread promotion and replication of the technology. This is not always easy to ensure in practice due to the urgent need to find solutions to pressing water problems, especially when an apparently appropriate solution is found.

8.8.9 Training, Quality Control, and Good Management

While many projects fail as a result of the use of inappropriate technologies and designs, frequently the technology may be sound but failure is due to a lack of training, quality control, or poor management. Specific problems stemming from this include poor workmanship, inadequate maintenance, and lack of the necessary skills, training, and supervision to ensure high-quality construction. Use of inappropriate materials, such as saline water or poorly graded sand cause structural weaknesses in the tank which may well act as obstacles to long-term project success.

In one major project in Kenya, for example, widespread tank cracking and failure resulted after a few years due to the "disappearance" of cement during construction, resulting in inadequate quantities being used in the building of large ferrocement tanks. It is probably fair to say that the checkered history of ferrocement tanks in Africa, and particularly problems with the construction of larger rainwater tanks in Botswana and Namibia stem mainly from insufficient attention having been given to thorough training, careful quality control, and project management.

8.8.10 Importance of Proper Operation and Maintenance

Manuals and literature relating to rainwater catchment system operation and maintenance generally recommend that regular system maintenance should be carried out. In reality, it seems, based on field observations, that regular cleaning of systems tends to be the exception rather than the rule.

Maintenance is also frequently neglected, often to the detriment of the system's life span.

The diagrams in Box 8.2 compare a good and bad roof catchment system. The poorly designed system is probably only about 10 % efficient. Less than a quarter of the roof area is being effectively utilized and only half of the storage capacity. Unfortunately, this is based on a real example observed in 1991 in Masunga, Botswana.

Leaking taps, blocked or broken gutters, and downpipes are very simple to maintain and repair but if left unattended, these frequently result in total system failure. Even such obvious measures, like closing a dripping tap properly so water loss is avoided, are sometimes not done, especially in the case of communal tanks. Unless specific training is provided and responsibilities allocated, it is probably safest to assume that very little effort will be made regarding operation and maintenance, particularly with communal systems, at least until they fail or breakdown completely. The situation regarding privately owned systems is somewhat better, but even here it sometimes helps raise awareness amongst system owners about the necessity and benefits to be derived from regular system cleaning and maintenance.

8.9 Gutters and Downpipes

Although gutters and downpipes are not the only method for delivering water from the catchment to the storage tank, they are by far the most common. Other methods used include the use of roof "glides" as in Bermuda, and cement channels are common in parts of rural China. Simple gutter troughs made of wood or sheet metal are also a common technique used for simple informal home-made systems at millions of poorer households across the developing world.

A carefully designed and constructed gutter system is essential for any roof catchment system to operate effectively. A properly fitted and maintained gutter–downpipe system is capable of diverting more than 90 % of all rainwater runoff into the storage tank (Ree 1976) even though the long-term collection efficiency is usually between 80 and 90 %.

Gutters and downpipes can be made of a variety of materials: metal, plastic, cement, wood, and bamboo. Typically, conventional 'off the shelf' metal or plastic gutters and downpipes will cost between 5 and 15 % of the total system cost, depending on local prices and conditions. All too often, both individuals and projects overlook the importance of guttering. This frequently results in only the runoff from part of the roof area being utilized.

The gutter and downpipe systems are crucial to any rainwater catchment system yet they are frequently the weak link, which result in poor system efficiency. Broken gutters often lead to little or no water reaching the tank. Regular gutter maintenance is, therefore, essential. Leaves and other debris in the gutter must be cleaned out and overhanging branches should be removed.

Box 8.2: Common problems with roof catchment design

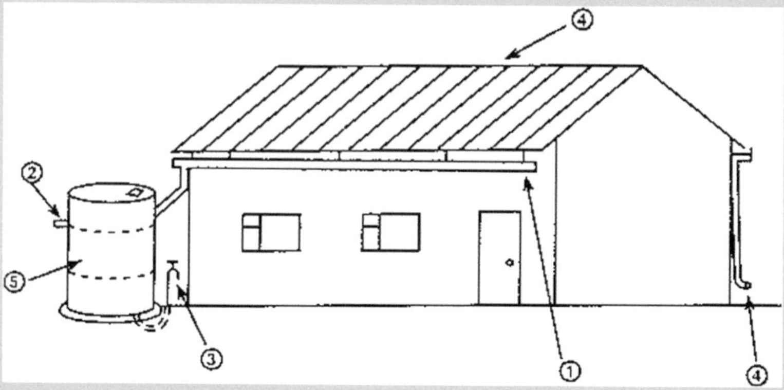

Example of poor system design

Many roof catchment systems are poorly designed. Common mistakes include:

1. Gutters which are horizontal or sloping away from the tanks
2. Overflow pipes placed well below the top of the tank
3. Outlet taps high above the base of the tank
4. Down pipes leading to waste
5. Only part of the roof area being used

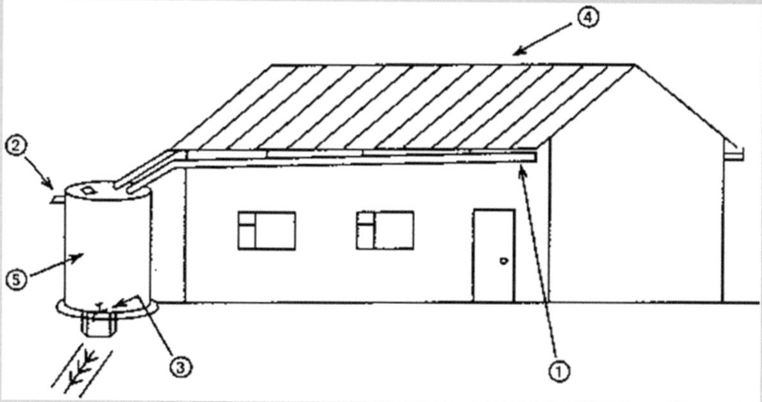

Example of good system design

Note Differences:

1. Gutter Slope
2. Height of Overflow
3. Height of Tap
4. Total Catchment Area Used
5. All Storage Volume Used.

8.9.1 Gutter and Downpipe Sizing

Gutter and downpipe sizing is a crucial element of the design of any system. Large quantities of runoff may be lost during heavy storms if gutters are too small and overflow. As a general guide to gutter dimensions for catchment areas of different sizes, a useful rule of thumb is to make sure that there is at least 1 cm^2 of gutter cross section for every 1 m^2 of roof area (Hasse 1989). See Sect. 9.8 in Chap. 9 for further guidance on sizing and types of guttering.

To avoid overflow during torrential downpours, it makes sense to provide a greater gutter capacity. The gutter must be of a sufficient size, in order to discharge water to the tank without any overflow in the gutter. The usual 10-cm (4″)-wide half-round gutter is generally not big enough for roofs larger than about 70 m^2. A 10 cm $\times$ 10 cm^2 gutter with a cross-sectional area of 100 cm^2 can be used for roof areas up to about 100 m^2 under most rainfall regimes.

For large roofs, such as at schools, the 14 cm $\times$ 14 cm V-shaped design described below, which has a cross-sectional area of 98 cm^2, is suitable for roof sections up to 50 m in length by 8 m in width (Fig. 8.12). When installed with a steeper gradient than 1:100 and used in conjunction with splash-guards, V-shaped gutters can cope with heavy downpours without large and unnecessary losses due to gutter overflow, splash, and spillage. A gradient of 1:100 also ensures less chance of gutter blockage from leaves or other debris as these are more easily flushed out. Under ideal conditions, a properly designed and installed gutter and downpipe system with splash guards can have a runoff coefficient in excess of 0.9 (90 %).

Downpipe cross sections are sometimes smaller than those of gutters as it is assumed that since they are normally vertical, water will pass through them faster than through gutters. In roof catchment systems, however, downpipes should have similar dimensions to gutters. This is because the downpipes are often not vertical and usually act as channels to convey water from the end of the gutter into the tank.

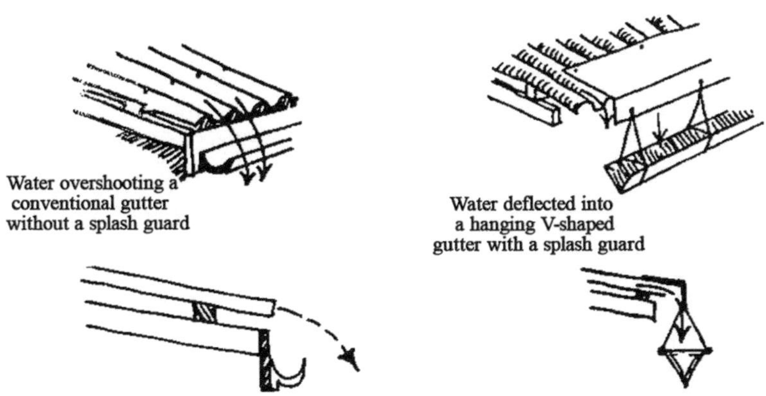

Fig. 8.12 Splash Guards—useful additions to use in conjunction with V-shaped gutters on long roofs to avoid water loss from over-shooting (Courtesy and adapted from Skinner 1990)

8.9.2 Splash Guards

During torrential downpours, large quantities of runoff can be lost due to gutter overflow and spillage (Fig. 8.12). This is particularly a problem on long roofs where, due to the slope of the gutter, it may hang many centimeters below the eaves of the roof. To overcome this problem, a device known as a splash guard, which was originally developed in Kenya, can be incorporated on corrugated iron roofs (Nissen-Petersen 1992). Splash guards consist of a long strip of sheet metal 30-cm wide, bent at an angle, and hung over the edge of the roof by 2–3 cm to ensure that all runoff from the roof enters the gutter. The splash-guard is nailed onto the roof and the lower half is hung vertically down from the edge of the roof. This simple device, which can be manufactured on site, serves two purposes:

(i) The gutter can be suspended from the splash-guard instead of being fitted in gutter-brackets nailed to a fascia-board, which becomes redundant.
(ii) The vertical flap of a splash-guard diverts all roof runoff into the V-shaped gutter hanging underneath it, preventing "over-shooting" or "under-cutting" of rainwater, which otherwise would lead to substantial losses.

An alternative way to deal with the problem of water over-shooting a conventional gutter is to use a specially designed extended V-shaped gutter or G-shaped gutter which wraps around the edge of the roof ensuring all runoff is directed into the gutter (diagrams of these designs can be seen in the next chapter Sect. 9.8).

8.10 Tank Inflow—with Self-Cleaning Mesh Screen

Coarse filters and screens are commonly used to exclude debris from entering storage tanks. A simple and appropriate design is the use of a self-cleaning screen. This consists of placing the end of the downpipe or down-gutter about 3 cm from the mesh screen in front of the inlet hole. The galvanized 5-mm mesh screen should slope at not less than 60° from the horizontal above the tank inlet.

Objects larger than 5 mm, such as stones, small branches, and leaves, are pushed down the gutter and downpipe by flowing water until they strike the mesh. Here the debris will be caught and roll downward off the screen while the water shoots through the mesh into the tank.

Any dust from a roof, which is finer than the mesh, will enter the tank along with the water and settle at the bottom of the tank. Since dust lying on roofs is sterilized by prolonged exposure to sunshine and it will settle on the tank floor below the draw-off pipe in-take, it should not adversely affect water quality once it has settled.

Guttersnipe

The guttersnipe or leaf-slide is another very simple device for removing leaves, insects, and other debris from entering the storage tank (Fig. 8.13) and works along the same principles as the self-cleaning tank inlet. The guttersnipe sits at

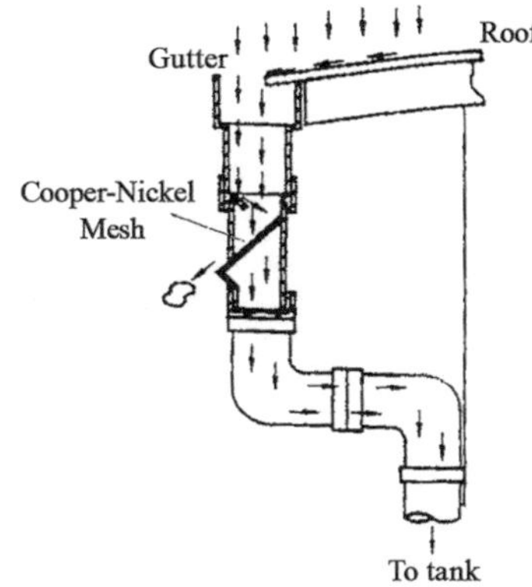

Fig. 8.13 Guttersnipe or Leafslide (Source and courtesy Finch 1994)

the top of the downpipe in PVC housing at least 15 cm below the gutter inlet, and consists of a stainless steel or copper-nickel mesh angled at 60° from the horizontal. This wire screen has 1-mm gaps between the wires, and has the size of about 18 cm by 9 cm for a standard gutter. The screen allows water to pass through it but excludes other material which is washed off in a self-cleaning mechanism. The maintenance required involves the cleaning of the screen once a month to remove any algae which may accumulate. Tests have revealed that use of guttersnipes may reduce bacteriological contamination of stored rainwater (Finch 1994).

8.11 Water Extraction Devices and Other Features

In order to withdraw water from the storage reservoir, some form of extraction device is needed. Normally this will be some form of water tap or pump and the extraction device is a vital link in the system. Broken or leaking taps, all too often, render systems useless for want of regular inspections and basic maintenance.

8.11.1 Taps

A properly functioning and well-maintained tap is a necessity for any surface catchment tank. A dripping or leaking tap can lose thousands of liters, quickly emptying most average sized rainwater tanks. Taps are most vulnerable to breakage on communal tanks particularly at schools where they are frequently used and occasionally abused by the children. Since a 20-year life expectancy is a reasonable assumption for a well-constructed and maintained water tank, a durable tap with a good life expectancy should be fitted, especially on communal tanks.

For communal tanks being used by large numbers of people, a lockable tap may be appropriate in order to control access and extraction rates from the tank. Sometimes tanks become empty simply because taps are left dripping or running.

Self-closing taps can help overcome this problem although they are more prone to breakage and maintenance problems. Privately owned household tanks generally suffer few problems with tap breakage and maintenance. While this is partly due to lower levels of usage, it is also an interesting reflection on human nature.

Often the water tap on a tank is built into the wall of a tank, where it is difficult to avoid seepage and impossible to draw water from that part of the tank situated below the level of the watertap—a wastage of storage capacity called "dead storage".

In other water tanks, the draw-off pipe is rightly placed in the concrete of the foundation but the tap is raised to about 60 cm above the floor of the tank to allow for a bucket to be placed under the tap. Again, this arrangement wastes a good portion of the tank volume on "dead storage" sometimes as much as 20 % of the tank volume. To avoid "dead storage" in the lower part of water tanks, the water tap must be positioned below the floor level of a tank. There are two ways of obtaining this: either the foundation of a tank can be elevated around 50 cm above the ground level by constructing a solid platform or the tap point can be situated below ground level.

8.12 Tank Overflow

Additional "dead storage" will be created at the top of water tanks, if the bottom of the overflow pipe is not placed at the maximum water level of a water tank. This means that in flat roofs made of reinforced concrete, the overflow pipe should be concreted into the base of the roof to avoid dead storage. In domes being used for storage, the overflow can either be placed at the level of the inlet for guttering, which determines the maximum water level or the gutter inlet can also be used as the overflow. In any case, the overflow should be situated vertically over the tap stand to force water overflowing to fall onto the concreted tap point excavation, from where it is drained to a soak-away pit without eroding the base of a water tank.

8.13 First Flush Systems

Although not absolutely essential for the provision of potable water in most circumstances, when effectively operated and maintained, first flush systems can significantly improve the quality of roof runoff.

If poorly operated and maintained, however, such systems may result in the loss of rainwater runoff, through unnecessary diversion or overflow and even the contamination of the supply.

In poor communities, where the provision of even a basic roof tank represents a substantial upgrading of the water supply, the addition of a first flush system will add some additional expense to the system and it may be worth considering it as a future upgrade.

In some locations, where roof surfaces are subjected to a significant amount of blown dirt and dust, or where particularly good quality water is required and proper operation and maintenance can be guaranteed, a first flush system can be very effective.

In a study by Yaziz et al. (1989), water quality analysis of the initial "foul flush" runoff from both a tile and galvanized iron roof in which the first, second, third, fourth, and fifth liter of runoff were sampled revealed high concentrations of most of the pollutants tested in the first liter with subsequent improvements in each of the following samples, with few exceptions. Fecal coliforms, for example, ranged from 4 to 41 per 100 mL in the first liter of runoff sampled but were absent entirely in samples of the fourth and fifth liters. The study also revealed that the rainfall intensity and number of dry days preceding a rainfall significantly affect runoff quality with higher pollution concentrations after long dry periods. Based on these findings, the minimum volume of foul flush ("first flush") which should be diverted for an average sized 'Australian' house was recommended to be 20–25 L by Cunliffe (1998).

The most effective first flush devices are often the simplest such as those used in northeast has been fitted as standard to thousands of tanks. Nevertheless, regular cleaning of the devices is needed.

To avoid the need of the manual resetting, draining and cleaning of the first flush system various self-cleaning systems have now been developed.

Based on this experience, it is recommended that if any kind of first flush device is to be considered, it should be simple, and should not require regular attention regarding its operation and maintenance.

Examples of such devices include:

- Self-cleaning first flush device.
- Self-cleaning gutter snipes sold commercially (see Fig. 8.13).
- Self-cleaning inlet mesh.
- Sedimentation chambers requiring only occasional cleaning.
- Movable downpipes for diverting the runoff from the season's first downpour.

The last device is appropriate in regions with distinct wet and dry seasons. This cleans the catchment and delivery system flushing away dust and other debris which may have accumulated in the dry season.

References

Cunliffe D. Guidance on the use of rainwater tanks, National Environmental Health Forum Monographs, Water Series 3, Public and Environmental Health Service, Department of Human Services, P.O. Box 6, Rundle Mall SA 5000, Australia, 1998.

Finch H. Development of the guttersnipe. In: Proceedings of the Tokyo international rainwater utilization conference. Sumida City, Tokyo, Japan, 1994. pp. 360–366.

Gould J, Nissen-Petersen E. Rainwater catchment systems for domestic water supply: design, construction and implementation. London: IT Publications; 1999. 300p.

Haebler RH, Waller DH. Water of rainwater collection systems in the Eastern Caribbean. In: Proceedings of the 3rd international conference on rain water cistern systems, Khon Kaen University, Thailand, 1987. F2, pp. 1–16.

Hasse R. Rainwater reservoirs above ground structures for roof catchment. Germany: Gate, Vieweg, Braunschweig/Wiesbaden; 1989 102p.

Latham B, Gould J. Lessons and field experience with rainwater collection systems in Africa and Asia. Aqua. 1986;4:183–9.

Lee MD, Visscher JT. Water Harvesting: a guide for planners and project managers. Technical Series Paper 30, IRC PO Box 93190, The Hague, Netherlands, 1992. 106p.

Nissen-Petersen E. How to build and install gutters with splash guards, a phot manual, ASAL Consultants/DANIDA, Kenya. www.morewaterforaridlands.org; 1992.

Ree W. Rooftop runoff for water supply, U.S. Dept. of Agriculture Report, ARS-S-133, Agricultural Research Service, US Dept. of Agriculture, 1976. 10p.

Skinner B. Community rainwater catchment. Unpublished Report, Water Engineering and Development Centre (WEDC), Loughborough University, UK, 1990. 109p.

Watt S. Ferrocement water tanks and their construction. London: IT Publications; 1978 118p.

Yaziz M, Gunting H, Sapiari N, Ghazali A. Variations in rainwater quality from roof catchments. Water Res. 1989;23:761–5.

Zhu Q, Liu C. Rainwater utilization for sustainable development of water resources in China. Paper presented at the Stockholm Water International Symposium, Stockholm, Sweden, 1998. 8p.

Chapter 9
Rainwater Harvesting Systems in the Humid Tropics

Terry Thomas

Keywords Roofwater harvesting design · Tank sizing · Gutter design

9.1 Climates of the Tropics

The Tropics comprise about 38 % of the earth's surface and contain over 40 % of the world's population (mostly in Asia). As a rough rule, the closer you are to the Equator, the wetter the climate—or more important for rainwater harvesting (RWH)—the more uniform the rainfall through the year.

A uniform rainfall favours RWH; however, it also favours surface and underground water supplies as well. Historically, the best-documented instances of RWH have been in semi-arid zones where all competing forms of water supply are 'difficult'.

We may divide the tropics into different rainfall zones—Uniform, Bimodal, Short dry season, Long dry season and Semi-arid. The strongest determinant of which zone a place lies in is its latitude, but other physical geographic factors also have influence, such as distance from the sea.

Below are monthly rainfall bar charts for five tropical locations (Fig. 9.1), chosen to illustrate the five zones listed above. These only show long-term monthly averages: they do not show the extent of year-by-year variations or any long-term trends due to, say, global warming. In general, the lower the rainfall at a location the more variable it is.

T. Thomas (✉)
DTU, School of Engineering, University of Warwick, Coventry, UK
e-mail: dtu@eng.warwick.ac.uk

Q. Zhu et al. (eds.), *Rainwater Harvesting for Agriculture and Water Supply*,
DOI 10.1007/978-981-287-964-6_9

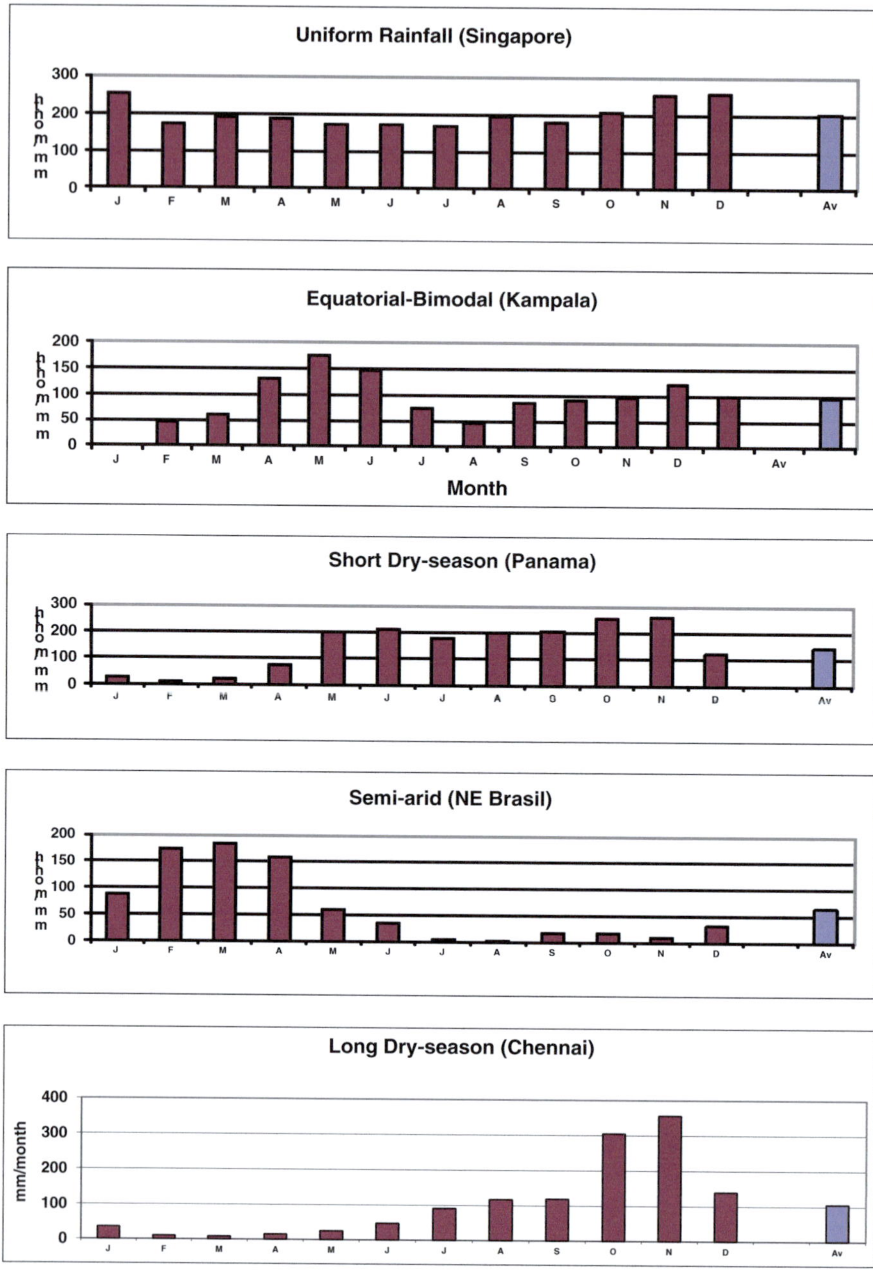

Fig. 9.1 Mean monthly rainfall bar graphs for 5 different tropical locations

Uniform: Rather few places have Uniform Rainfall (in contrast with most of Europe and N. America where the mean monthly precipitation is fairly constant round the year). Some islands, whose climate is dominated by sea breezes, have fairly uniform rainfall, as do locations in rain forests where much rainfall originates in recent local transpiration from plants. On the next page, Singapore, which is an island and only 1° from the Equator, represents this pattern. Singapore would be excellent for RWH were its population not living mainly in high blocks of flat with only a tiny roof area per inhabitant.

Bimodal: A commoner pattern close to the Equator is Bimodal Rainfall accruing as the sun passes overhead in March and September. This is represented by Uganda (e.g. Kampala, 1°N) and occurs right across Africa close to the Equator.

Short dry season: More than 5° from the Equator, however, this pattern is replaced by one of 'late summer' rains, centred on July in the Northern hemisphere, January in the Southern hemisphere. There is a single dry season but it is not intensely dry—few months have less than 30 mm rainfall. Panama, 9°N, represents this zone.

Long dry season: By 8° from the Equator in Africa and 15° in 'Monsoon' Asia, this single dry season has become more arid and long—often extending for over half the year. Chennai (Madras, 13°N) represents this pattern.

Semi-arid: Finally by 18° from the Equator in all but SE Asia the rainfall is falling and the dry season is extending to 8 months or more, as represented by the data from the dry zone of NE Brazil and by the Sahel in Africa.

It is, therefore, not easy to talk generally about RWH 'in the tropics'. Other features of the Tropics of interest to us are average incomes (generally below the world average) and the persistence of insect-vectored diseases such as malaria and dengue.

9.2 Styles of Roofwater Harvesting

The term 'RWH' is used for too wide a range of activities to be very useful. At one level all terrestrial water derives from precipitation from the sky, so all water engineering might be called 'RWH'. An immediately useful distinction is to restrict the term RWH to situations where water from the sky is captured and used close to where it fell—'close' might mean 'within 10 m'. Thus rain-fed agriculture would count as using RWH.

For household or institutional water we will normally restrict ourselves to water collected on relatively clean surfaces. The most suitable surface is a sloping roof of non-absorbent material: its height and slope discourage its use for other, potentially polluting, human activity. Flat roofs that are walked on, and courtyards, will generally generate run-off that is not potable until treated. Flat roofs have a poor reputation (for leaking) in the humid tropics and are usually more costly than sloping roofs.

For the rest of this chapter we will limit ourselves to considering roofwater harvesting.

The purpose of rainwater storage is clearly to retain water from the actual time of rainfall until when the consumer needs it. The time elapsed may be hours, days or weeks—the longer the interval, the larger the storage required. On this basis, we may usefully distinguish many different 'styles' of roofwater harvesting. Nine of these (S0–S8) are summarised below:

S0. *Opportunist* Domestic Rainwater Harvesting (DRWH) that employs no special equipment but yields a useful quantity of water on rainy days. This is already widely practised wherever roofing is suitable, generally taking the form of basins placed on the ground under roof edges during rainfall. In former times, *collecting from trees* was a common form of *Opportunist RWH.*

S1. *Informal DRWH* in which some of the equipment (short crude gutters, water stores up to say 400 l capacity per household) has been acquired specifically for the purpose. It gives a household convenient access to water throughout most of the wet months. One limit on the volume of water storage used is the low water quality. Water stored in open containers deteriorates, due to algal growth, after 3 or 4 days, particularly in the absence of any inlet screening.

S2. *Wet-season DRWH* uses a fairly small store to meet all household water needs for (in Uganda) 6–8 months of each year. It does not much enhance water security, since almost all dry season water must be obtained from other sources. It is intended to save labour, especially during the agriculturally busiest time of each year.

S3. *Potable DRWH* acknowledges roofwater to be the cleanest source locally available and water for drinking/cooking as the core of household water consumption. From 4 to 7 l per person per day are needed to meet this part of household water demand, a quantity that even a small roof and a fairly small storage tank can provide. Many rural households have access to convenient, but frequently polluted, water during the wet months, with which they can meet their non-potable needs. These sources may dry up in the dry months, to which the household response includes strongly reducing use of non-potable water during that season. We might consider two variants of *Potable DRWH*, namely *seasonal* and *perennial.* With the first (*seasonal potable*) the expected reliability of supply of potable water in the dry months is not high and other potable water will be then fetched from probably quite distant point sources. By contrast *perennial potable* DRWH systems aim to meet all potable water needs throughout most of an average year. Stores needed for perennial potable DRWH are somewhat larger than for *informal* DRWH, but their main difference is that they are designed for safe water storage, excluding light and having some form of inlet screening.

S4. **Example rules for 'Adaptive DRWH' (5-person household)**

When tank is over 2/3 full	**draw 105 l per day**
When tank is between 1/3 and 2/3 full	*draw 70 l per day*
When tank is under 1/3 full	**draw 35 l per day**

Adaptive DRWH we might think of as the standard form. A medium-sized store is combined with all-roof guttering to provide the bulk of household water consumption. However, it is managed so as to substitute security for quantity in the dry months. Thus, in wet months it meets all needs with a high availability and in dry months it provides potable water only with a fairly high availability. Dry season access to other water sources is still required, but neither the volume nor urgency from them is high.

S5. *Main-source DRWH* uses a large tank and roof to increase dry-season quantities and availability to a higher level than *Adaptive* DRWH can. It 'fails' sufficiently rarely that when the tank does run dry, expensive supplementation, such as bowsering water from a great distance, can be tolerated.

S6. *Sole-source DRWH* uses a very large tank in a context where there is no other water source (not even bowsering) available. This is unlikely to arise in Uganda.

S7. *Reserve* or *Emergencies DRWH* entails the storage of rainwater only to be used when other (normal) sources have failed. It is rarely economic unless serious emergencies are common—for example earthquakes in Japan, typhoons in Guam, flooding in Bangladesh. Uganda is not such a zone.

S8. *Supplementary* DRWH employs a buffer tank in combination with any erratic water source independent of rainfall. Urban houses with erratic or expensive piped supplies sometimes instal such a tank. Unfortunately the tank location appropriate for buffering a piped supply (namely an elevated 'header' tank) is too high up to be supplemented by roof run-off water. Thus multiple tanks and a pump may be required, increasing complexity and cost.

Styles S6 and S7 give a very poor return on investment. For example, it is almost always cheaper to import a little water each year than to design a RWH system capable of acting as the sole water source, S6. In Japan (earthquakes) and W Pacific (cyclones) there is some interest in employing *Emergencies DRWH* (S7). We will not discuss these last two styles any further. Styles S0 to S6 are represented by the 'ladder' diagram below, see Fig. 9.2.

Styles S1–S6 represent increasing system cost and increasing system performance. Within performance, there is some scope for exchanging 'security' for 'quantity', since high water security is achieved by sometimes **not** using available water now in order to save it for later. However, not using water in a wet season increases the probability of tank overflow and thereby reduces how much water can be captured and used in a year.

Daily rainfall exceeding 100 mm occur once or twice a year in the equatorial and monsoon tropics, so with all but the largest tank it is possible to experience overflow from a single day's rain. A more typical wet day has around 10 mm precipitation, which in isolation rarely causes tank overflow.

Institutional RWH is likely to start at rung 3 (*Potable-water RWH*). In the case of domestic RWH there are circumstances where a household might progress step-by-step up the ladder, periodically adding extra storage and hence extra performance.

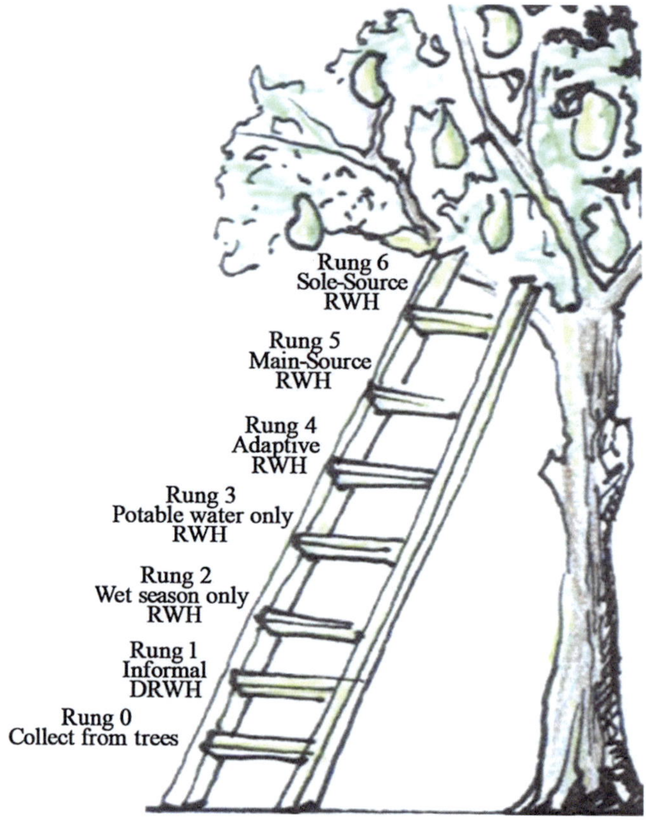

Fig. 9.2 Ladder illustrating rainwater harvesting styles

Actually the economics of RWH are not very favourable to piecemeal system upgrading; because it costs more to instal a given capacity in stages than to create it in one go. For example two 1000 l tanks are likely to cost about 30 % more than one 2000 l tank, due to size economies of scale in tank production.

The performance of the different styles, and their related storage requirements, are shown for a specific situation in Table 9.1. The 'situation' is an equatorial climate with bimodal rainfall (2 wet seasons a year).

Meanwhile, the popularity in 'the North' for the concept of autonomous or 'eco' housing is beginning to transfer to the Tropics. There have been some 'eco-houses' built in tropical Queensland (Australia) and there is considerable interest in India and Mexico. Roofwater harvesting, sits alongside water re-use, eco-sanitation, use of on-site materials, passive and active solar architecture in these situations.

In recent years, *Institutional RWH* has received more attention in Africa than *Domestic* RWH. By contrast, *Domestic* forms have been more common in Asia. Because of unfavourable tariffs for piped water, *Industrial RWH* has led the way in cities like Bangalore, India.

Table 9.1 Comparison of performance of different styles of DRWH

Form of DRWH	Required availability [wet (%)/ dry (%)]	Wet-season demand (lcd[a])	Dry-season demand (lcd[a])	Roof area $A = 8$ m^2/capita Rainfall = 1200 mm		
				Tank size needed (l/capita)	Water drawn (l/capita/year)	Tank usage[b] (times/year)
Opportunist	≈30 /0	14	0	14	1000	70
Informal	≈50/10	14	0	80	1700	21
Potable	98/95	7	7	200	2500	13
Wet-season	95/0	20	0	360	4500	13
Adaptive	98/90	<20	>7	360	5400	15
Main-source	98/90	20	20	1200	6800	5
Sole-source	100/98	20	20	2500	7200	3

[a]lcd = litres/capita/day
[b]Usage = Water drawn/tank size needed per capita

Meanwhile, the popularity in 'the North' for the concept of autonomous or 'eco' housing is beginning to transfer to the Tropics. There have been some 'eco-houses' built in tropical Queensland (Australia) and there is considerable interest in India and Mexico. Roofwater harvesting, sits alongside water re-use, eco-sanitation, use of on-site materials, passive and active solar architecture in these situations.

9.3 Historical Experience of Roofwater Harvesting

On a permeable surface like ploughed or vegetated soil, most rainfall percolates into the ground rather than 'runs off': run-off occurs only on steep slopes or during especially intense rainfall. For agricultural purposes, and for flood control, such percolation is desirable and is sometimes enhanced by special treatment of the ground (a type of 'RWH'). For domestic or institutional water supply, however, we wish to encourage run-off so that we can refill water-storage containers with it, and hence we prefer impermeable surfaces. Examples of impermeable surfaces are:

- Bare rock faces (high, unfortunately rare and usually polluted)
- Tree leaves and branches (high, often not very clean)
- Specially hardened or coated soil (low, humans and animals need to be excluded)
- Sloping roofs made of ceramic, metal, etc., but not grass (high, relatively clean)
- Flat roofs of concrete or hard mud (high, often not clean)

It is convenient if the bottom of the collecting surface is higher than the top of the intended storage vessel, so that water can flow from one to the other by gravity without pumping. Pumping during rainfall (intermittent but large flows) is awkward and costly and normally avoided. Thus, collection from a field or a courtyard invariable feeds an underground tank, whereas for the other surfaces named an above-ground tank can also be used.

It is tedious to clean water once it has become polluted (although long storage itself usually improves water quality via sedimentation and bacterial die off). Rainfall 'in the air' is generally clean, so for household or institutional water it is very attractive to intercept rain via a clean surface—or at least not a surface that is regularly walked on by humans or animals.

The most famous rock catchment is that of the often besieged peninsula Rock of Gibraltar at the entrance to the Mediterranean Sea.

Water for 50,000 people is obtained by storing run-off from the upper rock and storing it in blasted caverns within the Rock.

Harvesting from trees (including banana plants) is still practiced in several tropical countries, but usually by elderly people whose water demand is low. Some trees channel rainfall towards their own trunk where it can be collected by adding a spout to that trunk. Unfortunately, such trees also accumulate debris at the point where branches join together, the run-off coefficient of most trees is under 10 % but the rainfall interception area may exceed 100 m^2 so a useful flow is obtainable. Traditionally small vessels like pots are used for collection from trees, so these have to be frequently replaced during the rainstorm—an uncomfortable and cold procedure.

Collection from 'hardened' sloping fields is practised in some semi-arid regions (like W India) but usually outside the Tropics proper. In such areas of low rainfall, a large collection area per person is needed—much more than is likely to be available from roofing. Run-off from such fields during light rains may be negligible, due to evaporation or slight percolation, so the refilling of water stores takes place mainly during the occasional heavy rainfall. Collection from pavements, roads and courtyards is practiced in several places, including Gansu, China. The water is unlikely to be immediately potable and so needs long storage, treatment or application to non-potable uses.

Sloping roofs are characteristic of buildings in the humid tropics and have traditionally been of grass or palm. The growing scarcity of such materials, a high maintenance cost and problems with creatures living in the thatch have led to the steady replacement of such roofs by those of tile, asbestos-cement, corrugated iron (CI), aluminium etc. Some of these materials (especially CI) result in poor conditions inside the house—dry perhaps but very hot—or (e.g. tiles) require heavy strong roof structures. However, they are seen by householders to be 'modern', they require little maintenance, and they are coincidentally excellent for roofwater harvesting. So RWH has generally followed (rather than promoted) changes in roofing materials.

Since in many communities institutional buildings such as schools were the first to employ hard roofing, they were often the first to be used for RWH. This may have been unfortunate, because considerable and effective organisation is needed if the limited run-off from institutional roofs is to be equitable shared by community members. In Africa, in particular, there have been many institutional RWH programs and a high fraction of these have failed due to management conflicts.

As well as the changes in roofing, there have been changes in water containers. In the last 20 years, plastics have penetrated even rural areas in the form of bowls, buckets and jerry cans. The combination of hard roof and plastic bowl makes casual collection of roofwater fairly easy.

In those countries where sloping hard roofing has become common, for example Sri Lanka or Kenya, most rural households capture some roof run-off. However, this practice has received little official recognition because it does little to relieve dry-season water insecurity or because the choice of water storage (open-topped containers) does not result in safe water.

A group of interest in the pioneering of domestic RWH is that of richer householders. Often such householders wish to buy for themselves a better water supply than that generally on offer in their location. DRWH offers an option for to them to have more water, better water or more convenient water than their poorer neighbours. So in many tropical countries with good rainfall, such households have provided a market for the manufacturers of tanks—large galvanise—iron drums in Africa, brick tanks and mortar jars in S and SE Asia, underground-concrete in many countries and more recently large plastic tanks.

9.4 The Roofing Constraint

Inadequate roofing is a major constraint on the adoption of domestic RWH. There is the obvious problem of unsuitable roofing material, since collection off thatch is not usually practical (difficult construction, low-quality water, low run-off coefficient). However, time alone may remove this constraint. In Uganda, for example, the national fraction of households with *some* hard roofing has reached 60 % and is rising at 2 % per year, suggesting that in 20 years almost all households will have roofs suitable for DRWH.

However, a more difficult constraint is that many households with a suitable roof *type* do not have adequate roof hard roof *area*.

Although the annual fraction of run-off that can be used increases with increase in tank size, the required roof area can be thought of as fairly independent of size of storage, because starting with a nominal daily demand of say 20 l per person, we might crudely assume

Informal DRWH	captures 20 % of gutter flow and meets 20 % of annual demand
Adaptive DRWH	captures 65 % of gutter flow and meets 65 % of annual demand
Main-source DRWH	captures 92 % of gutter flow and meets 92 % of annual demand

Table 9.2 Required roof area per person for different annual rain

Annual rain (mm)	500	750	1000	1250	1500	1750	2000
Example location	Dodoma Tanzania	Cuzco Andes	Nairobi Kenya	Chennai India	Bangkok Thailand	Panama	Singapore
Roof area required for DRWH (m^2)	18	12	9	7.5	6	5	4.5

Thus, regardless of our style of DRWH, we require annual gutter flow to be not less than annual demand. Moreover, as gutter flow is typically 80–85 % of rainfall reaching the roof, we might say

"Roof area per person × annual rainfall should exceed 9000 l"

Table 9.2 for required roof area *per person* for the ready practice of DRWH.

These required areas are large in all but the wettest locations. A recent survey in Uganda, which averages about 1250 mm rainfall, suggested that although mean roof area per person (in the households *with* hard roofing) is about 8 m^2 and is just sufficient, a high fraction of households have under 6 m^2 per person. Moreover, this is a country of single-storey housing: any movement to two-storey construction will probably reduce roof area per person.

A similar survey established that the area of *institutional* roofing per inhabitant in that country is under 1 m^2 per person, and hence quite inadequate for widespread DRWH.

Thus if DRWH is to be treated as the normal source of water for all the homes in a community, then some provision for extra hard roofing will often be needed. Alternatively, roofwater can be reserved for potable uses and other surfaces than roofs used to collect rainwater for bathing, cleaning, livestock, etc.

9.5 Economic and Social Constraints

There are some basic facts underlying any economic assessment of roofwater harvesting.

One is that large tanks are cheaper *per litre of capacity* than small tanks, an 'economy of scale' similar to that found throughout industry. This suggests that large tanks have better value. However, this economy of scale in tank production is more than offset by the greater effectiveness, per litre, of small tanks. Doubling the size of a tank less than doubles its cost, but also less than doubles the annual water yield from it. Usually DRWH systems containing *small* tanks give better economics than systems containing large ones. Thus an agency with a fixed total sum to invest in roofwater harvesting will get a higher economic return by building many small systems than by building a few large ones.

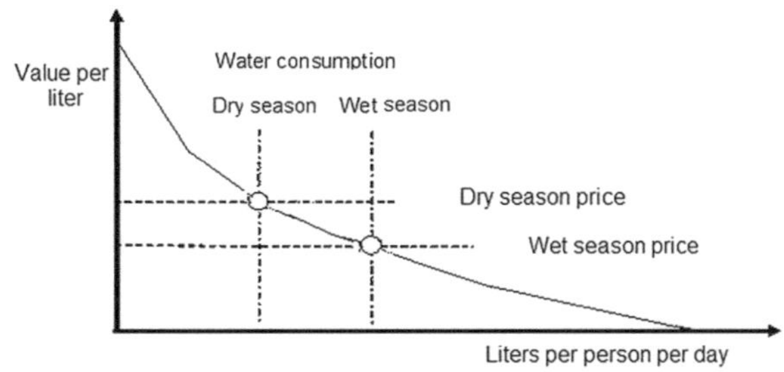

Fig. 9.3 Value per litre roofwater versus consumed water per day per person

Secondly, we must not consider all water as having the same value to a household. The first 3 l per day per person are of very high value—if they are absent the household may have to emergency relocate. By contrast a household may not be able to use more than 50 l per person per day, so that any litres beyond 50 per person have no value to it. Traditional economics would assign a 'consumer value' function to the household supply looking something like the graph below. And we would expect the water consumption by a household to vary with the 'price' of a litre of water (this price includes the perceived cost of the fetching time and any queuing time). For example in the dry season, the cost per litre is generally higher, by a factor of two or more, than the cost in the wet season; so most families use less water in the dry months, see Fig. 9.3.

This seasonal variation in consumption in line with the effective value of water is important for RWH, since a RW tank effectively absorbs low-value (wet season) water and returns some of it as high-value (dry season) water. If the ratio of value of dry-season water to wet-season water were very high—say 5:1—that would justify using a bigger tank than if it were just say 2:1.

There are two forms of economic assessment that particularly interest us. The first is 'return on investment in DRWH' and the second is 'comparison of the cost of RWH with that of other technologies delivering the same benefits'. We can look at these in turn, using data from one tropical country (Uganda) to illustrate the methods.

9.5.1 Is Investment in RWH Worthwhile in Economic Terms?

Of the many financial measures we might use to answer this question, the simplest and most suitable for DRWH is '*Payback Time*' (*PBT*) measured in months or years.

$$\text{PBT in years} = \text{Construction cost/Annual benefit} \ = C/B$$

However, not only do we need to calculate *PBT* for any given investment proposal, we also need a criterion for acceptability, for example

"For an investment in DRWH to be attractive to a household, its PBT should be under 18 months".

The investment in DRWH is primarily the cost *C* (e.g. in $) of constructing the system, since running costs are low and the life of the system is likely to much exceed its *PBT*. This investment cost depends on the technology used and in particular on the size of tank chosen. In Uganda RWH system costs are currently high for many reasons and may be expected to fall in the coming years; in countries like Thailand where the technique is long-established, costs are much lower.

The annual benefit *B* (e.g. in $/year) is the value of the water the system delivers 'to the door' in a typical year. We can, via modelling using past rainfall data, estimate the quantity of this water we expect the new system to deliver in the future.

We need in addition a value per litre for such water. Some water is actually traded—carriers sell water at the doorstep at a rate that varies with the seasons—but for most households the alternative to getting water from a DRWH system is to fetch it themselves from a free point source. In this latter case, the value of water is in effect the labour cost of fetching it, again something that may vary with the seasons.

As a gross simplification we might say that water in the dry season is worth about twice what it is in a wet season. The main reasons for this 'doubling' is the extra distance often travelled in the dry months and the longer queuing times normal at such times of year. In most Ugandan settlements, under 35 % of each year might be deemed 'dry', by contrast in much in countries further from the Equator the annual 'dry' fraction will be higher. For example in much of tropical India the dry period exceeds 60 % of each year and in semi-arid zones it may exceed 80 %. So any economic figures given here from Ugandan studies are poor guides to those in countries with lower annual rainfall and a single very long dry season.

For a median Ugandan household, making a number of reasonable assumptions about yields, roof size (40 m^2), household size (5) and collection time per litre, and on the basis of various recent surveys

- valuing water at 0.0025 $/l in the wet season and 0.005 $/l in the dry season;
- taking $100 as a benchmark cost for a RWH system with 2000 l storage;
- assuming a sensitivity of cost to storage volume of 0.6.

Payback times are given in Table 9.3.

Thus the best economic returns come from having a tank in the range 1000–2000 l. (The smaller tanks would score relatively better still had no premium on

Table 9.3 Payback times for different household tank sizes

Tank size (l)	500	1000	2000	4000	8000	16000
System cost ($)	45	66	100	150	230	350
Style of use	Informal	Potable/adapt	Adaptive	Adapt	Main	Sole[a]
Wet season yield (l/year)	8500	8500/15,000	19,000	23,000	23,300	23,700
Dry season yield (l/year)	1500	4500/3500	6000	7000	11,000	12,500
Value of yield ($/year)	29	43/55	78	93	113	122
Payback time (Years)	1.5	1.5/1.2	1.3	1.6	2.0	2.9

[a]Only in the case of *Sole-source RWH* is no use made of other (e.g. point) sources of water

dry-season water been assumed.) Of course everyone would like the higher dry-season yield that comes from a larger tank, but this liking is not economically justified. There is, however, also an economic *lower* limit on size, i.e. a size *below* which PBT starts to rise again. This is due to there being other costs, like guttering, that are fixed rather than varying with tank size and which, therefore, get relatively more important as the tank is made smaller.

9.5.2 Is RWH Cheaper than Rival Means of Attaining a Desired Service?

Here our first step is to define the 'desired service'. For study purposes we have adopted the hoped-for future standard of "average collection time per litre not to exceed 1 min" (corresponding to a mean distance threshold of 500 m from source) and a volumetric standard of 20 l/person/day. To reflect reality, the volumetric standard should be varied over the seasons, for example increased by 20 % in the wet season when water is cheap and reduced by 20 % in the dry seasons when it is expensive.

Our second step is to define what we are comparing in the chosen location, and we have chosen to compare the costs of the two alternatives of either:

(1) installing *Adaptive RWH* in all households over 500 m from an existing point source—which, in combination with *some* continuing use of that point source, is likely to achieve the '1 min per litre' target above for homesteads up to 2 km from point sources.
or
(2) increasing the number of point sources so that all households are within 500 m of one of them.

Detailed local studies to investigate this comparison have not been performed. They would require every household and every source in some selected location to be mapped. What is possible is to model larger areas on the assumption that point sources and households are uniformly spaced. However, we know that sources are where possible placed close to settlements and that settlement density may itself be enhanced by the presence of a source. A crude adjustment has, therefore, been made that within 500 m of a source (a zone whose area is $a = 0.78\ \text{km}^2$) the household densities are double those outside such a circle.

It is convenient to express all costs 'per average household'.

On this basis, the cost of alternative (i) above is the cost C_{RWH} of one DRWH system multiplied by the fraction of households over 500 m from a source. That fraction can be derived from the 'catchment' area A associated with each existing point source. [It can be shown that cost per household would be $C_{RWH} \times (A - a)/(A + a)$].

The corresponding cost of alternative (ii) is that of sufficient new point sources to reduce A to a, divided by the average number N of households per original point source. Thus, if one new borehole or other freely locatable point-source costs C_{PS} it can be shown that per household the upgrading would cost CPS $\times$ $(A - a)/(a \cdot N)$.

For Uganda as a whole, but excluding the driest zone (where only 3 % of households are located) there is one protected point source per (A=) 6 km^2. Assuming the suggested higher population densities near sources, this gives about 77 % of homesteads being over 500 m from a source. We may also assume for that country $C_{RWH} = \$100$ for a 2000 l RWH system, $C_{PS} = \$7500$ per borehole, $N = 80$ households per existing source.

Thus cost per homestead for the RWH option (i) is $77.

Cost per homestead for the point-source option (ii) is $627.

This indicates a major cost—saving using RWH rather than new point sources to achieve a reasonable standard of water access.

In other countries, the cost per 'adaptive' DRWH system may differ from the $100 value assumed here, if only because the 2000 l tank sufficient in Uganda would be insufficient in a country with less favourable rainfall distribution. Actually, Ugandan costs are high in global terms and size for size tanks are much cheaper in say SE Asia. Also the local cost of a point source may be less than the assumed $7500. Even so, it is likely that DRWH will be an attractive economic option for any organisation wishing to upgrade water supply to a higher standard of access.

It is unlikely that the sevenfold increase in the number of point sources assumed for option (ii) in Uganda would actually be built in the next decade, despite an over 30 % projected population increase in that time. Application of a less drastic program of reducing households per point source from the current value of 80 down to 40 (and thus not fully meeting the "1 min per litre" standard) would reduce option (ii) cost to $94 per homestead, which is still slightly more than the RWH cost.

Clearly lowering the costs of DRWH will increase its attraction to householders and its competitiveness with other modes. How this might be achieved will be discussed later.

9.5.3 Social Constraints

One can imagine a number of social constraints upon DRWH in general, or its use in an adaptive form, for example:

- A household water tank may be viewed as a too-conspicuous sign of economic superiority and attract criticism, vandalism or even poison.
- The effective operation of a limited water store requires its manager to retain authority over its use—against perhaps the pressures of other family members or of neighbours.

These constraints are likely to be culture specific.

9.6 Sizing Tanks

The right size for a tank is generally that which gives a good balance between its performance and its cost. That good balance depends upon local circumstance and in particular upon the availability of other, non-RWH domestic water sources. Total dependence on roofwater (*Sole-source DRWH*) is a very expensive option, to be used only in the face of necessity. Normally DRWH is operated as a partial supply meeting some (*Informal DRWH, Potable-only DRWH*) or most (*Adaptive DRWH, Main-source DRWH*) of household annual water needs.

The use of DRWH for 'productive' uses like irrigating gardens, watering livestock, brewing, making bricks etc., is rarely a serious option—roof areas and rainfall are usually barely sufficient to meet basic household water needs.

9.6.1 Theory of Tank Sizing

There is a large and rather confusing literature on sizing tanks. For *Sole-source DRWH* systems which have very large tanks, we need to know the annual rainfall over many past years in order to predict how well it will perform in the future—e.g. how often it will run dry. For *Informal DRWH*, *Adaptive DRWH* etc., that use only a small tank, we need to know *daily* rainfall over a few past years to be able to accurately predict future performance.

However, there is little point in trying to be very accurate in predicting such future performance as 'system reliability', because we do not exactly know what future rainfall will be and we also probably do not know very well how the household will try to use the system.

'Sizing a tank' means choosing a good size for it, but first we must decide what we mean by 'good'. That means we must agree on how to measure the performance of a RWH system. The three most straightforward measures are:

- 'reliability of supply' = fraction of days in a year the tank is not empty = (fraction of days the demand is fully met)
- 'demand satisfaction' = fraction of annual demand that the DRWH meets
- 'efficiency' = fraction of the yearly gutter flow that is actually drawn by the household

All three measures are fractions and it is convenient to express them as percentages. 'Satisfaction' is always higher than 'reliability', because there will be days when only part of the demand is satisfied: on such days the water drawn will count towards satisfaction but not towards the count of 'reliable' days. For economic assessment 'satisfaction' is a better measure than 'reliability', but for DRWH system users 'reliability' is easier to understand, especially if applied separately to the dry season and the wet season. We suggest that where only one measure can be used, that measure should be 'reliability'.

There are other more complicated measures, such a 'annual value of water harvested' that allow for the fact that dry-season water may be more valuable than wet-season water.

In the case of both 'reliability' and 'satisfaction', we need to know what the annual 'demand' put onto the RWH system is. This is easy if the demand is kept constant (e.g. 100 l/day = 3650 l/year), but not if the demand itself varies with how much water is in the tank (*Adaptive DRWH*). Generally, if we increase the demand, we will increase the efficiency (less water will overflow the tank) but we will reduce the reliability and the satisfaction.

Here is a model of the situation (Fig. 9.4):

It is easy to decide roof area (in square metres) but the other three inputs pose some problems.

Past rainfall pattern means ideally that we should be able to enter into the computer the daily rainfall for the last say 10 years. However, such data are rarely available for the actual site of our system and even for a nearby meteorological station the data may be hard to get or expensive. The only rainfall data that are fairly easy to get is *average annual rainfall*. Unfortunately this is little use for calculating a forecast of DRWH system performance. Much more useful are the twelve *average monthly rainfalls* and better still would be *actual monthly rainfalls* (over say 10 years). There is a way of using monthly rainfall to generate a sequence of artificial daily rainfalls to feed into the computer program that forecasts system performance. This technique is used in the Warwick University's RWH performance prediction service that you can access for free by going to the website:

http://www2.warwick.ac.uk/fac/sci/eng/research/structures/dtu/rwh/model

Tank size (in litres) is our main interest. Normally, therefore, we choose several tank sizes of interest and check the performance of each one.

Demand level and strategy is more complex. Often it is convenient to identify a 'standard daily demand' (for example household population X 20 l/person/day) and then a rule to describe how it might be varied with the season or with the amount of water remaining in the tank.

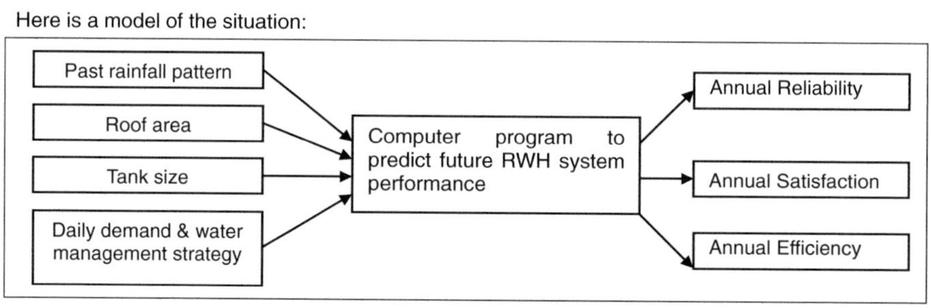

Fig. 9.4 Key inputs and outputs of the tank performance calculator computer program

9.6.2 Practical Tank Sizing

Very few people are comfortable with doing computer calculations to make a decision such as tank size. Certainly few householders or masons will be able to use such a method.

Therefore, we need 'design guides', prepared by rainwater specialists, but easy to use by RWH practitioners. The simplest guide is one in which the tank size is standardised to a single value, or perhaps to two values—'large tank' and 'small tank'. If, however, we are serious about matching the tank size to the particular household, we will have to generate more complex design aids. Unfortunately most people find it hard to 'read' tables or graphs, especially if they are very detailed. So we want to keep the variables in a design chart as few as possible.

There are two ways we might present information to a potential RWH user.

Either (a) we show them—for their particular roof size etc.—the expected performance of systems for several different tank sizes (and perhaps the cost of those systems) and leave them to choose which size they will buy

Or (b) we tell them which size to use (according to *our* judgement as to what is the right level of performance)

As each chart applies only to a place with a particular rainfall, it is best to prepare separate charts for each district rather than a library of charts covering many districts.

If we choose the option (a) above, namely we prepare *performance* tables that leave us with the following list of variables or choices:

- Roof area—we might choose 3 values.
- Tank size—we might choose 4 values.
- Demand level—we might offer 3 demand levels or alternatively 3 household sizes.
- Demand strategy—we might offer a choice of 2 strategies (fixed and adaptive).
- Performance measure (e.g. 'annual reliability' 'wet-season reliability' 'dry-season reliability' 'efficiency' and 'satisfaction')—from a user point of view 'wet-season reliability' and 'dry-season reliability' are the best ones to use.

Unfortunately this gives a table containing 144 cells ($3 \times 4 \times 3 \times 2$ input conditions and for each one 2 output measures). If we use graphs we would need a pair of graphs (one for wet-season, one for dry season) for each roof size—each graph would plot performance against tank size and have 6 lines on it.

If instead we choose option (b) above, and prepare tables of recommended tank sizes, we have fewer input variables:

- Roof area—we might choose 3 values.
- Demand level—we might offer 2 demand levels or alternatively 2 household sizes.
- Demand strategy—we might offer a choice of 2 strategies (fixed and adaptive).

Table 9.4 Recommended tank sizes in m^3 (Uganda, rainfall 1000–1200 mm/a)

Demand and choice criterion	Roof size	25 m^2			50 m^2			90 m^2		
(Reliabilities)	Family size	2	5	8	2	5	8	2	5	8
100 l/day Reliability = 98 %	Tank size (m^3) Sole-source RWH	3.0	RTS	RTS	1.5	10.0	RTS	1.1	4.0	10.0
100 l/day Reliability = 90 %	Tank size (m^3) Main-source RWH	1.3	RTS	RTS	0.8	6.0	RTS	0.6	2.0	6.0
35–100 l/day Reliability = 95 %	Tank size (m^3) Adaptive RWH	0.6	3.0	RTS	0.5	1.8	4.0	0.4	1.5	2.5
100 l/day Wet Rel. = 95 % Dry Rel. = 0 %	Tank size (m^3) Wet-season RWH	0.3	RTS	RTS	0.2	1.5	16.0	0.2	1.0	9.0
35 l/day Wet Rel. = 95 % Dry Rel. = 95 %	Tank size (m^3) Potable-only RWH	0.3	1.6	6.0	0.25	0.9	2.0	0.2	0.7	1.5

Note **RTS** means (roof too small)—the roof is too small and so even an enormous tank would not give the required service

There is now only one output instead of two or more.

The Table 9.4 is an example of such a design guide.

So in any RWH promotion program this problem of calculating and communication a recommended tank size needs to be addressed.

9.7 Technology Choice—Tanks

There is much that could be said about the selection of the type and size and method of delivery of DRWH tanks. Sizing was discussed in the last session and delivery will be addressed in the next one.

The key choices concerning tank type focus on location (below ground, partly below ground (PBG) and above ground), materials (metal, plastic, mortar, brick, mud), location of production (factory or on-site) and 'quality'. There are very many designs in use, some belonging to particular landscapes, and some reflecting the degree of usage of DRWH in the country. Many of these designs are unnecessarily expensive, so that a major concern of any DRWH promoter must be how to reduce tank costs by competition, R&D, mass-production, better quality control and avoidance of inappropriate standards. One sometimes sees water tanks much superior in construction (and cost) to the houses they serve—indeed sometimes it would be desirable for the family to abandon the house and move into the tank!

In sanitation there is the idea of progress up a 'sanitation ladder' from cheap and crude techniques (e.g. latrines) to costly but convenient ones (e.g. sewerage). In RWH we can think of both a *tank size ladder* as discussed in session T2 and a

Table 9.5 An example of a roofwater harvesting ladder

Type	Item	1000 l	2000 l	5000 l	10,000 l
Based on a low fixed demand of 50 l/day	Demand satisfaction (%)	61	68	79	94
	Max dry period (days)	163	151	113	51
Based on higher but variable demand[a] (Adaptive DRWH)	Demand satisfaction (%)	60	66	74	86
	Max dry period (days)	135	112	37	0
	Max low-use period (days)	83	98	159	147
Design 1: Pumpkin tank					
	Number of tanks	1000	680	410	280
	Labour contribution (per tank) (days)	6	8	13	19
	HH cash contribution	0	0	0	0
Design 2: Dome tank					
	Number of tanks	2000	1360	820	560
	Labour contribution (per tank) (days)	6	8	13	19
	HH cash contribution	0	0	0	0
Design 3: Mud tank					
	Number of tanks	3000	2040	1230	840
	Labour contribution (per tank) (days)	9	14	24	35
	HH cash contribution	0	0	0	0

[a]Variable demand: if tank is more than 2/3 full, 100 l/day; if it is less than 1/3 full, 20 l per day; otherwise, 50 l per day

tank quality ladder. An example of a quality ladder is as on the following page. Three tank designs of different qualities (or at least of different appearances) and different sizes are compared. Of interest is the different number of tanks that a fixed budget (of ca $20000) could buy. The lowest of the three qualities (the mud tank) will need more maintenance than the others. It achieves its low cost by separating the 'strength' and 'water-tightness' functions of a water store and using different materials to address each.

In order for a community to make an informed choice among technologies, they will need information about how different sizes systems behave, as well as the costs and trade-offs involved in different designs. A finished ladder presents this information in two sections. The top rows are generic descriptions of tanks of various sizes. These figures should not change regardless of tank type. The following rows are dedicated to tank type. They should include:

- A picture of the tank (ideally, the community should also have access to examples of the actual tanks).
- Cost, if cost recovery is being sought, or the number that can be built if they are to be subsidised (or given free) from a fixed budget.
- Any cash contribution and any HH unskilled labour contribution.

The table can either be presented "as is" or can be altered during discussions of trade-offs with the community, e.g. if the household cash contribution were raised, would the community be able to afford better or larger tanks, and what impact would this have on the overall water picture (Table 9.5).

9.8 Technology Choice—Gutters

Guttering on a very-low-cost roofwater harvesting system can take up a substantial amount of the cost, so its optimisation is important here. Typical state-of-the-art gutters in developing countries tend to be quite expensive with a typical 10-m length costing from $16 to $40. Some work has been done in East Africa with V-shaped gutters which have a typical cost of $13 for a similar length. Research at Warwick on optimising gutter size based on carrying capacity suggests that a V-shaped gutter of only 7.5-cm width is sufficient to carry water from a domestic roof in all but the most severe downpours and will deliver more than 90 % of the water it catches. Such a small gutter should cost less than $5 for a 10-m run.

In general we are looking for a gutter that

- Is small (uses little material) and hence cheap
- Is efficient (loses, by overshoot or overflow, under 5 % of the water running off the roof)
- Is easy to fix to the desired profile and does not need an (expensive) fascia board

We want to hang gutters along a profile that strikes a good compromise between *intercepting* run-off and *conveying* run-off. Experimentation and calculations suggests that to lay the first half of a gutter horizontal and the second half at a slope of 1 % is a good solution. (for gutters longer than 8 m, slopes of 0, 0.5 and 1 % would be suitable for the first, second and third 'third' of the length) (Table 9.6).

Water interception is a slightly more difficult issue. Water often has to fall some distance from the roof to the gutter and is thrown from the roof different distances depending on the intensity of the downpour. It can also be blown by wind in unexpected directions. Solutions for limiting wind-blown loss include (Fig 9.5):

The first configuration (a) is a complete solution that captures the water at the end of the roof and directs it into the gutter below. Such gutters are also very quick to instal as the slope is determined by a variable manufactured length of vertical support (between A and B) so no adjustment is necessary. Cleaning is also simple, as the inside edge is open for inserting a brush all the way along its length. Problems with the gutter appear when the length to be guttered is longer than 5 m or when thick roofs need to be accommodated. Under these circumstances the vertical support becomes very long and can flex causing the gutter to spill. This can be alleviated by using support wires with the loss of some ease of cleaning, however, as the vertical support can use a substantial amount of material, the gutter starts to become expensive at over $15 for 10 m.

The second configuration (b) uses the concept of an "upstand", where one side of the gutter stands proud of the other, effectively raising the catchment height of the gutter. In the design the usual square gutter has been simplified to a V-shape and the upstand is merely an extension of one arm of the V. This extends the catchment of the gutter upwards and moves the centre of the catchment out from the roof edge, better matching the profile of water flowing from a roof. The gutter is very cheap (less than $5 for a 10-m run) and can be applied to any sized roof without the need for a fascia board. Like all suspended gutters, the design does need adjustment to maintain the slope and suffers from guy wires obstructing cleaning.

The fixing of gutters to the proper vertical profile (and the correct distance 'out' from the roof edge) is difficult, especially when the roof edge itself is not straight

Table 9.6 Recommended sizes for 'U'-shaped gutters based on roof area ('roof area' is area drained by one gutter)

Roof area (m^2)	13	17	21	25	29	34	40	46	54	66
Gutter width (mm)	55	60	65	70	75	80	85	90	95	100

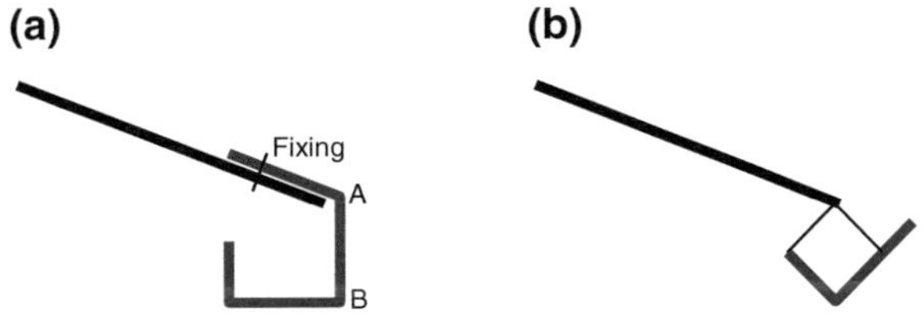

Fig. 9.5 Gutter profiles. **a** G-shaped gutter. **b** Extended V-shape

or horizontal. The job might be reserved for an artisan who has been taught where the best location lies, has developed the skill to attach gutters in that right place and who has the tooling to define a level (within 0.2 %).

Gutter failings commonly include

- oversizing
- mis-location (including sloping the wrong way)
- actual absence
- leaking joints
- twisting
- bad/no connection to downpipes

Any RWH promotion program should, therefore, specifically address the selection, sizing, responsibility for providing and the proper fixing of gutters. None of these has been addressed properly in many past programs.

9.9 Different Ways of 'Delivering' DRWH

Already 'historically' several modes of NGO delivery have been used. These may be classified as

- subsistence
- private-sector
- NGO/communal
- Government (covered in Sect. 9.10)

Subsistence provision of DRWH—the do-it-yourself (DIY) approach—is the norm for opportunist or informal DRWH. The investment and skill levels are low, but the performance is also quite low. To function well, a subsistence technology must be simple enough to copy and be supported by some level of material supplies. In some societies, DIY construction is supported by various aids to greater quality such as the existence of popular handbooks (Europe) or some overlap with artisanal practice (Africa). Shortage of competent builders, taxation and wage differentials have driven rich countries in a DIY direction for the last 30 years, so that many householders in them possess toolkits far in excess of specialist builders in poorer countries. In the tropics the trend seems to be the other way, i.e. the replacement of self-help by the employment of artisans for activities like building, bicycle repair and grain milling. *Opportunist DRWH* is such a 'low' technology that is has propagated by imitation. *Informal DRWH* is more complex and challenging. Higher forms of DRWH seem too complex to be reliably performed entirely on a DIY basis. To date there seem to have been no programs to directly encourage DIY DRWH by popularising/demonstrating good practices, issuing construction guides as pamphlets or posters or encouraging the provision of key components.

Private-sector provision of DRWH normally involves both builders ('masons', 'fundis') and manufacturers. Naturally the activity is orientated to satisfying those with purchasing power. Specialist producers of gutters and some sorts of tank exist in several countries. Specialist installers—whose knowledge includes correct sizing and placing of components—are less common, despite some evidence of demand, so that the building is often done by generalist builders with little appreciation of DRWH. In a class of their own are tank-builders—masons who have the confidence and perhaps the skill to produce economical and effective tanks. In Sri Lanka for example, many masons and some householders build domestic brick RW tanks; unfortunately these are usually open-topped and hence yield low-quality water.

There is an interesting tension between enhancing production quality of tanks (and hence reducing materials use) by manufacturing rather than building and avoiding difficult transport problems. For example consider the following range of tank designs and their suitability for on-site or for factory production.

Large underground tank and cover—must be made on-site (i.e. by masons)
Brick above-ground tank—must be made on-site
Large mortar jar—must be made on-site unless terrain is easy for transport of 500 kg loads
Medium/small jars—may be made on-site or transported from a small local factory
Galvanised iron tanks—normally transported from a small local factory
Plastic tanks—transported from a large centralised factory

The new legislation in India forcing widespread inclusion of DRWH in larger new housing will surely lead to the development of specialist builders. Perhaps as guttering becomes widespread its correct construction will become a normal part of a roofer's trade. One sometimes finds isolated examples of specialist DRWH services, ranging from turnkey packages for rich houses to masons who specialise in restoring rusty galvanised iron tanks.

There seems to be considerable scope for providing external assistance to specialist installers and builders, both via production training and by initial marketing on their behalf (justified on grounds of enterprise promotion and income generation). In theory some sort of accreditation, inspection or franchising of such specialists could improve their performance, their prosperity and their customers' satisfaction with DRWH.

NGO/Communal Provision of DRWH have been popular in Africa, despite DRWH, unlike the development of shared point sources, being intrinsically a household activity. The assumption is usually that DRWH is for the poor, that substantial subsidy is required, that it offers good opportunities for empowerment of marginal people and that some form of community control must be incorporated into programs. At its best such NGO/CBO promotions leave behind an ongoing capacity to build DRWH systems—some of the church programs have achieved this. At worst a high training, consultation and supervision cost is incurred—out of all proportion to the number of systems finally built. A popular scheme has been to involve women's groups in managing 'revolving' funds for the self-building of DRWH systems—thus in conjunction with regular savings, every household in a defined club eventually receives a system, built by group effort. Such construction can be fitted well into the routines and constraints of rural women. The success of such schemes is very variable—often the chosen group (privileged by training and initial funding) becomes closed and secretive, impeding rather than encouraging the take-off of DRWH in their community. NGOs themselves find it hard to 'let go' of a new technology and are naturally attracted to techniques requiring ongoing subsidy—whose distribution they can of course manage.

There has recently been a move towards government and international aid agencies using NGOs as delivery agents for larger numbers of DRWH systems, perhaps in conjunction with specialised relief programs (RWH for refugees and IDPs, for the disabled in Cambodia, for AIDS victims in E Africa etc.)

9.10 Incorporating DRWH into Publicly Funded Rural Water Supply

The era of "government will provide water, free (if you wait for long enough)" is drawing to an end in the tropics, although in countries like Sri Lanka publicly linking rural water supply to payment is still politically contentious. Governments, often with the support of overseas aid agencies do feel obliged—or politically rewarded—to address severe water stress. However, DRWH poses particular problems for incorporation into government programs. The principal of these problems are:

- Dislike of putting publicly owned infrastructure on private property;
- Ignorance of most water engineers of how to design and supervise DRWH systems and preference for 'communal' (=large scale or complex) technologies;
- The bias-to-the-rich of DRWH unless special measures are taken, such as funding roof construction;
- Lack of competent contractors to tender for DRWH contracts and difficulty of NGOs (which hold much of the experience) engaging in tendering;
- Tendency to think in terms of single-source supply, despite multiple-sourcing being a norm in much of the tropics;
- Difficulty in monitoring system delivery or water quality.

So the piloting of different forms of government delivery of DRWH is needed if the technique is to be normalised into public water supply practice and in particular ways of addressing the problem of anti-poor bias must be found.

Chapter 10
Rainwater Quality Management

John Gould

Keywords Rainwater quality · Rainwater and health · Rainwater contamination

10.1 Introduction

People have consumed rainwater stored temporarily in natural surface depressions, artificial reservoirs and household containers since ancient times. In almost all regions and traditional cultures, rainwater is universally recognized as a pure and clean source of water (Gould and Nissen-Petersen 1999).

In recent decades, however, there has been a growing concern regarding the quality of small-scale drinking water supplies around the world. Since rainwater has tended to be used for potable purposes mainly by individual households, or small scattered rural communities especially in developing countries, the issue of the safety of rainwater supplies has until recently been largely ignored.

The quality of rainwater used for domestic supply is of vital importance because in most cases, it is used for drinking untreated. The issue of water quality is a complex and sometimes controversial one, partly because rainwater supplies often do not meet WHO drinking water standards (the definitive global water quality standard) especially with respect to bacteriological water quality. The same, however, is also true for many improved rural water sources in much of the developing world.

Just because water quality does not meet some arbitrary national or international standard, it does not automatically mean the water is harmful to drink. In the

J. Gould (✉)
Lincoln University, Canterbury, New Zealand
e-mail: johnegould@gmail.com

Q. Zhu et al. (eds.), *Rainwater Harvesting for Agriculture and Water Supply*,
DOI 10.1007/978-981-287-964-6_10

1980s, drinking water quality guidelines, more appropriate for rural conditions in developing countries, were proposed and widely adopted. These less strict guidelines allowed water with mean *Escherichia coli* counts of up to 10 per 100 ml to be accepted for drinking. This is compared to the WHO-recommended limit of 0 per 100 ml (WHO 1993). When the quality of stored rainwater samples is judged according to this less strict but more realistic criterion, the number of rainwater supplies meeting this standard for water acceptable for drinking is greatly increased.

In 1993, bacteriological water quality standards for potable rainwater were formally proposed (Krishna 1993). It was recommended that standards should be achievable and less stringent in tropical regions and developing countries. Faecal coliforms were proposed as the most appropriate indicator of cistern water quality, and the following three-tier classification was suggested as a useful guide to cistern water quality originating from rooftop run-off.

Class I 0 faecal coliforms/100 ml;
Class II 1–10 faecal coliforms/100 ml;
Class III >10 faecal coliforms/100 ml.

In this classification, Class I represent the highest and ideal water quality, Class II represents water of marginal quality, and Class III represents water unacceptable for drinking purposes.

Compared with most unprotected traditional water sources, drinking rainwater from well-maintained roof catchments usually represents a considerable improvement, even if it is untreated and normally it is quite safe to drink.

In answer to the question "Is rainwater safe to drink", the official line of the Australian government (National Environmental Health Forum) can be summarized by the following statement from Cunliffe (1998).

> Providing the rainwater is clear has little taste or smell and is from a well-maintained system it is probably safe and unlikely to cause any illness for most users.

Only immuno-compromised individuals might be at risk such as the very young, very old, those with cancer, diabete, organ transplant or HIV-positive status for whom disinfection and thorough boiling of rainwater is recommended before consumption.

The quality of water in ground catchment tanks, however, is generally low due to contamination of the catchment area by children and animals and is not recommended for drinking unless it is first boiled or filtered. The water extracted directly from ground tanks is, however, suitable for most non-consumptive purposes without treatment.

People used to drinking rain water with small amounts of bacteriological contamination tend to build up resistance and generally suffer no ill effects, although a visitor not used to drinking untreated water may suffer from diarrhoea (gastroenteritis) or even dysentery when drinking the same water.

10.2 Quality Issues for Collected Rainwater

There is no simple answer to the question "Is rainwater safe to drink?" No water supply is 100 % safe all the time, and the issue is really one of what is an acceptable level of risk based on cultural and socio-economic standards and the quality of alternative water supplies.

The level of risk associated with drinking rainwater is a product of the concentration of pathogens/toxins present, the level of exposure and impact of the infective agent/toxin, and the vulnerability of the individual or population exposed.

Local circumstances and in particular environmental factors are also critical, such as the degree of atmospheric pollution, the type of construction materials and level of maintenance of the rainwater catchment system. In rural areas where atmospheric pollution is not generally a problem, several simple steps to reduce contamination of rainwater supplies can be taken. Most of these require little or no extra cost and are outlined below.

10.3 Sources of Contamination

10.3.1 Contamination by the Atmosphere

In industrialized urban areas, atmospheric pollution often renders rainwater unsafe to drink. Heavy metals such as lead are potential hazards especially in areas of high traffic density or in the vicinity of heavy industries (Yaziz et al. 1989).

Chemicals such as organochlorines and organophosphates used as pesticides and herbicides in agriculture can contaminate rainwater, and the studies in the USA have revealed their presence (Richards et al. 1987).

Despite the numerous sources of atmospheric pollution, in most parts of the world, especially in rural and island locations, levels of contamination of rainfall are low. Most contamination of rainwater occurs after contact with the catchment surface (roof or ground) and during subsequent delivery and storage.

10.3.2 Contamination by the Catchment

Generally, rainwater collected from a clean, well-designed and maintained roof catchment in a rural area is safe to drink unless lead paints or lead flashings have been used in the construction of the roof.

Water collected from roofs in industrialized urban areas or in the close proximity of smelters or other sources of serious air pollution should also *not* be consumed.

Reports of disease outbreaks linked to roof water sources are rare, although there have been a few reports of gastrointestinal illnesses linked to large quantities of bird or animal droppings on the roof. If trees or branches overhang roofs, these should be removed.

Rainwater collected from ground or rock catchment systems, however, tends to be more heavily contaminated by debris on the catchment surface making it unsuitable for drinking unless first boiled or treated. The water is, however, suitable for most non-consumptive purposes without treatment.

10.3.3 Contamination During Storage

Immediately following heavy rainfall, the quality of water in the tank may be lowered due to any debris washed into the tank or stirred up from the bottom, which may take some time to settle out. It is, therefore, appropriate to avoid drinking water directly from the tank for a few days.

Contrary to popular myth, however, rather than becoming stale with extended storage, rainwater quality usually improves. This is because bacteria and pathogens gradually die off during the first several days of storage (Watt 1978; Michaelides 1986, 1989; Skinner 1990). For this process to occur, however, it is essential that both light and organic matter are excluded from the tank. Any light getting into the tank will allow algal growth, and the presence of organic matter will provide nutrients enabling bacteria and other micro-organisms to survive. In such situations, the stored water may become stale and unpalatable. Measures to exclude both light and organic debris (leaves, moss, etc.) such as covers and screens will also help to keep small mammals, reptiles, frogs, birds and insects out of the tank. Even with the best efforts, some matter and small organisms will get into tanks from time to time, so regular inspections and occasional cleaning and maintenance of the systems are essential. If all reasonable measures are taken, however, the water quality should not normally deteriorate, and because the stored water is normally cool, clear and sweet, it is often preferred over other sources.

One potential problem which can occur during storage can be caused if rainwater becomes acidic due to organic material decaying in the tank or because of atmospheric pollution. These concerns do not relate to any direct threat posed by rainwater with low pH values but are due to the indirect effects of this more "aggressive water". When pH is depressed, rainwater becomes more active in leaching out metals and other constituents from storage tanks, taps, fittings and sludge deposits on the tank floor. In Hawaii, this has been shown as the source of lead and other metals found in storage tanks as a result of acidic rain resulting from volcanic gaseous releases (Sharpe and Young 1982).

10.3.4 Contamination After Storage

The reality in much of the developing world today is such that even where water supplies are clean and meet WHO standards at source, they are normally contaminated during collection and transportation to the household. The water may be further contaminated in the home through transfer to another storage container or even contamination from the cup or vessel used for drinking (Wirojanagud et al. 1989). While this secondary contamination problem should never be used as an argument for not trying to achieve the highest possible water quality at source, it does put the issue of water quality standards into context and highlights the crucial importance of hygiene education.

10.4 Health Risks

10.4.1 Microbiological Contamination

Rather than comparing bacteriological water quality data against some general standard guidelines, a more direct source of evidence implicating a rainwater supply as a potential health risk comes from identifying the presence of specific pathogens.

There are many references to pathogens including *Salmonella, Legionella-like spp, Clostridiumperfringens, Aeromonas, Vibrio parahaemolyticus, Campylobacter, Cryptosporidium, and Giardia* having been isolated from rainwater samples (Cunliffe 1998). Proving direct causation between microbial contamination and health is, however, more difficult especially when contamination levels are low.

There are only a few studies citing proven links between disease outbreaks and rainwater sources is perhaps not surprising given this difficulty and the fact that many individual cases go unreported or result in no further investigation (Simmons and Heyworth 1999). The lack of reported cases is mainly due to the fact that most rainwater supplies are used by single families, thus reducing the likelihood of large numbers of people being affected in any single outbreak.

Case studies—Microbiological contamination of rainwater

One of the first well-documented cases records an outbreak of gastrointestinal illness including diarrhoea, headaches, fever and vomiting among 63 individuals from a group of 83, mainly children 5–19 years of age, who attended a rural camp in Trinidad, West Indies (Koplan et al. 1978).

The probable cause for the outbreak was postulated as *Salmonella arechevalata* contained in animal or bird excrement on the camp roof and washed into the rainwater tank, from which water was used for drinking.

Other studies that have established links between rainwater consumption and illness include outbreaks of:

- *salmonellosis* in New Zealand (Simmons and Smith 1997);
- *campylobacteriosis* in Australia (Brodribb et al. 1995);
- *giardiasis and cryptosporidosis* in Australia (Lester 1992).

In Queensland, Australia, 28 cases of gastroenteritis among 200 workers at a construction site, Salmonella Saintpaul, were isolated from both individuals and rainwater samples (Taylor et al. 2000). Animal access was suggested as being the source of the contamination with several live frogs being found in one of the suspect tanks.

In South Australia an investigation into the relationship between tank rainwater consumption and gastroenteritis involved a survey of 9500 four-year-old children (Heyworth et al. 1998; Heyworth 2001, 2004). This study found that children drinking tank rainwater were not at a greater risk of gastroenteritis than children drinking public mains water. One important limitation of this study was that the majority of participants had drunk tank rainwater for at least one year. Hence, an alternative explanation to there being no increased risk associated with tank water was that acquired immunity to common pathogens may have resulted from long-term exposure to tank rainwater. A number of research studies in New Zealand has also alluded to the potential risk of consuming rainwater due to microbiological contamination (Abbott 2004a, b, 2006) and also of the effectiveness of reducing this risk using First Flush devices.

10.4.2 Chemical Contamination

Potentially, the adverse health implications of the long-term consumption of rainwater containing elevated levels of heavy metals such as lead pose a serious health threat.

Case studies—Chemical contamination of Rainwater

Evidence from Ohio, in an area with serious atmospheric pollution and acidic rainfall, suggested that elevated lead and cadmium levels in cistern sediment and water posed a potentially serious health risk.

A recent pilot study of 25 potable household rainwater supplies around Auckland in New Zealand found lead exceeding national drinking water standards in 12 % of the tanks surveyed (Simmons and Smith 1997). Lead levels exceeding 3.5 times WHO drinking water standards have also been noted in Selangor, Malaysia (Yaziz et al. 1989).

Investigations in Halifax, Nova Scotia, also revealed high lead concentrations in run-off water collected from an old roof with considerable amounts of lead flashing from rainwater with pH 4. Evidence of the potential health dangers of excessive lead levels in stored rainwater also comes from a study in Port Pirie, an industrial port in South Australia and location of one of the world's biggest smelters. This study revealed a correlation between blood lead levels in children under 7 and lead in tank waters, one source of which may have been highly leaded roof paint. The effect of acidic (pH 3) water and the presence of leaf litter in the tank was also shown to increase the rate of dissolution of lead from tank sludge by up to 50 times.

10.4.3 Insect Vectors

Rainwater tanks can provide breeding sites for various mosquito larvae. Where containers are open or lack secure covers or screens, they are far more vulnerable to infestation. This is of particular concern in tropical areas where they are vectors of serious diseases such as malaria, yellow fever, dengue fever and filariasis.

It should be noted that if all light and organic material (nutrients) are excluded from the tank, even if mosquito eggs hatch in the tank, the larvae will die off before reaching maturity, so no adult mosquitoes will result.

Several approaches to mosquito control have been tried with some success. These include:

- adding small amounts (5 ml per 1000 l) of domestic kerosene to stored rainwater;
- biological control using fish and dragonfly larvae to consume mosquito larvae.

Although insecticides, such as DDT, are sometimes sprayed on open water breeding sites, these should <u>never</u> be applied to stored rainwater for consumption.

10.4.4 Tank Failure and Collapse

Large surface tanks pose a potential threat to life and property if they fail or collapse. This small but potentially serious risk should be considered when determining the siting of the tank.

10.4.5 Risk of Drowning Especially for Children in Sub-surface Tanks

Drowning of young children and livestock only poses a risk in areas where uncovered ground tanks exist. Any risk to human life can be significantly mitigated by incorporating a simple ladder or steps in the tank design to facilitate an easy exit.

10.5 Health Benefits

On the positive side, we should also note the undoubted health benefits for women by spending less time collecting water—benefits such as fewer accidents to unattended infants, better nutrition, less female back injury and of course the hygiene benefits of greater water consumption which introducing RWH sometimes brings.

10.6 Protecting Rainwater Quality

Methods to protect or improve rainwater quality include appropriate system design, sound operation and maintenance, the use of first flush devices and treatment. Treatment is mainly appropriate as a remedial action if contamination is suspected.

10.6.1 Appropriate System Design

The best initial step to protecting water quality is to ensure good system design. Water quality will generally improve during storage provided light and living organisms are excluded from the tank, and fresh inflows do not stir up any sediment. The design should include (Fig. 10.1):

- A clean impervious roof made from smooth, clean non-toxic material. Overhanging branches above the catchment surface should be removed.
- Taps should be at least 5 cm above the tank floor (more if debris accumulation rates are high).
- Wire or nylon mesh should cover all inlets to prevent any insects, frogs, toads, snakes, small mammals or birds from entering the tank. The tank must be covered, and all light excluded to prevent growth of algae, etc.
- A leaf slide (guttersnipe) and first flush device should be fitted to intercept water before it enters the tank for removing leaves and other debris.

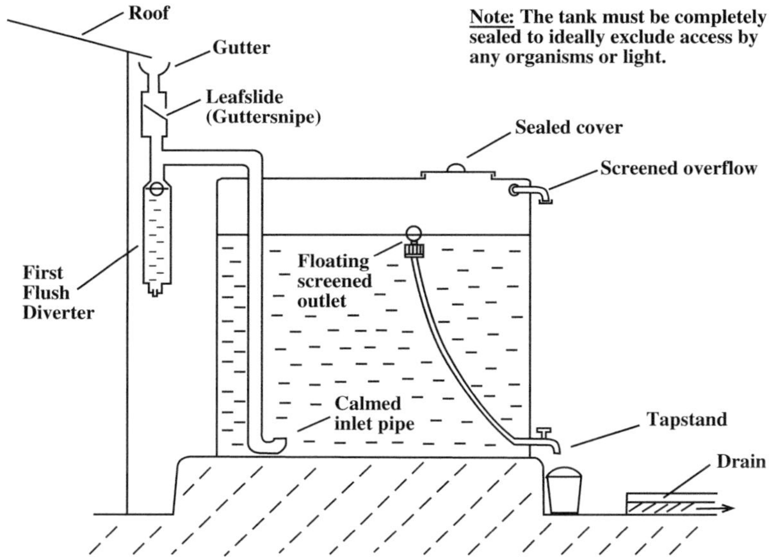

Fig. 10.1 Idealized tank design incorporating water quality protection measures (Drawing courtesy John Gould 2015)

10.6.2 Filtration, Treatment and First Flush Systems

Although not absolutely essential for the provision of potable water, when effectively operated and maintained, first flush and filter systems can significantly improve the quality of roof run-off. If poorly operated and maintained, however, such systems may result in the loss of rainwater run-off, through unnecessary diversion or overflow and even the contamination of the supply. If any kind of first flush (foul flush) device is to be considered, it should be extremely simple, and should not require regular attention regarding its operation and maintenance. Examples of such devices include:

- First flush diverter with self-closing ball valve (Figs. 7.2 and 10.2);
- Self-cleaning gutter snipes—sold commercially (Fig. 8.13);

In a study by Yaziz et al. (1989), water quality analysis was conducted on the initial "first flush" run-off from both a tile and galvanized iron roof in which the first, second, third, fourth and fifth litres of run-off were sampled. This revealed high concentrations of most of the pollutants tested in the first litre with subsequent improvements in each of the following samples, with few exceptions. Faecal coliforms, for example, ranged from 4–41 per 100 ml in the first litre of run-off sampled but were absent entirely in samples of the fourth and fifth litres.

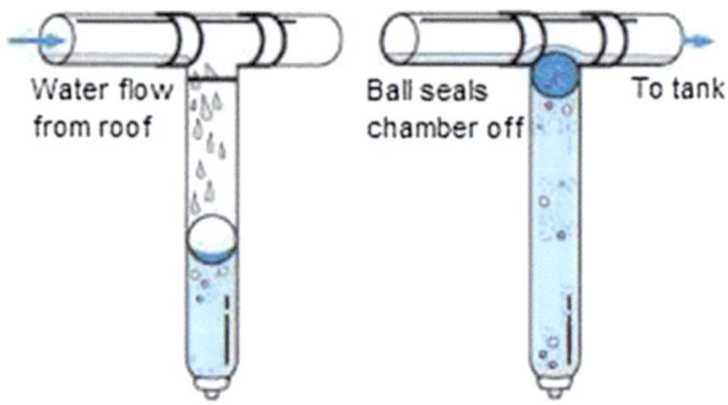

Fig. 10.2 Rainwater tank first flush diverter (*Source* www.rainharvesting.com.au)

10.6.3 Treatment

Treatment of stored rainwater only makes sense if it is done properly and if hygienic collection and use of the water will ensure it does not suffer from recontamination. There are several types of treatment possible, the most common being chlorination, boiling, filtration and exposure to ultraviolet or natural sunlight.

- Chlorination is most appropriately used to treat rainwater if contamination is suspected due to the rainwater being coloured or smelling bad. It should only be done if the rainwater is the sole source of supply and the tank should first be thoroughly inspected to try to ascertain the cause of any contamination.
- Boiling water thoroughly for at least 1 min normally ensures it is free from harmful bacteria and pathogens. Since it is not practical to treat water regularly in this way, it is again only usually appropriate as an emergency measure.
- Filtration can be used both to prevent material from entering the storage tank, during extraction of water from the tank or prior to consumption. Potentially, if properly maintained, filters will improve water quality.
- Direct sunlight can also be used to kill many of the harmful bacteria in water by exposing it in clear glass or plastic bottles for several hours. Although, feasible in some circumstances, the water must be clear, the weather fine and the water cooled overnight before consumption. This process is known as the Solar Disinfection (SODIS) method and is explained in detail at www.sodis.ch.
- A solar-powered ultraviolet unit is able to process 1.5 litres of water per minute developed by Joklik (1995). It was evaluated on rainwater cistern water in Hawaii and found to be 99.9 % effective in removing indicator bacteria (Fujioka et al. 1995). However, effective ultraviolet sterilization may be too expensive for widespread use.

A wide range of commercially available and systems for improving roof rainwater harvesting can be purchased at a price, and these include:

- Gutter meshes,
- Leaf and debris screens,

- First flush diverters,
- Tank vacuum systems and kits,
- Calmed inlet water pipes,
- Floating valve out-take systems,
- Two or more tanks in series.

Examples of a range of such devices can be found at www.rainharvesting.com.au.

10.7 Conclusion

Untreated roof run-off has been widely used for drinking purposes for many years with very few reports of serious health problems. Compared with most unprotected traditional water sources, drinking rainwater from well-maintained roof catchments usually represents a considerable improvement and even if it is untreated it is generally safe to drink. Rainwater collected from ground catchment systems is generally subjected to high levels of microbial contamination, and its consumption without treatment is not recommended.

A number of pathogens harmful to health and disease outbreaks linked to rainwater sources have been documented. While serious, these outbreaks are all isolated and comparatively rare incidents. Nevertheless, they do provide an important warning regarding the potential hazards associated with drinking rainwater from sources which are poorly located, constructed or maintained. Pathogens will die off after several days if all light and organic material are excluded from the tank.

Clearly, in certain circumstances such as in areas of severe air pollution or where lead flashing, lead-based paints or other potentially toxic building materials have been used, and chemical contamination of rainwater can pose a health threat.

Generally, serious chemical contamination of rainwater is rare. Nevertheless, extra vigilance may be advisable with respect to very old roof catchment systems, which may have been constructed at times when building codes and regulations relating to the use of lead-based paints, lead fittings and other hazardous materials were far more lax than today.

Keeping leaves and organic material from accumulating in the tank and gutters will also help to prevent the stored rainwater from becoming too acidic and potentially leaching lead or other heavy metals from the tank walls, fittings or sludge deposits into the water.

Essential measures to protect quality of water collected from any roof catchment system include good system design and regular system inspection and maintenance (Michaelides 1986, 1989). First flush devices can also be used and, if these are properly maintained and operated, can greatly improve the quality of the initial roof wash entering the tank. Treatment should be used only if rainwater contamination is suspected and no alternative potable water source is available.

References

Abbott SE, Caughley BP, Bell SJ. Microbiological health risks of roof-collected rainwater—a review. In: Proceedings of national environmental health conference war memorial centre, Napier. NZ Institute of Environmental Health; 2004a.

Abbott SE, Caughley BP, Bell SJ. Microbiological health risks of roof-collected rainwater. Water Wastes New Zealand. 2004b;134:26–27.

Abbott SE, Douwes J, Caughley BP. A survey of the microbiological quality of roof-collected rainwater of private dwellings in New Zealand. New Zealand J Environ Health. 2006;29:6–16.

Brodribb R, Webster P, Farrell D. Recurrent *Campylobacter fetus* subspecies bacteraemia in a febrile neutropaenicpatient linked to tank water. Commun Dis Intell. 1995;19:312–3.

Cunliffe D. Guidance on the use of rainwater tanks. National Environmental Health Forum Monographs, Water Series 3, Public and Environmental Health Service, Department of Human Services, P.O. Box 6, Rundle Mall SA 5000, Australia; 1998.

Fujioka R, Rijal G, Ling B. A solar powered UV system to disinfect cistern waters. In: Proceedings of the 7th international conference on rain water cistern systems. Beijing, China; 1995. Section 9, pp 48–57.

Gould J, Nissen-Petersen E. Rainwater catchment systems for domestic water supply: design, construction and implementation. London: IT Publications; 1999 300.

Heyworth J. The diary study of gastroenteritis and tank rainwater consumption in young children in South Australia. In: Proceedings of the 10th international conference on rain water catchment systems. Mannheim, Germany; September 2001. Paper 3.3, pp 1–9.

Heyworth J. An epidemiologic study of childhood gastroenteritis and consumption of tank rainwater, an untreated water supply in South Australia. PhD thesis, Flinders University Australia; 2004.

Heyworth JS, Maynard EJ, Cunliffe D. Who drinks what: potable water consumption in South Australia. Water. 1998;25:9–13.

Joklik O. Potabilization of rainwater. In: Proceedings of the 7th international conference on rain water cistern systems. Beijing, China; 1995. Section 9, pp 33–47.

Koplan J, Deen R, Swanston W, Tota B. Contaminated roof-collected rainwater as a possible cause of an outbreak of salmonellosis. J Hyg Cambridge. 1978;8:303–9.

Krishna J. Water quality standards for rainwater cistern systems. In: Proceedings of the 6th international conference on rain water cistern systems. Nairobi, Kenya; 1993. pp 389–392.

Lester R. A mixed outbreak of cryptosporidiosis and giardiasis. Update. 1992;1(1):14–5 Australian Government, Department of Health.

Michaelides G. Investigations into the quality of roof-harvested rainwater for domestic use in developing countries. PhD Research Thesis. Scotland: University of Dundee; 1986.

Michaelides G. Investigation into the quality of roof-harvested rainwater for domestic use in developing countries: a PhD research study. In: Proceedings of the 4th international rainwater cistern systems conference. Manila, Philippines; 1989. E2, pp 1–12.

Richards R, Kranmer J, Baker D, Krieger K. Pesticides in rainwater in the northeastern United States. Nature. 1987;327(6118):129–31.

Sharpe W, Young E. Occurrence of heavy meals in rural roof-catchment cistern systems. In: Proceedings of the international conference on rain water cistern systems. Hawaii, Honolulu; 1982. pp 249–256.

Simmons G, Heyworth J. Assessing the microbial health risks of potable rainwater. In: Proceedings of the 9th international rainwater catchment systems conference. Petrolina, Brazil; 1999.

Simmons G, Smith J. Roof water a probable source of Salmonella infections. New Zealand Public Health Rep. 1997:4(1):5.

Skinner B. *Community rainwater catchment.* Unpublished Report. U.K.: Water Engineering and Development Centre (WEDC), Loughborough University; 1990. p 109.

Taylor R, Sloan D, Cooper T, Morton B, Hunter I. A waterborne outbreak of Salmonella Saintpaul. Commun Dis Intell. 2000;24:336–40.

Watt S. Ferrocement water tanks and their construction. London: IT Publications; 1978. p. 118.

WHO. Guidelines for drinking-water quality, 2nd ed, vol 1. Geneva: World Health Organization; 1993. p 188.

Wirojanagud W, Hovichitr V, et al. Evaluation of rainwater quality: heavy metals and pathogens. Ottawa: IDRC; 1989. p. 103.

Yaziz M, Gunting H, Sapiari N, Ghazali A. Variations in rainwater quality from roof catchments. Water Res. 1989;23:761–5.

Chapter 11
Runoff Farming

Zhijun Chen

Keywords Runoff farming · Flood farming · Microcatchments · Bunds

11.1 Introduction

This chapter is intended to provide technicians and extension workers with systematic introduction on the application of run-off farming technologies in agriculture production.

The focus of this section is on simple, field scale systems for improved production of crops, trees and rangeland species in drought-prone areas. The purpose is to assist in selecting suitable runoff harvesting techniques and designing efficient runoff farming systems.

Systems and experiences outlined and described below are drawn from examples around the world, especially arid and semi-arid areas such as Sub-Saharan Africa, where the basic problems—low and erratic rainfall, high rates of runoff, and unreliable food production—are similar.

Selection criteria and detailed technical designs for various systems, as well as information on field layout and construction, are given below.

Note: This chapter is based on material taken with permission from an FAO publication (AGL/MISC/17/91) by Critchley W. and Siegert K. 1991, Water Harvesting : A Manual for the Design and Construction of Water Harvesting Schemes for Plant Production, FAO, Rome 133p.

Z. Chen (✉)
Formerly based at FAO Regional Office for Asia and the Pacific, Bangkok, Thailand
e-mail: zhijun.chen@fao.org

Q. Zhu et al. (eds.), *Rainwater Harvesting for Agriculture and Water Supply*,
DOI 10.1007/978-981-287-964-6_11

11.1.1 Definition

As land pressure rises, more and more marginal areas in the world are being used for agriculture. Much of this land is located in the arid or semi-arid belts where rain falls irregularly and much of the precious water is soon lost as surface runoff. Recent droughts have highlighted the risks to human beings and livestock.

While existing water resources limit the development of regular irrigation, which has been considered to be the most obvious response to drought, conventional water resources development are getting more and more difficult and costly; there is now increasing interest in a low-cost alternative—generally referred to as "water harvesting."

There are globally three levels of activities for improving more efficient rainwater use in agriculture production:

- Keep the rainwater on the field by preventing runoff and evaporation. This is achieved by non-tillage, mulching with dead or live materials, etc. This is what is commonly known as "Conservation Farming." It works as long as there is enough rainfall to grow a crop.
- If rainfall is not enough, then there is a need to harvest runoff water from a larger non-cultivated (runoff or catchment) area to a small cultivated (run-on) area. Where water is directly conveyed to a cropping area and stored in the soil profile for immediate uptake by the crops, we call it "Runoff Farming" (see Fig. 11.1).
- Finally runoff water can also be collected in reservoirs on the ground, under the ground or behind dams. This water can be used later for "Irrigated Farming" or other purposes. This way of storing and using water is much more expensive than runoff farming as water needs to be mobilized again and an irrigation system is needed. Further more, it requires certain hydrological and geological conditions.

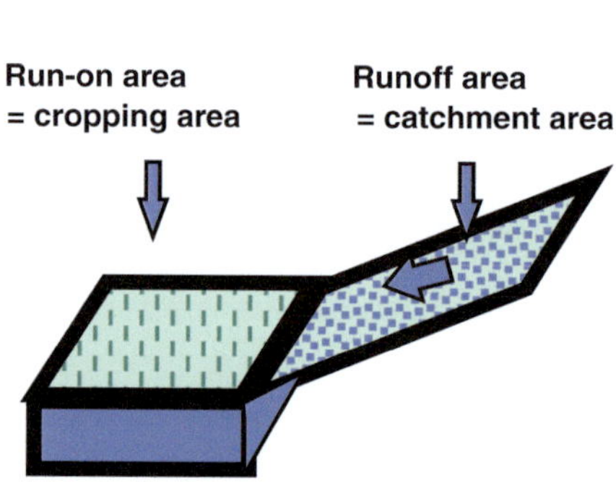

Fig. 11.1 Definition of runoff farming (Adapted with courtesy, FAO 1991)

11.1.2 Description

Runoff farming can be considered as a rudimentary form of irrigated agriculture. The major differences are that with runoff farming the water is provided by natural runoff from catchments areas rather than artificial irrigation systems; water is stored in the soil of cultivated land rather than reservoirs, ponds and cisterns; the farmers have no control over timing since runoff can only be harvested when it rains.

(1) **Suitable Areas**

Runoff farming systems are usually suitable in arid and semi-arid areas, with an average annual rainfall of at least 100 mm in the cold season and 250 mm in the hot season (Table 11.1).

(2) **Main Benefits**

Immediate benefits: in arid and semi-arid areas, where other water resources are not available or are uneconomic, runoff farming can contribute to achieve the immediate productive goal of increasing food production, either by extending areas under cultivation or increasing the productivity of arable and grazing land that suffers from inadequate rainfall (high variability). In regions where crops are entirely rainfed, a reduction of 50 % in the seasonal rainfall, for example, may result in a total crop failure. If, however, the available rain can be concentrated on a smaller area, reasonable yields will still be received. Of course in a year of severe drought there may be no runoff to collect, but an efficient water harvesting system will improve plant growth in the majority of years.

Long-term benefits: rehabilitation of degraded land through re-afforestation and increased vegetation cover and soil conservation (reduction of soil erosion). In the case of floodwater farming systems, other benefits include enhancement of groundwater recharge and reduction of damage caused by flash floods to rural structures.

Table 11.1 Suitable areas for runoff farming (FAO 1991)

Rainfall/ETP[a]	Zone	Annual rainfall (mm)	Suitability	Length growing period (day)
<0.05	Hyper-arid	<100	Too dry for runoff farming	0
0.05–0.2	Arid	100 < x < 200	Suitable for runoff farming	1–59
0.2–0.5	Semi-arid	200 < x < 400	Suitable for runoff farming	60–119
0.5–0.65	Dry sub-humid	400 < x < 800	Suitable for runoff storage	120–179
>0.65	Humid	>800	Too wet for runoff farming	>179

[a]Evapotranspiration

(3) **Limitations**

A major limitation is the risks of climatic uncertainty. Hence there is no guarantee of regular high yields. Other limitations remain in technical development, conflicts between upstream and downstream users and difficulty of project implementation etc.

(4) **Integration with Soil and Nutrient Management**

In many places, only some 15–30 % of the natural occurring rainfall is productively used by the crops. Most of the water evaporates from bare soil (30–50 %) or is lost through runoff (10–25 %) or deep percolation (10–30 %). Despite the above situation, recent experience shows that, even in drought—prone environments, with improved rainwater management strategies, there are no agro-hydrological limitations to doubling or even quadrupling on–farm staple food crops from its current low yields (Fig. 11.2).

To improve the performance of runoff farming systems, interdisciplinary approaches are required to (1) increase the amount of water made available to the crops to satisfy their crop water requirements over time (2) maximize water infiltration and water holding capacity of the soils, as well as reduce evaporation losses and (3) increase crop access to water and improve plant water uptake capacity.

In this context, integrated land, water and plant nutrient management is needed, combining runoff farming systems with improved soil moisture management (e.g., manure, mulching, and zero-tillage) and/or crop husbandry (e.g., selection of crop variety and species—*length of growing period, root depth,* etc, agronomic practices—*sowing date,* etc, weed and pest management).

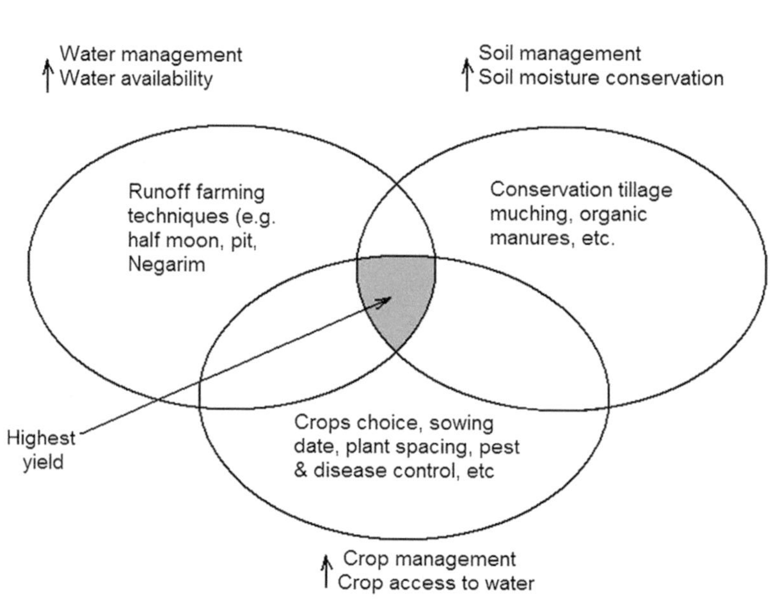

Fig. 11.2 Integrated water-soil-crop management strategies (Adapted with courtesy FAO 1991)

11.2 History and Development

11.2.1 Historical Perspectives

Various forms of runoff farming have been used traditionally throughout the centuries. Some of the very earliest agriculture, in the Middle East, was based on techniques such as diversion of "wadi" flow (spate flow from normally dry watercourses) onto agricultural fields. In the Negev Desert of Israel, runoff farming systems dating back 4000 years or more have been discovered. These schemes involved the clearing of hillsides of vegetation to increase runoff, which was then directed to fields on the plains.

Floodwater farming has been practiced in the desert areas of Arizona and northwest New Mexico for at least the last 1000 years. The Hopi Indians on the Colorado Plateau, cultivated fields situated at the mouth of ephemeral streams. Microcatchments techniques used for tree growing were discovered by travelers in southern Tunisia in the nineteenth century. In the "Khadin" system of India, floodwater is impounded behind earth bunds, and crops then planted into the residual moisture when the water infiltrates into the soil.

Traditional, small-scale runoff farming systems have been used in Sub-Saharan Africa. Simple stone lines are used, for example, in some West African countries, notably Burkina Faso, and earth bunding systems are found in Eastern Sudan and the Central Rangelands of Somalia.

11.2.2 Recent Developments

A growing awareness of the potential of runoff farming for improved crop production arose in the 1970s and 1980s, with the widespread droughts in Africa leaving a trail of crop failures.

A number of runoff farming projects have been set up in Sub-Saharan Africa during the past decade. Their objectives have been to combat the effects of drought by improving plant production (usually annual food crops), and in certain areas rehabilitating abandoned and degraded land. However, few of the projects have succeeded in combining technical efficiency with low cost and acceptability to the local farmers. This is partially not only due to the lack of technologies but also often due to the selection of an inappropriate approach with regard to the prevailing socio-economic conditions.

11.2.3 Future Directions

Appropriate systems should ideally evolve from the experience of traditional techniques—where these exist. They should also be based on lessons learned from the shortcomings of previous projects. Above all it is necessary that the systems are appreciated by the communities where they are introduced.

Water harvesting technology is especially relevant to the semi-arid and arid areas where the problems of environmental degradation, drought, and population pressures are most evident. It is an important component of the package of remedies for these problem zones, and there is no doubt that implementation of runoff farming techniques will expand.

11.2.4 Classification and Categories

In general, two types of water harvesting technologies could be used for runoff farming systems:

- Rainwater harvesting
- Floodwater harvesting

Various runoff farming systems can be classified into three basic types whose main characteristics are summarized as follows:

(1) **Microcatchment Systems** (Also referred to as within-field system)

- Overland flow harvested from short catchment length
- Catchment length usually between 1 and 30 m
- Runoff stored in soil profile
- Ratio catchment: cultivated area usually 1:1–3:1
- Normally no provision for overflow
- Plant growth is even

Typical Examples:

- Negarim Microcatchments (for trees)—see Fig. 11.3.
- Contour Bunds (for trees)
- Contour Ridges (for crops)
- Semi-Circular Bunds (for rangeland and fodder)

(2) **Macrocatchment Systems** (Rainwater harvesting, also referred to as external system)

- Overland flow or rill flow harvested
- Runoff stored in soil profile
- Catchment usually 30–200 m in length
- Ratio catchment: cultivated area usually 2:1–10:1
- Provision for overflow of excess water
- Uneven plant growth unless land leveled

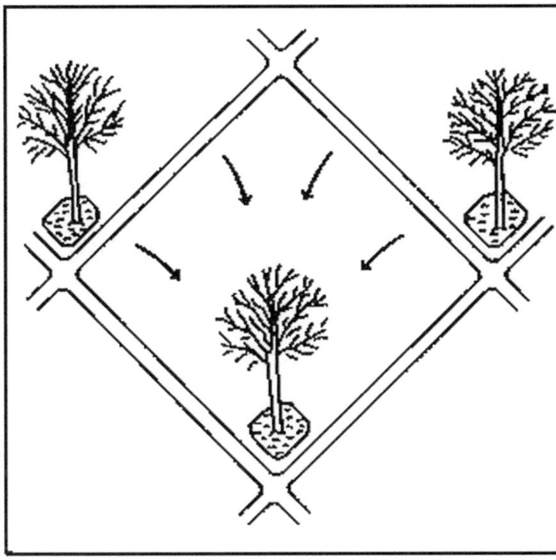

Fig. 11.3 Microcatchment system: Negarim microcatchment for trees (Courtesy FAO 1991)

Typical Examples:

- Trapezoidal Bunds (for crops)—see Fig. 11.4
- Contour Stone Bunds (for crops)

(3) **Floodwater Farming** (Floodwater harvesting)

Floodwater harvesting also referred to as "Water Spreading" and sometimes "Spate Irrigation"—see Fig. 11.5

- Turbulent channel flow harvested either (a) by diversion or (b) by spreading within channel bed/valley floor
- Runoff stored in soil profile

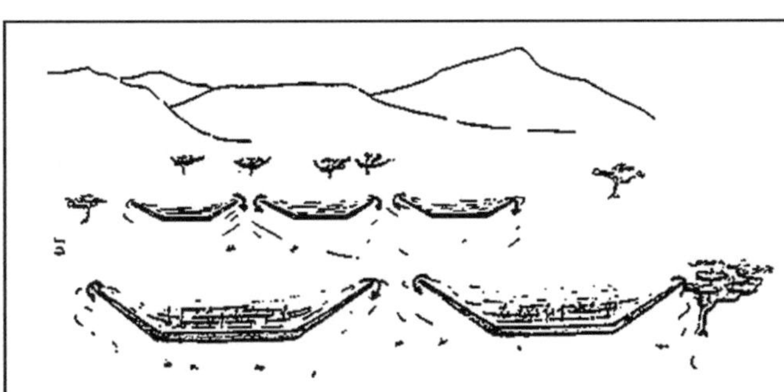

Fig. 11.4 Macro-catchment system: trapezoidal bunds for crops (Courtesy FAO 1991)

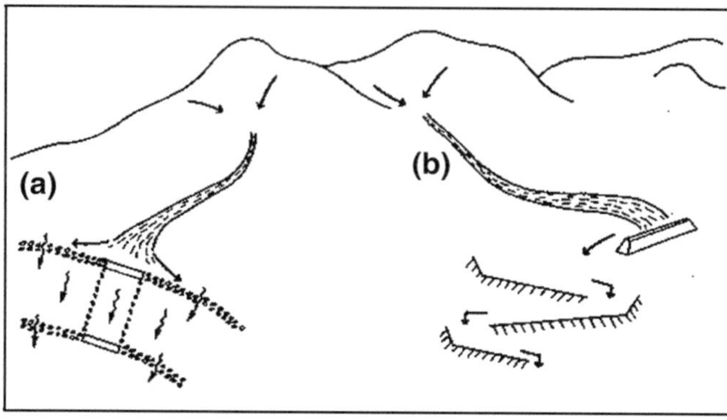

Fig. 11.5 Floodwater farming systems: **a** spreading within channel bed; **b** diversion system (Courtesy FAO 1991)

- Catchment long (may be several kilometers)
- Ratio catchment: cultivated area above 10:1
- Provision for overflow of excess water

Typical Examples:

- Permeable Rock Dams (for crops)
- Water Spreading Bunds (for crops)

11.2.5 Overview of Main Runoff Farming Systems

Overviews of the main runoff farming systems are described in detail in Table 11.2 (FAO 1991). The eight techniques presented and explained in the table are not the only systems currently utilized and known. But they do represent the major range of runoff farming techniques for different situations and productive uses. In a number of cases, the system which is described here is the most typical example of a technique for which a number of variations exist—trapezoidal bunds are a case in point.

11.3 Soil and Water Requirements

11.3.1 Soil Requirements for Runoff Farming

Ideally the soil in the catchment area should have a low infiltration rate and high runoff coefficient while the soil in the cultivated area should be a deep, fertile loam with low runoff coefficient, high infiltration rate, and high water storing capacity. Where the conditions for the cultivated and catchment areas conflict, the

Table 11.2 Summary chart of main runoff farming techniques

Item	Classification	Main uses	Description	Where appropriate	Limitations	Example
Negarim microcatchments	Microcatchment (short slope catchment) technique	Trees and grass	Closed grid of diamond shapes or open-ended "V" s formed by small earth ridges, with infiltration pits	For tree planting in situations where land is uneven or only a few tree are planted	Not easily mechanised therefore limited to small scale. Not easy to cultivate between tree lines	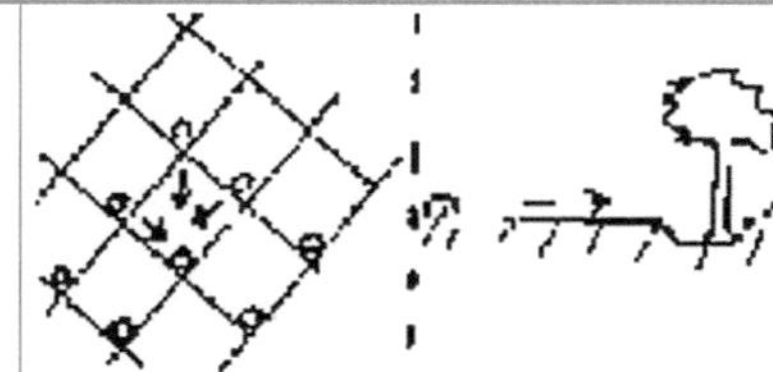
Contour bunds	Microcatchment (short slope catchment) technique	Trees and grass	Earth bunds on contour spaced at 5–10 m apart with furrow upslope and cross-ties	For tree planting on a large scale especially when mechanised	Not suitable for uneven terrain	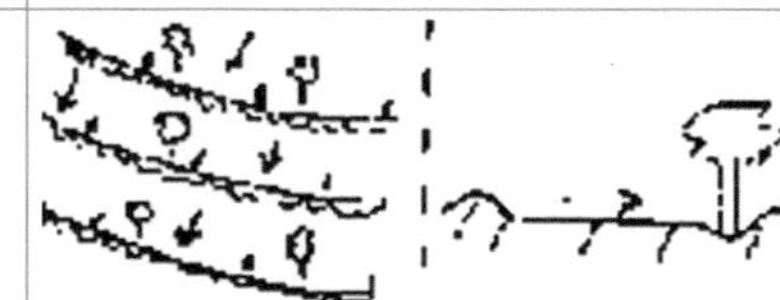
Semi circular bunds	Microcatchment (short slope catchment) technique	Rangeland and fodder (also trees)	Semi-circular shaped earth bunds with tips on contour. In a series with bunds in staggered formation	Useful for grass reseeding, fodder or tree planting in degraded rangeland	Cannot be mechanized therefore limited to areas with available hand labor	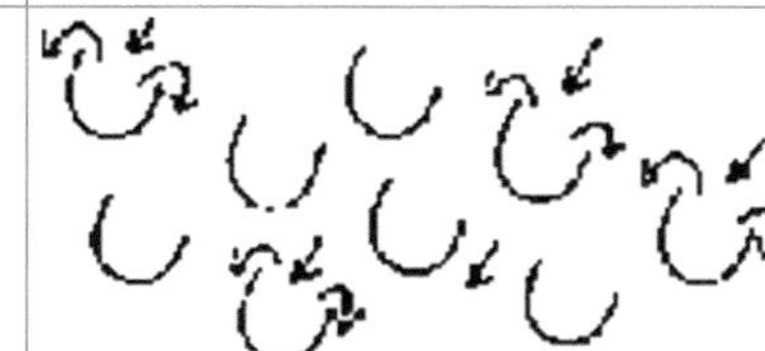
Contour ridges	Microcatchment (short slope catchment) technique	Crops	Small earth ridges on contour at 1.5–5 m apart with furrow upslope and cross-ties Uncultivated catchment between ridges	For crop production in semi-arid areas especially where soil fertile and easy to work	Requires new technique of land preparation and planting, therefore may be problem with acceptance	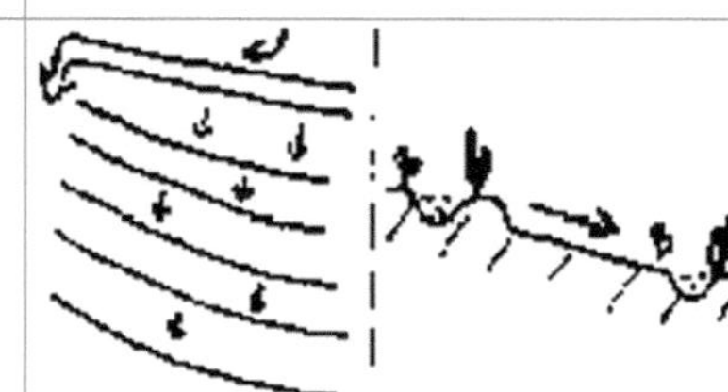

(continued)

Table 11.2 (continued)

Item	Classification	Main uses	Description	Where appropriate	Limitations	Example
Trapezoidal bunds	External catchment (long slope catchment) technique	Crops	Trapezoidal shaped earth bunds capturing runoff from external catchment and overflowing around wingtips	Widely suitable (in a variety of designs) for crop production in arid and semi-arid areas	Labor-intensive and uneven depth of runoff within plot	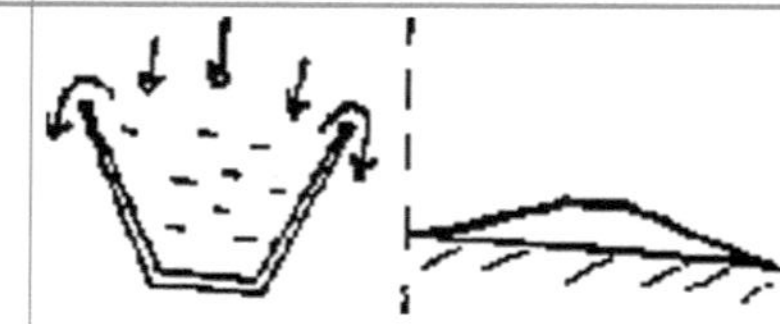
Contour stone bunds	External catchment (long slope catchment) technique	Crops	Small stone bunds constructed on the contour at spacing of 15–35 m apart slowing and filtering runoff	Versatile system for crop production in a wide variety of situations. Easily constructed by resource-poor farmers	Only possible where abundant loose stone available	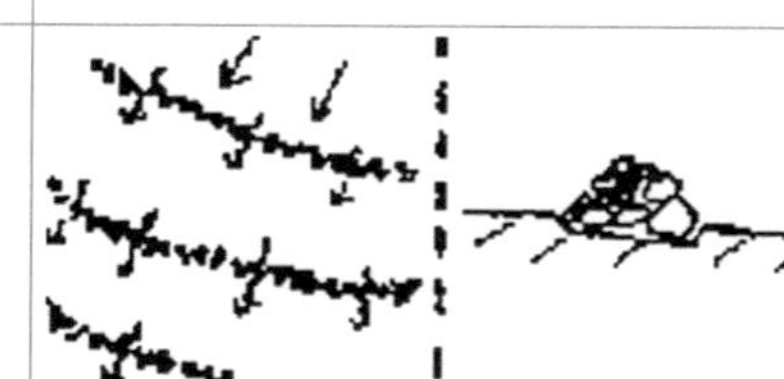
Permeable rock dams	Floodwater farming technique	Crops	Long low rock dams across valleys slowing and spreading floodwater as well as healing gullies	Suitable for situation where gently sloping valleys are becoming gullies and better water spreading is required	Very site-specific and needs considerable stone as well as provision of transport	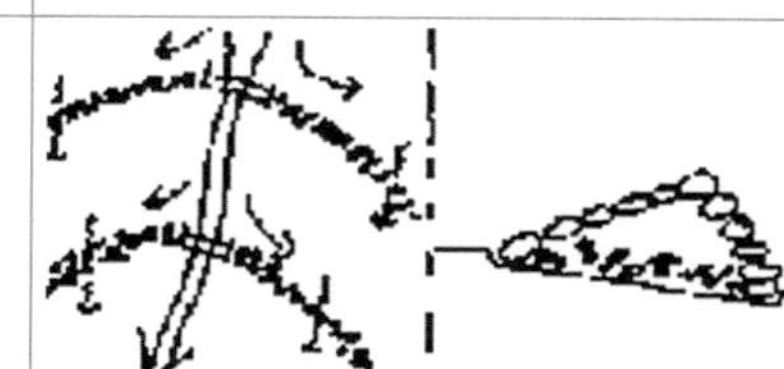
Water spreading bunds	Floodwater farming technique	Crops and rangeland	Earth bunds set at a gradient, with a "dogleg" shape, spreading diverted floodwater	For arid areas where water is diverted from watercourse onto crop or fodder block	Does not impound much water and maintenance high in early stages after construction	

requirements of the cultivated area should always take precedence. After all, soil for runoff farming should be suitable for construction. In detail, the suitability of soil for runoff farming depends on following aspects.

(1) **Texture**

To improve the performance of runoff farming systems, interdisciplinary approaches are required to (a) increase the amount of water made available to the crops to satisfy their crop water requirements over time; (b) maximize water infiltration and water holding capacity of the soils, as well as reduce evaporation losses; and (c) increase crop access to water and improve plant water uptake capacity.

In this context, integrated land, water and plant nutrient management is needed, combining runoff farming systems with improved soil moisture management (e.g., manure, mulching, and zero-tillage) and/or crop husbandry (e.g., selection of crop variety and species—*length of growing period, root depth,* etc, agronomic practices—*sowing date,* etc, weed and pest management).

(2) **Structure**

A good soil structure is usually associated with loamy soil and a relatively high content of organic matter. Inevitably, under hot climatic conditions, organic matter levels are often low, due to the rapid rates of decomposition. The application of organic materials such as crop residues and animal manure is helpful in improving the structure.

(3) **Depth**

The depth of soil is particularly important where runoff farming systems are proposed. Deep soils have the capacity to store the harvested runoff as well as providing a greater amount of total nutrients for plant growth. Soils of less than one meter deep are poorly suited to runoff farming. Two meters depth or more is ideal, though rarely found in practice.

(4) **Fertility**

In many of the areas where runoff farming may be introduced, lack of moisture and low soil fertility are the major constraints to plant growth. Some areas in Sub-Saharan Africa, for example, may be limited by low soil fertility as much as by lack of moisture. Nitrogen and phosphorus are usually the elements most deficient in these soils. While it is often not possible to avoid poor soils in areas under runoff farming system development, attention should be given to the maintenance of fertility levels.

(5) **Salinity/Sodicity**

Sodic soils, which have a high exchangeable sodium percentage, and saline soil, which have excess soluble salts, should be avoided for runoff farming systems. These soils can reduce moisture availability directly, or indirectly, as well as exerting direct harmful influence on plant growth.

(6) **Infiltration Rate**

A very low infiltration rate can be detrimental to runoff farming because of the possibility of waterlogging in the cultivated area. On the other hand, a low infiltration rate leads to high runoff, which is desirable for the catchment area. The soils of the cropped area however should be sufficiently permeable to allow adequate moisture to the crop root zone without causing waterlogging problems. Therefore, the requirements of the cultivated area should always take precedence.

Crust formation is a special problem of arid and semi-arid areas, leading to high runoff and low infiltration rates. Soil compaction as a result of heavy traffic either from machinery or grazing animals could also result in lower infiltration rates.

(7) **Available Water Capacity (AWC)**

The capacity of soils to hold, and to release adequate levels of moisture to plants are vital to runoff farming. AWC is a measure of this parameter, and is expressed as the depth of water in mm readily available to plants after a soil has been thoroughly wetted to "field capacity." AWC values for loams vary from 100 to 200 mm/m. Not only is the AWC important, but the depth of the soil is critical also. In runoff farming which pond runoff, it is vital that this water can be held by the soil and made available to the plants.

(8) **Constructional Characteristics**

The ability of a soil to form resilient earth bunds (where these are a component of the runoff farming system) is very important, and often overlooked. Generally the soils which should particularly be avoided are those which crack on drying, namely those which contain a high proportion of montmorillonite clay (especially vertisols or "black cotton soils"), and those which form erodible bunds, namely very sandy soils, or soils with very poor structure.

11.3.2 Water Requirements for Runoff Farming

For the design of runoff farming systems, it is necessary to assess the water requirement of the crops, trees, rangeland, and fodder.

(1) **Water Requirements of Crops**

In the absence of any measured climatic data, it is often adequate to use estimates of water requirements for common crops (Table 11.3).

Normal procedure for determining crop water needs includes three steps:

(i) Determination of ET_o

Three basic methods could be used to determine ET_o:

(a) Pan evaporation method

Table 11.3 Indicative values of crop water needs (FAO 1991)

Crop	Crop water need (mm/total growing period)
Beans	300–500
Citrus	900–1200
Cotton	700–1300
Groundnut	500–700
Maize	500–800
Sorghum/millet	450–650
Soybean	450–700
Sunflower	600–1000

Table 11.4 Indicative values of ET_o (mm/day) (FAO 1991)

Climatic zone	Mean daily temperature		
	15°	15–25 °C	25°
Desert/arid	4–6	7–8	9–10
Semi-arid	4–5	6–7	8–9
Sub-humid	3–4	5–6	7–8
Humid	1–2	3–4	5–6

$$ET_o = E_{\text{pan}} \times K_{\text{pan}}$$

(b) The Blaney-Criddle Method

$$ET_o = p(0.46T_{\text{mean}} + 8)$$

(c) Indicative values of ET_o

Table 11.4 contains approximate values for ET_o which may be used in the absence of measured or calculated figures.

(ii) Determination of K_c

Table 11.5 contains crop factors for the most commonly crops grown under runoff farming.

Table 11.5 also contains the number of days which each crop takes over a given growth stage. However, the length of the different crop stages will vary according to the variety and the climatic conditions where the crop is grown. In the semi-arid/arid areas where runoff farming is practiced crops will often mature faster than the figures in Table 11.5 (FAO 1991).

(iii) Determination of ET_{crop}

$$ET_{\text{crop}} = K_c \times ET_o.$$

Since the values for ET_o are normally measured or calculated on a daily basis (mm/day), an average value for the total growing season has to be determined and

Table 11.5 Crop factors (*Kc*)

Crop	Initial stage	(Days)	Crop dev. stage	(Days)	Mid-season stage	(Days)	Late season	(Days)	Season average
Cotton	0.45	(30)	0.75	(50)	1.15	(55)	0.75	(45)	0.82
Maize	0.40	(20)	0.80	(35)	1.15	(40)	0.70	(30)	0.82
Millet	0.35	(15)	0.70	(25)	1.10	(40)	0.65	(25)	0.79
Sorghum	0.35	(20)	0.75	(30)	1.10	(40)	0.65	(30)	0.78
Grain/small	0.35	(20)	0.75	(30)	1.10	(60)	0.65	(40)	0.78
Legumes	0.45	(15)	0.75	(25)	1.10	(35)	0.50	(15)	0.79
Groundnuts	0.45	(25)	0.75	(35)	1.05	(45)	0.70	(25)	0.79

then multiplied with the average seasonal crop factor K_c as given in the last column of Table 11.5.

While conventional irrigation strives to maximize the crop yields by applying the optimal amount of water required by the crops at well determined intervals, this is not possible with runoff farming techniques. As already discussed, the farmer has no influence on the occurrence of the rains neither in time nor in the amount of rainfall. Bearing the above in mind, it is therefore a common practice to only determine the total amount of water which the crop requires over the whole growing season.

(2) **Water Requirements of Trees**

(i) Multipurpose trees

In general, the water requirements for trees are more difficult to determine than for crops. Trees are relatively sensitive to moisture stress during the establishment stage compared with their ability to withstand drought once their root systems are fully developed. There is no accurate information available on the response of these species, in terms of yields, to different irrigation/water regimes. Table 11.6 gives some basic data of multipurpose trees often planted in semi-arid

Table 11.6 Naturally preferred climatic zones of multipurpose trees (FAO 1991)

	Semi-arid/marginal 500–900 mm rain	Arid/semi-arid 150–500 mm rain	Tolerance to temporary waterlogging
Acacia albida	Yes	Yes	Yes
A. nilotica	Yes	Yes	Yes
A. saligna	No	Yes	Yes
A. Senegal	Yes	Yes	No
A. seyal	Yes	Yes	Yes
A. tortilis	Yes	Yes	No
Albizia lebbeck	Yes	No	No
Azadirachta indica	Yes	No	Some
Balanites aegyptiaca	Yes	Yes	Yes
Cassia siamea	Yes	No	No
Casuarina equisetifolia	Yes	No	Some
Colophospermum mopane	Yes	Yes	Yes
Cordeauxia edulis	No	Yes	?
Cordia sinensis	No	Yes	?
Delonix elata	Yes	No	?
Eucalyptus camaldulensis	Yes	Yes	Yes
Prosopis chilensis	Yes	Yes	Some
Prosopis cineraria	Yes	Yes	Yes
Prosopis juliflora	Yes	Yes	Yes
Ziziphus mauritiana	Yes	Yes	Yes

Table 11.7 Fruit tree water requirements (FAO 1991)

Species	Seasonal water requirement (mm)	Place
Apricots	550	Israel
Peaches	700	Israel
Pomegranate	265	Israel
Jujube (*Zixiphus mauritiana*)	550–750	India

areas. The critical stage for most trees is in the first two years of seedling/sapling establishment.

(ii) Fruit trees

There are some known values of water requirements for fruit trees under runoff farming systems—most of the figures have been derived from Israel. Table 11.7 contains the water requirements for some fruit trees.

(3) **Water Requirements of Rangeland and Fodder**

Water requirements for rangeland and fodder species grown in semi-arid/arid areas under runoff farming systems are usually not calculated.

The objective is to improve performance, within economic constraints, and to ensure the survival of the plants from season to season, rather than fully satisfying water requirements.

11.4 Ratio Between Runoff and Run-on Areas

11.4.1 Rainfall

Runoff is generated by rainstorms, and its occurrence and quantity are dependent on the characteristics of the rainfall event, i.e., intensity, duration, and distribution.

(1) **Intensity**

Studies carried out in Saudi Arabia suggest that, on average, around 50 % of all rain occurs at intensities in excess of 20 mm/hour and 20–30 % occurs at intensities in excess of 40 mm/h. This relationship appears to be independent of the long-term average rainfall at a particular location.

(2) **Variability**

Planning and management of runoff farming in arid and semi-arid zones present difficulties which are due less to the limited amount of rainfall than to the inherent degree of variability associated with it.

In temperate climates, the standard deviation of annual rainfall is about 10–20 % and in 13 years out of 20, annual amounts are between 75 and 125 % of the mean. In arid and semi-arid climates, the ratio of maximum to minimum annual amounts is much greater and the annual rainfall distribution becomes increasingly skewed with increasing aridity. With mean annual rainfalls of 200–300 mm, the rainfall in 19 years out of 20 typically ranges from 40 to 200 % of the mean and for 100 mm/year, 30–350 % of the mean. At more arid locations, it is not uncommon to experience several consecutive years with no rainfall.

(3) **Probability**

For a runoff farming planner, the most difficult task is therefore to select the appropriate "design" rainfall according to which the ratio of catchment to cultivated area will be determined.

From rainfall probability curve, it is possible to obtain the magnitude of the rain corresponding to a given probability.

11.4.2 Runoff

(1) **Surface Runoff Process**

When rain falls, the first drops of water are intercepted by the leaves and stems of the vegetation. This is usually referred to as interception storage. As the rain continues, water reaching the ground surface infiltrates into the soil until it reaches a stage where the rate of rainfall (intensity) exceeds the infiltration capacity of the soil. Thereafter, surface puddles, ditches, and other depressions are filled (depression storage), after which runoff is generated.

The infiltration capacity of the soil depends on its texture and structure, as well as on the antecedent soil moisture content (previous rainfall or dry season). The initial capacity (of a dry soil) is high but, as the storm continues, it decreases until it reaches a steady value termed as final infiltration rate.

The process of runoff generation continues as long as the rainfall intensity exceeds the actual infiltration capacity of the soil but it stops as soon as the rate of rainfall drops below the actual rate of infiltration.

Apart from rainfall characteristics such as intensity, duration, and distribution, there are a number of site (or catchment)-specific factors which have a direct bearing on the occurrence and volume of runoff, including soil type, vegetation, slope, and catchments size.

(2) **Runoff Coefficient**

The runoff coefficient from an individual rainstorm is defined as runoff divided by the corresponding rainfall both expressed as depth over catchment area (mm):

$$K = \frac{\text{Runoff(mm)}}{\text{Rainfall(mm)}}.$$

Actual measurements should be carried out until a representative range is obtained. Some experts recommend that at least 2 years should be spent to measure rainfall and runoff data before any larger construction program starts.

When analyzing the measured data, it will be noted that a certain amount of rainfall is always required before any runoff occurs. This amount, usually referred to as threshold rainfall, represents the initial losses due to interception and depression storage as well as to meet the initially high infiltration losses.

The threshold rainfall depends on the physical characteristics of the area and varies from catchment to catchment. In areas with only sparse vegetation and where the land is very regularly shaped, the threshold rainfall may be only in the range of 3 mm while in other catchments this value can easily exceed 12 mm, particularly where the prevailing soils have a high infiltration capacity. The fact that the threshold rainfall has first to be surpassed explains why not every rainstorm produces runoff. This is important to know when assessing the annual runoff-coefficient of a catchment area.

The knowledge of runoff from individual storms as described before is essential to assess the runoff behavior of a catchment area and to obtain an indication both of runoff-peaks which the structure of a runoff farming system must withstand and of the needed capacity for temporary surface storage of runoff, for example the size of an infiltration pit in a microcatchment system.

However, to determine the ratio of catchment to cultivated area, it is necessary to assess either the annual (for perennial crops) or the seasonal runoff coefficient. This is defined as the total runoff observed in a year (or season) divided by the total rainfall in the same year (or season).

$$k = \text{Yearly(seasonal) total runoff(mm)}/\text{Yearly(seasonal) total runoff(mm)}.$$

The annual (seasonal) runoff coefficient differs from the runoff coefficients derived from individual storms as it takes into account also those rainfall events which did not produce any runoff. The annual (seasonal) runoff-coefficient is therefore always smaller than the arithmetic mean of runoff coefficients derived from individual runoff-producing storms.

(3) **Runoff Plots**

Runoff plots are used to measure surface runoff under controlled conditions. The plots should be established directly in the project area. Their physical characteristics, such as soil type, slope, and vegetation must be representative of the sites where runoff farming systems are planned.

The size of a plot should ideally be as large as the estimated size of the catchment planned for the runoff farming project. This is not always possible mainly due to the problem of storing the accumulated runoff. A minimum size of 3–4 m in width and 10–12 m in length is recommended. Smaller dimensions should be avoided, since the results obtained from very small plots are rather misleading.

Care must be taken to avoid sites with special problems such as rills, cracks, or gullies crossing the plot. These would drastically affect the results which would not be representative for the whole area. The gradient along the plot should be regular and free of local depressions. During construction of the plot, care must be taken not to disturb or change the natural conditions of the plot such as destroying the vegetation or compacting the soil. It is advisable to construct several plots in

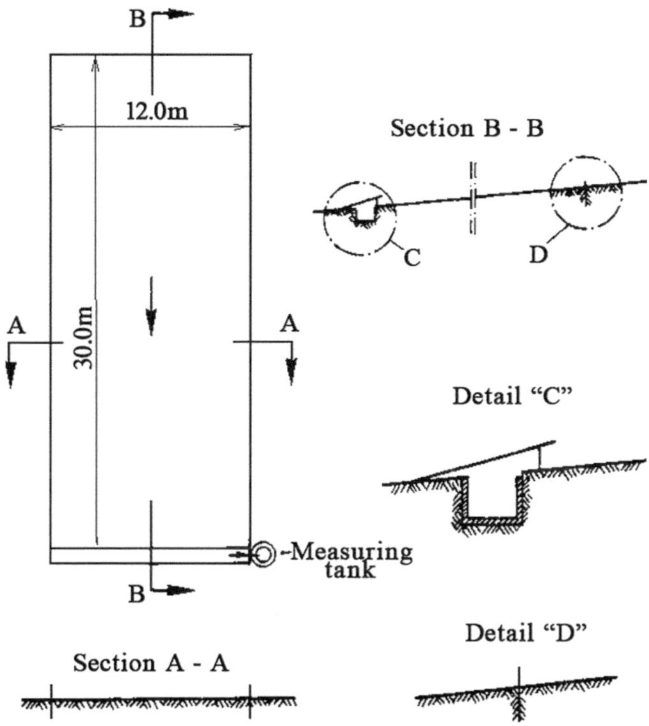

Fig. 11.6 Standard layout of a runoff plot (Courtesy FAO 1991)

series in the project area which would permit comparison of the measured runoff volumes and to judge on the representative character of the selected plot sites.

Around the plots metal sheets or wooden planks must be driven into the soil with at least 15 cm of height above ground to stop water flowing from outside into the plot and vice versa (see Fig. 11.6). A rain gauge must be installed near to the plot. At the lower end of the plot, a gutter is required to collect the runoff. The gutter should have a gradient of 1 % toward the collection tank. The soil around the gutter should be backfilled and compacted. The joint between the gutter and the lower side of the plot may be cemented to form an apron in order to allow a smooth flow of water from the plot into the gutter. The collection tank may be constructed from stone masonry, brick or concrete blocks, but a buried barrel will also meet the requirements. The tank should be covered and thus be protected against evaporation and rainfall. The storage capacity of the tank depends on the size of the plot but should be large enough to collect water also from extreme rain storms. Following every storm (or every day at a specific time), the volume of water collected in the rain gauge and in the runoff tank must be measured. Thereafter the gauge and tank must be completely emptied. Any silt which may have deposited in the tank and in the gutter must be cleared.

11.4.3 Ratio Between Runoff and Run-on Areas

Each runoff framing system consists of a runoff (catchment) and a run-on (cultivated) area. The relationship between the two, in terms of size, determines by what factor the rainfall will be "multiplied." For an appropriate design of a system, it is recommended to determine the ratio between catchment (C) and cultivated (CA) area.

Many successful runoff farming systems have been established by merely estimating the ratio between catchment and cultivated area. This may indeed be the only possible approach where basic data such as rainfall, runoff, and crop water requirements are not known. However, calculation of the ratio will certainly result in a more efficient and effective system provided the basic data are available and accurate.

Nevertheless, it should be noted that calculations are always based on parameters with high variability. Rainfall and runoff are characteristically erratic in regions where runoff farming is practiced. It is, therefore, sometimes necessary to modify an original design in the light of experience, and often it will be useful to incorporate safety measures, such as cut-off drains, to avoid damage in years when rainfall exceeds the design rainfall.

(1) **Crop Production Systems**

The calculation of C:CA ratio is primarily useful for runoff farming systems where crops are intended to be grown. The calculation is based on the concept that the design must comply with the rule:

$$\text{Water harvested} = \text{Extra water required.}$$

The amount of water harvested from the catchment area is a function of the amount of runoff created by the rainfall on the area. This runoff, for a defined time scale, is calculated by multiplying a "design" rainfall with a runoff coefficient. As not all runoff can be efficiently utilized (because of deep percolation losses, etc.), it must be additionally multiplied with an efficiency factor.

$$\begin{aligned}\text{Water harvested} = {} & \text{Catchment area} \times \text{Design rainfall} \\ & \times \text{Runoff coefficient} \times \text{Efficiency factor.}\end{aligned}$$

The amount of water required is obtained by multiplying the size of the cultivated area with the net crop water requirements which is the total water requirement less the assumed "design" rainfall.

$$\text{Extra water required} = \text{Cultivated area} \times (\text{Crop water requirement} - \text{Design rainfall}).$$

By substitution in original equation

$$\text{Water harvested} = \text{Extra water required,}$$

we obtain as follows:

$$\begin{aligned}&\text{Catchment area} \times \text{Design rainfall} \times \text{Runoff coefficient} \times \text{Efficiency factor} \\ &\quad = \text{cultivated area} \times (\text{Crop water requirement} - \text{Design rainfall}).\end{aligned}$$

If this formula is rearranged we finally obtain

$$\begin{aligned}&(\text{Crop water requirement} - \text{design rainfall})\\&\div(\text{Design rainfall} \times \text{Runoff coefficent} \times \text{Efficiency factor})\\&= \text{Catchment area} \div \text{Cultivated area.}\end{aligned}$$

(i) Crop Water Requirement: Crop water requirement depends on the kind of crop and the climate of the place where it is grown. Estimates as given previously should be used when precise data are not available.

(ii) Design Rainfall: The design rainfall is set by calculations or estimates. It is the amount of seasonal rain at which, or above which, the system is designed to provide enough runoff to meet the crop water requirement. If the rainfall is below the "design rainfall," there is a risk of crop failure due to moisture stress. When rainfall is above the "design," then runoff will be in surplus and may overtop the bunds.

Design rainfall is calculated at a certain probability of occurrence. If, for example, it is set at a 67 % probability, it will be met or exceeded (on average) in two years out of three and the harvested rain will satisfy the crop water requirements also in two out of three years.

A conservative design would be based on a higher probability (which means a lower design rainfall), in order to make the system more "reliable" and thus to meet the crop water requirements more frequently. However, the associated risk would be a more frequent flooding of the system in years where rainfall exceeds the design rainfall.

(iii) Runoff Coefficient: This is the proportion of rainfall which flows along the ground as surface runoff. It depends on the degree of slope, soil type, vegetation cover, antecedent soil moisture, rainfall intensity, and duration. The coefficient ranges usually between 0.1 and 0.5. When measured data are not available, the coefficient may be estimated from experience. However, this method should be avoided whenever possible.

(iv) Efficiency Factor: This factor takes into account the inefficiency of uneven distribution of the water within the field as well as losses due to evaporation and deep percolation. Where the cultivated area is leveled and smooth the efficiency is higher. Microcatchment systems have higher efficiencies as water is usually less deeply ponded. Selection of the factor is left to the discretion of the designer based on his experience and of the actual technique selected. Normally the factor ranges between 0.5 and 0.75. Following are three examples on how to calculate the ratio *C*:CA for crop production systems.

(2) **Systems for Trees**

The ratio between catchment and cultivated area is difficult to determine for systems where trees are intended to be grown. As already discussed, only rough estimates are available for the water requirements of the indigenous, multi-purpose species commonly planted in runoff farming systems. Furthermore, trees are almost exclusively grown in microcatchment systems where it is difficult to determine which proportion of the total area is actually exploited by the root zone bearing in mind the different stages of root development over the years before a seedling has grown into a mature tree.

In view of the above, it is therefore considered sufficient to estimate only the total size of the microcatchment (MC), that is the catchment and cultivated area (infiltration pit) together, for which the following formula can be used:

$$MC = RA \times \frac{WR - DR}{DR - K - EFF},$$

where
MC = total size of microcatchment (m^2),
RA = area exploited by root system (m^2),
WR = water requirement (annual) (mm),
DR = design rainfall (annual) (mm),
K = runoff coefficient (annual), and
EFF = efficiency factor.

As a rule of thumb, it can be assumed that the area to be exploited by the root system is equal to the area of the canopy of the tree, and for multipurpose trees in the arid/semi-arid regions, the size of the microcatchment per tree (catchment and cultivated area together) should range between 10 and 100 m^2, depending on the aridity of the area and the species grown. Flexibility can be introduced by planting more than one tree seedling within the system and removing surplus seedlings at a later stage if necessary.

(3) **Systems for Rangeland and Fodder**

In most cases, it is not necessary to calculate the ratio C:CA for systems implementing fodder production and/or rangeland rehabilitation. As a general guideline, a ratio of 2:1–3:1 for microcatchments (which are normally used) is appropriate (Table 11.8).

11.5 Microcatchment Systems

11.5.1 Negarim Microcatchments

(1) **Background**

Negarim microcatchments are diamond-shaped basins surrounded by small earth bunds with an infiltration pit in the lowest corner of each (Fig. 11.7). Runoff is collected from within the basin and stored in the infiltration pit. Microcatchments are mainly used for growing trees or bushes. This technique is appropriate for small-scale tree planting in any area which has a moisture deficit. Besides harvesting water for the trees, it simultaneously conserves soil. Negarim microcatchments

Table 11.8 Dimensions of catchments and cultivated areas for tree microcatchments (Courtesy FAO 1991)

Species	Country	Catchment area (m^2)	Cultivated area (m^2)	Source
Ziziphus mauritiana	Rajastan, India	31.5–72	36	Sharma et al. (1986)
Pomegranate	Negev, Israel	160	16	Shanan and Tadmore (1979)
Almonds	Negev, Israel	250	10	Ben-Asher (1988)
Fodder/fuelwood spp.	Baringo, Kenya	10–20[a]		Critchley and Reij (1989)
Fodder spp.	Turkana, Kenya	93.75[a]	6.25[b]	Barrow (in Rocheleu et al. 1988)
Fuelwood/multi-purpose spp.	Guesselbodi forest, Niger	64[a]		Critchley and Reij (1989)
Fuelwood/multi-purpose spp.	Keita valley, Niger	12	1.8	Critchley and Reij (1989)

[a]No breakdown given between catchment and cultivated area/infiltration pit
[b]In a number of cases two trees planted within the same system

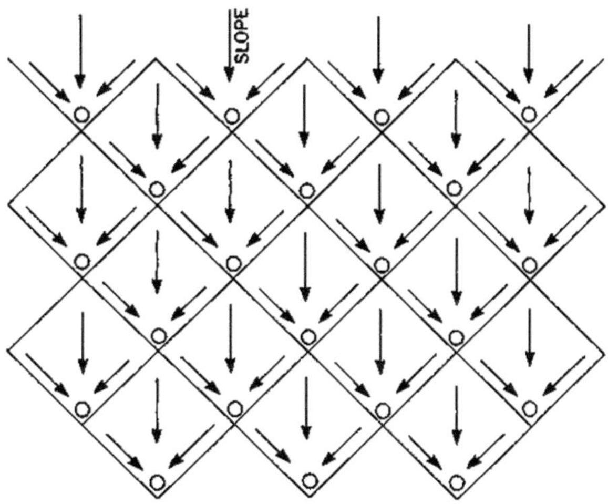

Fig. 11.7 Negarim microcatchments—field layout (Courtesy FAO 1991)

are neat and precise, and relatively easy to construct. Around 500 per hectare can be constructed on land sloping up to 2° and more on land sloping up to 5°. See Table 11.9 (Courtesy FAO 1991).

Although the first reports of such microcatchments are from southern Tunisia, the technique has been developed in the Negev desert of Israel. The word "Negarim" is derived from the Hebrew word for runoff—"Neger." Negarim microcatchments are the most well known form of all runoff farming systems.

Israel has the most widespread and best developed Negarim microcatchments, mostly located on research farms in the Negev Desert, where rainfall is as low as

Table 11.9 Earthwork/stonework for various runoff farming systems

	Earthwork (m^3/ha treated)							Stonework (m^3/ha treated)
System name and number	Negarim micro-catchments (trees)	Contour bunds (trees)	Semi circular bunds (grass)	Contour ridges (crops)	Trapezoidal bunds (crops)	Water spreading bunds (crops)	Contour stone bunds (crops)	Permeable rock dams (crops)
Slope %	(1)	(2)	(3)	(4)	(5)	(8)	(6)	(7)
0.5	500	240	105	480	370	305	40	70
1.0	500	360	105	480	670	455	40	140
1.5	500	360	105	480	970	N/R[a]	40	208
2.0	500	360	210	480	N/R[a]	N/R[a]	55	280
5.0	835	360	210	480	N/R[a]	N/R[a]	55	N/R[a]

[a]Not recommended

Note

1 Typical dimensions are assumed for each system: for greater detail see relevant chapters

2 For microcatchment systems (1, 2, 3 and 4), the whole area covered (cultivated and within-field catchment) is taken as "treated"

3 Labor rates for earthworks: Larger structures (e.g., trapezoidal bunds) may take 50 % more labor per unit volume of earthworks than smaller structures (e.g., Negarim microcatchments) because of increased earthmoving required. Typical rates per person/day range from 1.0 to 3.0 m^3

4 Labor rates for stoneworks: Typical labor rates achieved are 0.5 m^3 per person/day for construction. Transport of stone increases this figure considerably

100–150 mm per annum. However, the technique, and variations of it, is widely used in other semi-arid and arid areas, especially in North and Sub-Saharan Africa. Because it is a well-proven technique, it is often one of the first to be tested by new projects.

(2) **Technical Details**

(i) Suitability

Negarim microcatchments are mainly used for tree growing in arid and semi-arid areas.

- Rainfall: can be as low as 150 mm per annum.
- Soils: should be at least 1.5 m but preferably 2 m deep in order to ensure adequate root development and storage of the water harvested.
- Slopes: from flat up to 5.0 %.
- Topography: need not be even—if uneven a block of microcatchments should be subdivided.

(ii) Overall configuration

Each microcatchment consists of a catchment area and an infiltration pit (cultivated area). The shape of each unit is normally square, but the appearance from above is of a network of diamond shapes with infiltration pits in the lowest corners (Fig. 11.7).

(iii) Limitations

While Negarim microcatchments are well suited for hand construction, they cannot easily be mechanized. Once the trees are planted, it is not possible to operate and cultivate with machines between the tree lines.

(iv) Microcatchment size

The area of each unit is either determined on the basis of a calculation of the plant (tree) water requirement or, more usually, an estimate of this.

Size of microcatchments (per unit) normally ranges between 10 and 100 m^2 depending on the species of tree to be planted but larger sizes are also feasible, particularly when more than one tree will be grown within one unit.

(v) Design of bunds

The bund height is primarily dependent on the prevailing ground slope and the selected size of the microcatchment. It is recommended to construct bunds with a height of at least 25 cm in order to avoid the risk of over-topping and subsequent damage.

Where the ground slope exceeds 2.0 %, the bund height near the infiltration pit must be increased. Table 11.10 gives recommended figures for different sizes and ground slopes.

The top of the bund should be at least 25 cm wide and side slopes should be at least in the range of 1:1 in order to reduce soil erosion during rainstorms.

Table 11.10 Bund heights (cm) on higher ground slopes (Courtesy FAO 1991)

Size unit microcatchment (m^2)	Ground slope			
	2 %	3 %	4 %	5 %
3 × 3	Even bund height			
4 × 4	Of 25 cm			30
5 × 5			30	35
6 × 6			35	45
8 × 8		35	45	55
10 × 12	30	45	55	
12 × 12	35	50	Not recommended	
15 × 15	45			

Note These heights define the maximum height of the bund (below the pit). Excavation/total bund volume remain constant for a given microcatchment size

Table 11.11 Quantities of earthworks for Negarim microcatchments (Courtesy FAO 1991)

(1)	(2)	(3)	(4)	(5)	(6)
Size unit microcatch-ment (m^2)	Size infiltration pit (m)	Ground slopes suitable for 25 cm bund	Volume earthwork per unit[b] (m^2)	No. units per ha	Earthworks (m^3/ha)
Sides (×) area	Sides,(y) depth	Height[a] (%)			
3 × 3 = 9	1.4 × 1.4 × 0.4	up to 5	0.75	1110	835
4 × 4 = 16	1.6 × 1.6 × 0.4	up to 4	1.00	625	625
5 × 5 = 25	1.8 × 1.8 × 0.4	up to 3	1.25	400	500
6 × 6 = 36	1.9 × 1.9 × 0.4	up to 3	1.50	275	415
8 × 8 = 64	2.2 × 2.2 × 0.4	up to 2	2.00	155	310
10 × 10 = 100	2.5 × 2.5 × 0.4	up to 1	2.50	100	250
12 × 12 = 144	2.8 × 2.8 × 0.4	up to 1	3.25	70	230
15 × 15 = 225	3.0 × 3.0 × 0.4	up to 1	3.50	45	160

[a]These ground slopes allow construction of a bund of 25 cm height throughout its length. Above these gradients the bund should be constructed relatively higher at the bottom (below the pit) and lower upslope. Table 11.10 gives the height of the bund below the pit for given microcatchment sizes

[b]Calculation of earthworks per unit includes only two of the sides around the catchment: the other two sides are included in the microcatchment above. Does not include earthworks required for diversion ditch (which is 62.5 m^3 for each 100 m length)

Whenever possible, the bunds should be provided with a grass cover since this is the best protection against erosion.

(vi) Quantities of Earthworks

Table 11.11 further gives required quantities of earthworks for different layouts. Quantities per unit include only the infiltration pit and two sides of the catchment, while the other two bunds are included in the microcatchment above. When a

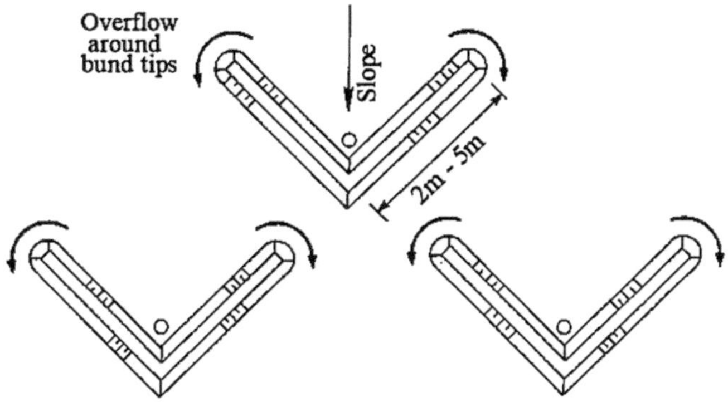

Fig. 11.8 "V"-shaped microcatchments (Courtesy FAO 1991)

diversion ditch is required additional earthworks of 62.5 m^3 per 100 m length of ditch will be needed.

(vii) Size of infiltration pit

Table 11.11 presents recommended values for pit dimensions. A maximum depth of 40 cm should not be exceeded in order to avoid water losses through deep percolation and to reduce the workload for excavation. Excavated soil from the pit should be used for construction of the bunds.

(viii) Design Variations

A common variation is to build microcatchments as single, open-ended structures in "V" or semi-circular shape (see Fig. 11.8). The advantage is that surplus water can flow around the tips of the bunds. However, the storage capacity is less than that of a closed system. These types of bunds are particularly useful on broken terrain, and for small numbers of trees around homesteads.

(3) **Maintenance**

Maintenance will be required for repair of damage to bunds, which may occur if storms are heavy soon after construction when the bunds are not yet fully consolidated. The site should be inspected after each significant rainfall as breakages can have a "domino" effect if left unrepaired.

(4) **Husbandry**

Tree seedlings of at least 30 cm height should be planted immediately after the first rain of the season. It is recommended that two seedlings are planted in each microcatchment—one in the bottom of the pit (which would survive even in a dry year) and one on a step at the back of the pit, see Fig. 11.9. If both plants survive, the weaker can be removed after the beginning of the second season. For some species, seeds can be planted directly. This eliminates the cost of a nursery.

Manure or compost should be applied to the planting pit to improve fertility and water-holding capacity. If grasses and herbs are allowed to develop in the catchment area, the runoff will be reduced to some extent; however, the fodder

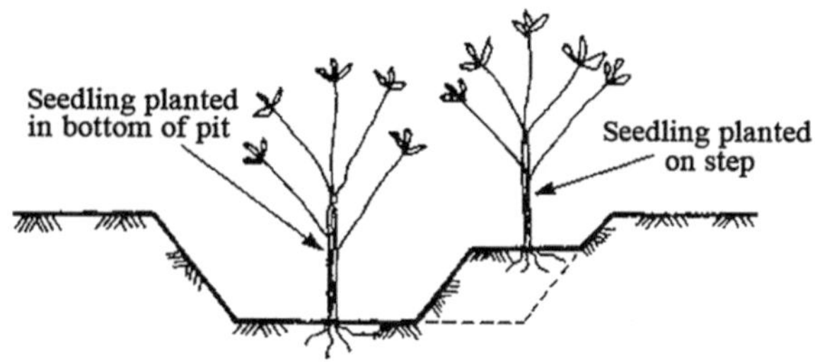

Fig. 11.9 Planting site for seedling (Courtesy FAO 1991)

obtained gives a rapid return to the investment in construction. Regular weeding is necessary in the vicinity of the planting pit.

11.5.2 Contour Bunds for Trees

(1) **Background**

Contour bunds for trees are a simplified form of microcatchments (Fig. 11.10). Construction can be mechanized and the technique is therefore suitable for implementation on a larger scale. As its name indicates, the bunds follow the contour, at close spacing, and by provision of small earth ties the system is divided into individual microcatchments. Whether mechanized or not, this system is more economical than Negarim microcatchment, particularly for large scale implementation on even land—since less earth has to be moved. A second advantage of contour bunds is their suitability to the cultivation of crops or fodder between the bunds. As with other forms of microcatchment runoff farming techniques, the yield of runoff is high, and when designed correctly, there is no loss of runoff out of the system.

Contour bunding for tree planting is not yet as common as Negarim microcatchments. Examples of its application come from Baringo District, Kenya.

(2) **Technical Details**

(i) Suitability

Contour bunds for tree planting can be used under the following conditions:

- Rainfall: 200–750 mm; from semi-arid to arid areas.
- Soils: Must be at least 1.5 m and preferably 2 m deep to ensure adequate root development and water storage.

Fig. 11.10 Contour bunds for trees (Courtesy FAO 1991)

- Slopes: from flat up to 5.0 %.
- Topography: must be even, without gullies or rills.

(ii) Limitations

Contour bunds are not suitable for uneven or eroded land as overtopping of excess water with subsequent breakage may occur at low spots.

(iii) Overall Configuration

The overall layout consists of a series of parallel, or almost parallel, earth bunds approximately on the contour at a spacing of between 5 and 10 m. The bunds are formed with soil excavated from an adjacent parallel furrow on their upslope side. Small earth ties perpendicular to the bund on the upslope side subdivide the system into microcatchments. Infiltration pits are excavated in the junction between ties and bunds. A diversion ditch protects the system where necessary.

(iv) Unit Microcatchment Size

The size of microcatchment per tree is estimated in the same way as for Negarim microcatchments. However, the system is more flexible, because the microcatchment size can be easily altered by adding or removing cross-ties within the fixed spacing of the bunds. Common sizes of microcatchments are around 10–50 m^2 for each tree, see Fig. 11.11.

(v) Bund and Infiltration Pit Design

Bund heights vary, but are in the order of 20–40 cm depending on the prevailing slope. As bunds are often made by machine the actual shape of the bund depends on the type of machine; whether for example a disc plough or a motor grader is used. It is recommended that the bund should not be less than 25 cm in height. Base width must be at least 75 cm. The configuration of the furrow upslope of the bund depends on the method of construction.

Bunds should be spaced at either 5 or 10 m apart. Cross-ties should be at least 2 m long at spacing of 2–10 m. The exact size of each microcatchment is thus defined. It is recommended to provide 10 m spacing between the bunds on slopes of up to 0.5 % and 5 m on steeper slopes. A common size of microcatchment for

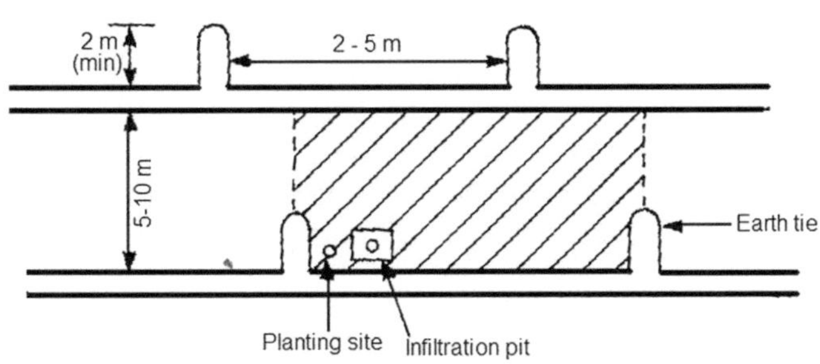

Fig. 11.11 Microcatchment unit (Courtesy FAO 1991)

Table 11.12 Quantities of earthworks (Courtesy FAO 1991)

Size unit microcatchment			Volume earthworks per unit (m^3)	No. units per ha	Earthworks m^3/ha
Bund spacing (m)	Tie spacing (m)	Area (m^2)			
5	2	10	0.5	1000	500
5	5	25	0.9	400	360
5	10	50	1.5	200	300
10	2.5	25	0.6	400	240
10	5	50	0.9	200	180

multipurpose trees is 25 m^2. This corresponds to 10 m bund spacing with ties at 2.5 m spacing or 5 m bund-spacing with ties at 5 m spacing. Excavated soil from the infiltration pit is used to form the ties. The pit is excavated at the junction of the bund and the cross-tie. A pit size of 80 cm × 80 cm and 40 cm deep is usually sufficient.

(vi) Quantities of Earthwork

Table 11.12 gives quantities of earthworks required for various layouts of contour bunds. The bund height assumed is 25 cm with 75 cm base.

(vii) Maintenance

As with Negarim microcatchments, maintenance will in most cases be limited to repair of damage to bunds early in the first season. It is essential that any breaches—which are unlikely unless the scheme crosses existing rills—are repaired immediately and the repaired section compacted. Damage is frequently caused if animals invade the plots. Grass should be allowed to develop on the bunds, thus assisting consolidation with their roots.

(viii) Husbandry

The majority of the husbandry factors noted under Negarim microcatchments also apply to this system: there are, however, certain differences.

Tree seedlings, of at least 30 cm height, should be planted immediately after the first runoff has been harvested. The seedlings are planted in the space between the infiltration pit and the cross-tie. It is advisable to plant an extra seedling in the bottom of the pit for the eventuality of a very dry year. Manure or compost can be added to the planting pit to improve fertility and water holding capacity.

One important advantage of contour bunds for tree establishment is that oxen or mechanized cultivation can take place between the bunds, allowing crops or fodder to be produced before the trees becomes productive. However, this has the disadvantage of reducing the amount of runoff reaching the trees.

11.5.3 Small Semi-circular Bunds

(1) **Background**

Small semi-circular bunds are earth embankments in the shape of a semi-circle with the tips of the bunds on the contour. They are used mainly for rangeland rehabilitation or fodder production. This technique is also useful for growing trees and shrubs and, in some cases, has been used for growing crops.

Small semi-circular bunds are recommended as a quick and easy method of improving rangelands in semi-arid areas. They are more efficient in terms of impounded area to bund volume than other equivalent structures—such as trapezoidal bunds for example. Surprisingly, this technique has never been used traditionally.

(2) **Technical Details**

(i) Suitability

Small semi-circular bunds for rangeland improvement and fodder production can be used under the following conditions:

- Rainfall: 200–750 mm: from arid to semi-arid areas.
- Soils: all soils which are not too shallow or saline.
- Slopes: below 2 %, but with modified bund designs up to 5 %.
- Topography: even topography required

The main limitation of semi-circular bunds is that construction cannot easily be mechanized.

(ii) Overall configuration

Semi-circular bunds with radii of 6 m are constructed in staggered lines with runoff producing catchments between structures.

It is a short slope catchment technique, and is not designed to use runoff from outside the treated area, or to accommodate overflow.

(iii) Catchment:cultivated area ratio

Small semicircular bunds have a *C*:CA ratio of only 1.4:1, and does not require provision for overflow. A detailed calculation is not required. The reasons for applying low ratios are that already adapted rangeland and fodder plants in semi-arid and arid areas need only a small amount of extra moisture to respond significantly with higher yields. Larger ratios would require bigger and more expensive structures, with a higher risk of breaching.

(iv) Bund design

Suitable for slopes of 1 % or less, consists of a series of small semi-circular bunds with radii of 6 m (see Fig. 11.12). Each bund has a constant cross section over the whole length of 19 m. The recommended bund height is 25 cm with side slopes of 1:1 which result in a base width of 75 cm at a selected top width of 25 cm.

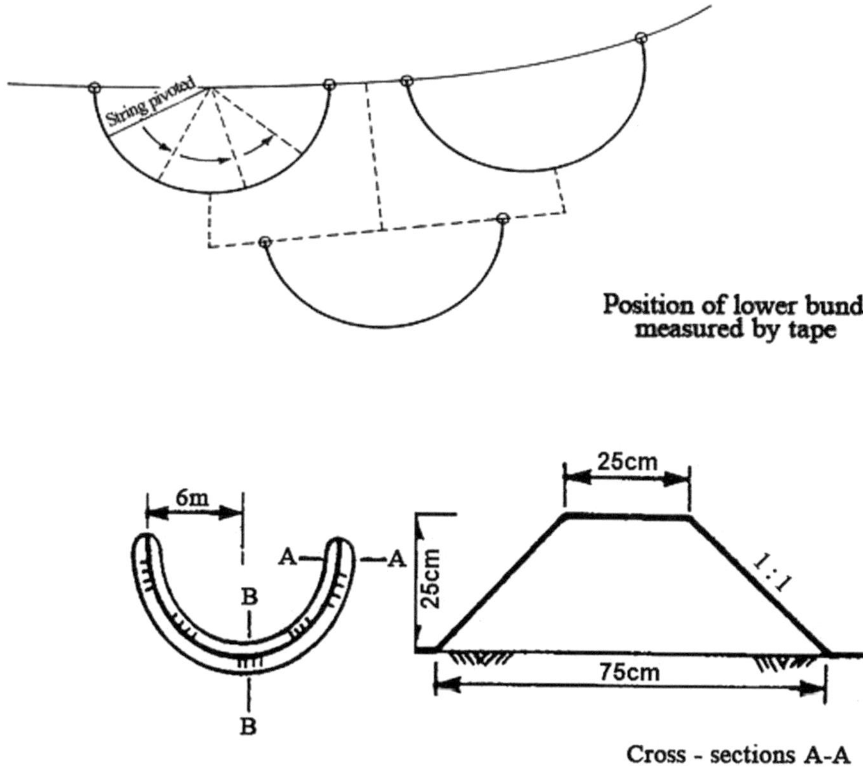

Fig. 11.12 Small semi-circular bunds (Courtesy FAO 1991)

The tips of each bund are set on the contour, and the distance between the tips of adjacent bunds in the same row is 3 m. Bunds in the row below are staggered, thus allowing the collection of runoff from the area between the bunds above. The distance between the two rows, from the base of bunds in the first line to tips of bunds in the second, is 3 m. At this spacing 70–75 bunds per hectare are required.

(v) Quantities of earthworks

Table 11.13 gives quantities of earthworks required for different layouts.

It should be noted that where a diversion ditch is required (Design "a" only), 62.5 m^3 for each 100 m of length has to be added to the figures in column 6.

(vi) Design variations

Semi-circular bunds can be constructed in a variety of sizes, with a range of both radii and bund dimensions. Small radii are common when semi-circular bunds are used for tree growing and production of crops. A recommended radius for these smaller structures is 2–3 m, with bunds of about 25 cm in height.

Table 11.13 Quantities of earthworks for semi-circular bunds (Courtesy FAO 1991)

Land slope (%)	Radius (m)	Length of bund (m)	Impounded area per bund (m^2)	Earthworks per bund (m^3)	Bunds per ha	Earthworks per ha (m^3)
	(1)	(2)	(3)	(4)	(5)	(6)
Design “a” up to 1.0	6	19	57	2.4	73	175
Design “b” up to 2.0	20	63	630	26.4	4	105
4.0	10	31	160	13.2	16	210

(3) **Maintenance**

As with all earthen structures, the most critical period for semi-circular bunds is when rainstorms occur just after construction, since at this time the bunds are not yet fully consolidated. Any breakages must be repaired immediately. If damage occurs, it is recommended that a diversion ditch is provided if not already constructed. Semi-circular bunds which are used for fodder production normally need repairs of initial breaches only. This is because in the course of time, a dense network of the perennial grasses will protect the bunds against erosion and damage. The situation is different if animals have access into the bunded area and are allowed to graze. In this case, regular inspections and maintenance (repair) of bund damages will be necessary. See Fig. 11.13.

(4) **Husbandry**

It may be possible to allow the already existing vegetation to develop—provided it consists of desirable species or perennial rootstocks. In most cases, however, it will be more appropriate to re-seed with seed from outside. Local collection of perennial grass seed from useful species can also be appropriate provided the seed is taken from “virgin land.” Together with grass, trees and shrub seedlings may be planted within the bunds.

11.5.4 Contour Ridges for Crops

(1) **Background**

Contour ridges, sometimes called contour furrows or micro-watersheds, are used for crop production. This is again a microcatchment technique. Ridges follow the contour at a spacing of usually 1–2 m. Runoff is collected from the uncultivated strip between ridges and stored in a furrow just above the ridges. Crops are planted on both sides of the furrow. The system is simple to construct—by hand or by machine—and can be even less labor intensive than the conventional tilling of a plot.

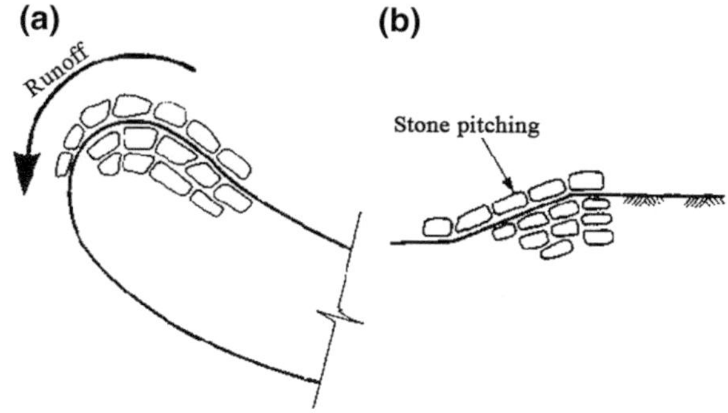

Fig. 11.13 Protection of wingtips. **a** Plon, **b** Side view (Courtesy FAO 1991)

The yield of runoff from the very short catchment lengths is extremely efficient and when designed and constructed correctly there should be no loss of runoff out of the system. Another advantage is an even crop growth due to the fact that each plant has approximately the same contributing catchment area.

Contour ridges for crops are not yet a widespread technique. They are being tested for crop production on various projects in Africa.

(2) **Technical Details**

(i) Suitability

Contour ridges for crop production can be used under the following conditions:

- Rainfall: 350–750 mm.
- Soils: all soils which are suitable for agriculture. Heavy and compacted soils may be a constraint to construction of ridges by hand.
- Slopes: from flat up to 5.0 %.
- Topography: must be even—areas with rills or undulations should be avoided.

(ii) Limitations

Contour ridges are limited to areas with relatively high rainfall, as the amount of harvested runoff is comparatively small due to the small catchment area.

(iii) Overall configuration

The overall layout consists of parallel, or almost parallel, earth ridges approximately on the contour at a spacing of between one and two meters. Soil is excavated and placed down slope to form a ridge, and the excavated furrow above the ridge collects runoff from the catchment strip between ridges. Small earth ties in the furrow are provided every few meters to ensure an even storage of runoff. A diversion ditch may be necessary to protect the system against runoff from outside. See Fig. 11.14.

(iv) Catchment:cultivated area ratio

The cultivated area is not easy to define. It is a common practice to assume a 50 cm strip with the furrow at its center. Crops are planted within this zone, and

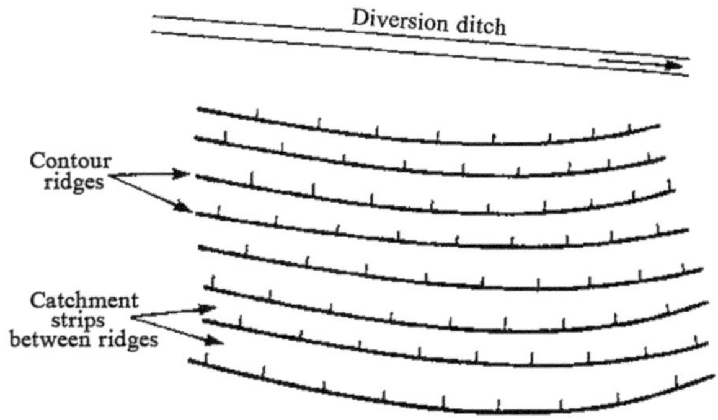

Fig. 11.14 Contour ridges: field layout (Courtesy FAO 1991)

use the runoff concentrated in the furrow. Thus for a typical distance of 1.5 m between ridges, the *C*:CA ratio is 2:1; that is a catchment strip of one meter and a cultivated strip of half a meter. A distance of 2 m between ridges would give a 3:1 ratio. The *C*:CA ratio can be adjusted by increasing or decreasing the distance between the ridges.

The calculation of the catchment: cultivated area ratio follows the general procedure described previously. In practice a spacing of 1.5–2.0 m between ridges (*C*:CA ratios of 2:1 and 3:1, respectively) is generally recommended for annual crops in semi-arid areas.

(v) Ridge design

Ridges need only be as high as necessary to prevent overtopping by runoff. As the runoff is harvested only from a small strip between the ridges, a height of 15–20 cm is sufficient. If bunds are spaced at more than 2 m, the ridge height must be increased. See Fig. 11.15.

(vi) Quantities and labor

Quantities of earthwork for different contour ridge spacings and ridge heights are given in Table 11.14. It should be noted that the construction of the ridges already includes land preparation—no further cultivation is required. Where a diversion ditch is necessary, an additional 62.5 m^3 for each 100 m of length of ditch has to be added.

(vii) Design variations

Design variations developed in Israel are the "runoff strips" ("Shananim") and "strip collectors" (Fig. 11.16). A series of wide, but shallow ridges and furrows,

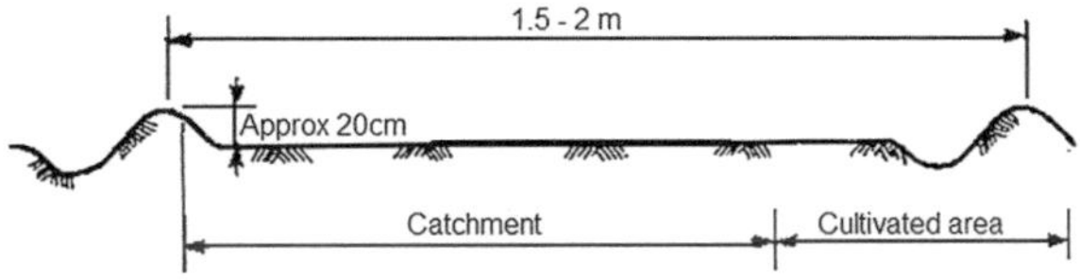

Fig. 11.15 Contour ridge dimensions (Courtesy FAO 1991)

Table 11.14 Quantities of earthworks for contour ridges (Courtesy FAO 1991)

Ridge spacing (m)	Ridge and tie height (cm)	Earthworks per ha (m^3)
1.5	15	270
1.5	20	480
2.0	20	360

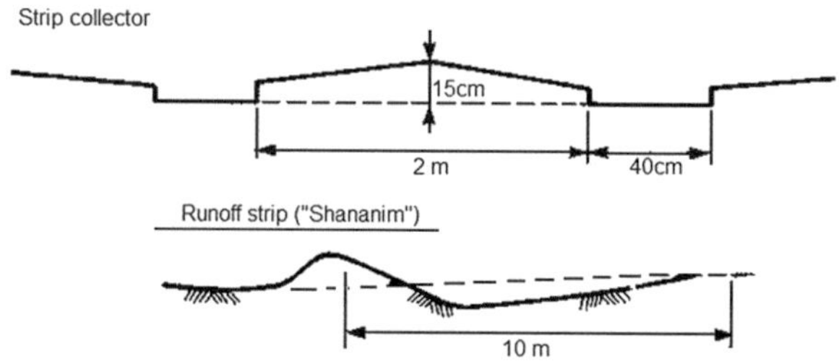

Fig. 11.16 Design variations (Courtesy FAO 1991)

are formed by means of a blade grader. The space in between the ridges can be several meters (for strip a collector, the space is usually between 2 and 5 m).

(3) **Maintenance**

If contour ridges are correctly laid out and built, it is unlikely that there will be any overtopping and breaching. Nevertheless if breaches do occur, the ridges or ties must be repaired immediately. The uncultivated catchment area between the ridges should be kept free of vegetation to ensure that the optimum amount of run-off flows into the furrows.

At the end of each season the ridges need to be rebuilt to their original height. After two or three seasons, depending on the fertility status of the soils, it may be necessary to move the ridges down slope by approximately a meter or more, which will result in a fresh supply of nutrients to the plants.

(4) **Husbandry**

The main crop (usually a cereal) is seeded into the upslope side of the ridge between the top of the ridge and the furrow. At this point, the plants have a greater depth of top soil. An intercrop, usually a legume, can be planted in front of the furrow. It is recommended that the plant population of the cereal crop be reduced to approximately 65 % of the standard for conventional rained cultivation. The reduced number of plants thus has more moisture available in years of low rainfall. See Fig. 11.17

Weeding must be carried out regularly around the plants and within the catchment strip.

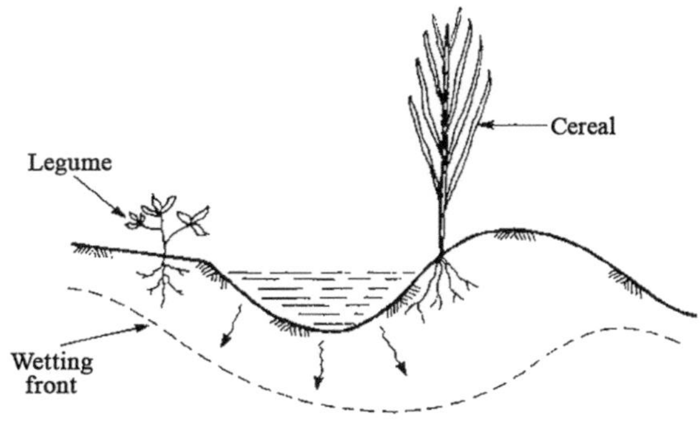

Fig. 11.17 Planting configuration (Courtesy FAO 1991)

11.6 Macrocatchment Systems

11.6.1 Trapezoidal Bunds

(1) **Background**

Trapezoidal bunds are used to enclose larger areas (up to 1 ha) and to impound larger quantities of runoff which is harvested from an external or "long slope" catchment. The name is derived from the layout of the structure which has the form of a trapezoid—a base bund connected to two side bunds or wingwalls which extend upslope at an angle of usually 135°. Crops are planted within the enclosed area. Overflow discharges around the tips of the wingwalls.

The general layout, consisting of a base bund connected to wingwalls is a common traditional technique in parts of Africa. The concept is similar to the semi-circular bund technique: in this case, three sides of a plot are enclosed by bunds while the fourth (upslope) side is left open to allow runoff to enter the field. The simplicity of design and construction and the minimum maintenance required are the main advantages of this technique. This section is based on the design and layout of trapezoidal bunds implemented in Turkana District in northern Kenya.

(2) **Technical Details**

(i) Suitability

Trapezoidal bunds (Fig. 11.18) can be used for growing crops, trees, and grass. Their most common application is for crop production under the following site conditions:

- Rainfall: 250–500 mm; arid to semi-arid areas.
- Soils: agricultural soils with good constructional properties i.e., significant (non-cracking) clay content.
- Slopes: from 0.25 to 1.5 %, but most suitable below 0.5 %.
- Topography: area within bunds should be even.

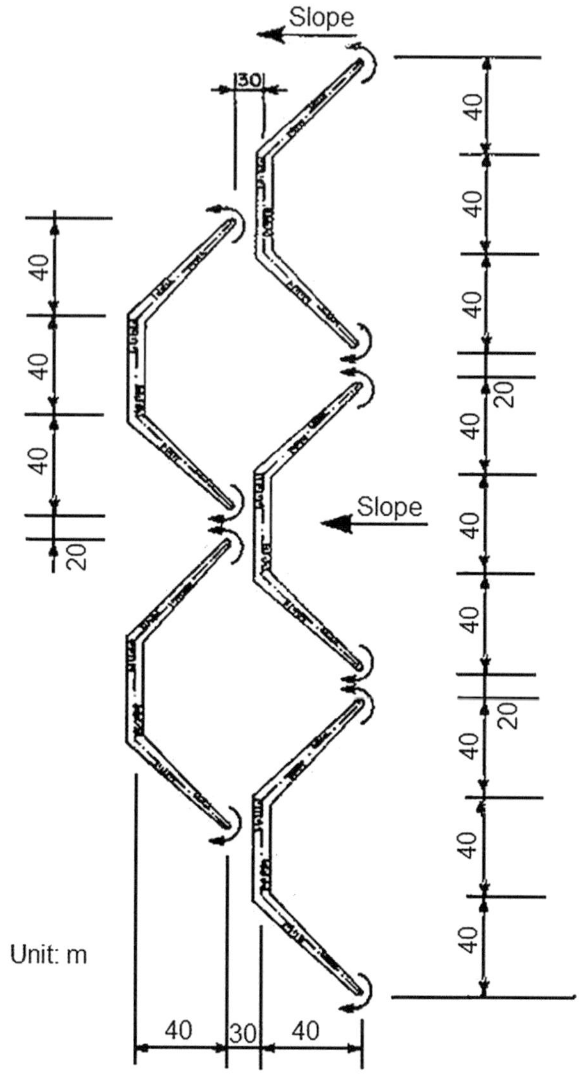

Fig. 11.18 Trapezoidal bunds: field layout for 1 % ground slope (Courtesy FAO 1991)

(ii) Limitations

This technique is limited to low ground slopes. Construction of trapezoidal bunds on slopes steeper than 1.5 % is technically feasible, but involves prohibitively large quantities of earthwork.

(iii) Overall configuration

Each unit of trapezoidal bunds consists of a base bund connected to two wingwalls which extend upslope at an angle of 135°. The size of the enclosed area depends on the slope and can vary from 0.1 to 1 ha. Trapezoidal bunds may be constructed

Table 11.15 Quantities of earthworks for trapezoidal bund (Courtesy FAO 1991)

Slope (%)	Length of base bund (m)	Length of wingwall (m)	Distance between tips (m)	Earthworks per bund (m^3)	Cultivated area per bund (m^2)	Earthworks per ha cultivated (m^3)
0.5	40	114	200	355	9600	370
1.0	40	57	120	220	3200	670
1.5	40	38	94	175	1800	970

Note Where diversion ditches or collection arms are required these add 62.5 m^3 for each 100 m length

as single units, or in sets. When several trapezoidal bunds are built in a set, they are arranged in a staggered configuration; units in lower lines intersect overflow from the bunds above.

A common distance between the tips of adjacent bunds within one row is 20 m with 30 m spacing between the tips of the lower row and the base bunds of the upper row (see Fig. 11.18). The planner is of course free to select other layouts to best fit into the site conditions. The staggered configuration as shown in Fig. 11.18 should always be followed. It is not recommended to build more than two rows of trapezoidal bunds since those in a third or fourth row receive significantly less runoff. Recommended dimensions for one unit of trapezoidal bunds are given in Table 11.15.

(iv) Catchment:cultivated area (C:CA) ratio

The basic methodology of determining C:CA ratio is given previously, for the case where it is necessary to determine the necessary catchment size for a required cultivated area. It is sometimes more appropriate to approach the problem the other way round, and determine the area and number of bunds which can be cultivated from an existing catchment.

Example Calculate the number of trapezoidal bunds needed to utilize the runoff from a catchment area of 20 ha under the following conditions:

Slope	1 %
Crop water requirement	475 mm per season
Design rainfall	250 mm per season
Runoff coefficient	0.25
Efficiency factor	0.50

$$\frac{C}{\text{CA}} = \frac{475 - 250}{250 \times 0.5 \times 0.25} = \frac{225}{31.25} = 7.2.$$

But $C = 20$ ha

Thus CA $= \frac{20}{7.2} = 2.8$ ha.

From Fig. 11.18 the area available for cultivation within one trapezoidal bund on a 1 % slope is 3200 $m^2 = 0.32$ ha.

Therefore, number of bunds required: $N = 2.8/0.32 = 8$

In common with all runoff farming techniques which rely on external catchments, the C:CA ratio is based on seasonal rainfall reliability in a year of relatively low rainfall. In years of high rainfall, and particularly under storm conditions resulting in excessive inflow, damage can be caused to crops and to the bunds themselves. This is particularly the case for bunds on steeper slopes and for those with high C:CA ratios. These results in recommendation of a maximum C:CA ratio of 10:1, although ratios of up to 30:1 are sometimes used. Where the use of a large catchment is unavoidable a temporary diversion ditch is advisable to prevent excessive inflow of runoff. Conversely, in situations where the catchment is not of adequate size, interception ditches can be excavated to lead runoff from adjacent catchments to the bunds.

(v) Bund design

The criteria used in the design of bunds in Turkana were as follows:

Length of base bund	40 m
Angle between base and side bunds	135°
Maximum bund height	0.60 m
Minimum bund height (at tips)	0.20 m

The configuration of the bunds is dependent upon the land slope, and is determined by the designed maximum flooded depth of 40 cm at the base bund. Consequently as the gradient becomes steeper the wing walls extend less far upslope as is illustrated in Fig. 11.18. The greater the slope above 0.5 %, the less efficient the model becomes because of increasing earthwork requirements per cultivated hectare (see Table 11.15).

Bund cross sections are based on a 1 m crest width and 4:1 (horizontal:vertical) side slopes.

(vi) Dimensions and Quantities of Earthworks

Table 11.15 gives details of dimensions and earthworks quantities in the Turkana model for different slopes. Earthworks are also quoted per hectare of cultivated area.

(vii) Design variations

The configurations and design criteria outlined above apply to bunds installed in the Turkana District of Kenya. Considerable variations are possible dependent on climatic, physical, and socio-economic conditions.

Traditional forms of runoff farming (Fig. 11.19), similar to trapezoidal bunds, are found in the clay plains of Eastern Sudan and also in Somalia. In Sudan, the layout of the bunds is rectangular with the wing walls extending upslope at right angles to the base bund. In Northwest Somalia a development project has constructed banana-shaped bunds by bulldozer.

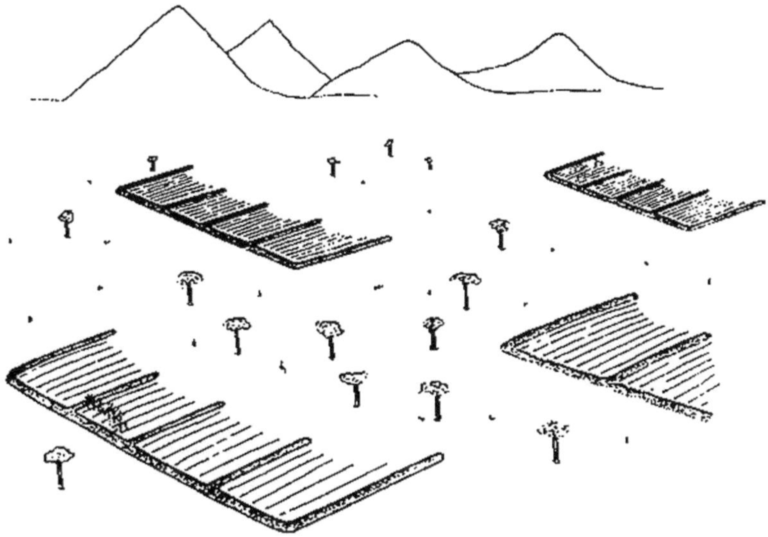

Fig. 11.19 "Teras" system, Eastern Sudan (Courtesy FAO 1991)

(3) **Maintenance**

If there are breaches in the bund, these must be repaired immediately, and the earth compacted thoroughly. Breaches are often caused by poor construction, or because the catchment area is producing damaging amounts of runoff—or both. It is advisable to construct a diversion ditch to protect the repaired bund.

Holes burrowed by rodents can be another cause of breaching. These should be filled in whenever spotted. Allowing natural vegetation to grow on the bunds leads to improved consolidation by the plant roots. Repairs to the wing tips will frequently be needed when overflow has occurred. These should be built up, and extra stone pitching provided if required.

(4) **Husbandry**

Trapezoidal bunds are normally used for production of annual crops in dry areas. The most common crops are cereals, and of these sorghum and bulrush (pearl) millet are by far the most usual. Sorghum is particularly appropriate for such systems because it is both drought tolerant and withstands temporary waterlogging. In the trapezoidal bund, water tends to be unevenly distributed because of the slope, and ponding often occurs near the base bund. Likewise the upper part may be relatively dry. Sorghum can tolerate both these situations.

Planting is carried out in the normal way, after ordinary cultivation of the soil within the bund. It is usual to plough parallel to the base bund, so that the small furrows formed by ploughing will locally accumulate some water. In the driest areas planting is sometimes delayed until a runoff event has saturated the soil within the bund, and germination/establishment is guaranteed. It is also possible to make use of out-of-season showers by planting a quick maturing legume, such as cowpea or tepary beans. Another useful technique is to plant curcurbits like gourds or watermelons on the bottom bund if water ponds deeply.

11.6.2 Contour Stone Bunds

(1) **Background**

Contour stone bunds are used to slow down and filter runoff, thereby increasing infiltration and capturing sediment. The water and sediment harvested lead directly to improved crop performance. This technique is well suited to small-scale application on farmer's fields and, given an adequate supply of stones, can be implemented quickly and cheaply. Making bunds—or merely lines—of stones is a traditional practice in parts of Sahelian West Africa, notably in Burkina Faso. Improved construction and alignment along the contour makes the technique considerably more effective. The great advantage of systems based on stone is that there is no need for spillways, where potentially damaging flows are concentrated. The filtering effect of the semi-permeable barrier along its full length gives a better spread of runoff than earth bunds are able to do. Furthermore, stone bunds require much less maintenance.

Stone bunding techniques for runoff farming (as opposed to stone bunding for hillside terracing, a much more widespread technique) is best developed in Yatenga Province of Burkina Faso. It has proved an effective technique, which is popular and quickly mastered by villagers.

(2) **Technical Details**

(i) Suitability

Stone bunds for crop production can be used under the following conditions:

- Rainfall: 200–750 mm; from arid to semi-arid areas.
- Soils: agricultural soils.
- Slopes: preferably below 2 %.
- Topography: need not be completely even.
- Stone availability: must be good local supply of stone.

(ii) Overall configuration

Stone bunds follow the contour, or the approximate contour, across fields or grazing land. The spacing between bunds ranges normally between 15 and 30 m depending largely on the amount of stone and labor available. There is no need for diversion ditches or provision of spillways.

(iii) Catchment:cultivated area ratio

Contour stone bunds are a long slope technique relying on an external catchment. Theoretical catchment:cultivated area (*C*:CA) ratios can be calculated using the formula given in Sect. 11.4. Initially it is advisable to be conservative in estimation of areas which can be cultivated from any catchment. The area can be extended either downslope or upslope in subsequent cropping seasons, if appropriate.

(iv) Bund design

Although simple stone lines can be partially effective, an initial minimum bund height of 25 cm is recommended, with a base width of 35–40 cm. The bund should be set into a shallow trench, of 5–10 cm depth, which helps to prevent undermining by runoff. As explained in the construction details, it is important to incorporate a mixture of large and small stones. A common error is to use only large stones, which allow runoff to flow freely through the gaps in-between. The bund should be constructed according to the "reverse filter" principle—with smaller stones placed upstream of the larger ones to facilitate rapid siltation.

Bund spacings of 20 m for slopes of less than 1 %, and 15 m for slopes of 1–2 %, are recommended.

(v) Quantities and labor

Table 11.16 gives details of the quantities of stone involved in bunding and the associated labor.

(vi) Design variations

Where there is not enough stone readily available, stone lines can be used to form the framework of a system. Grass, or other vegetative material, is then planted immediately behind the lines and forms, over a period of time, a "living barrier" which has a similar effect to a stone bund. Alternatively, earth contour bunds can be constructed, with stone spillways set into them.

Table 11.16 Quantities and labor requirements for contour stone bunds (Courtesy FAO 1991)

	Bund size cross-section (m^2)	Bund spacing = 15 m		Bund spacing = 20 m	
		Vol. stone m^3/ha	Person days/ha	Vol. stone m^3/ha	Person days/ha
Stones available in field	Small (0.05)	35	70	25	50
	Medium (0.08)	55	110	40	80
Stones transported locally (wheelbarrow etc.)	Small (cross section 0.05 m^2)	35	105	25	75
	Medium (cross section 0.08)	55	165	40	120

Note Labor requirements are very sensitive to availability of stone. The productivity figures quoted above are based on experience where suitable sized stones were available in-field (productivity 0.5 m^3 person-day) or in the immediate locality (productivity 0.33 m^3 person-day). These rates of productivity would decrease significantly if stone has to be transported over greater distances and/or is of too large a size and has to be broken

(3) **Maintenance**

During heavy runoff events stone bunds may be overtopped and some stones dislodged. These should be replaced. A more common requirement is to plug any small gaps with small stones or gravel where runoff forms a tunnel through.

Eventually stone bunds silt-up, and their water harvesting efficiency is lost. It normally takes 3 seasons or more to happen, and occurs more rapidly where bunds are wider apart, and on steeper slopes. Bunds should be built up in these circumstances with less tightly packed stones, to reduce siltation, while maintaining the effect of slowing runoff.

Alternatively grasses can be planted alongside the bund. *Andropogon guyanus* is the best grass for this purpose in West Africa. It can be seeded, and the mature grass is used for weaving into mats. The grass supplements the stone bund and effectively increases its height.

(4) **Husbandry**

Stone bunds in West Africa are often used for rehabilitation of infertile and degraded land. In this context it is recommended that the bunds be supported by a further technique—that of planting pits or "zai." These pits, which are usually about 0.9 m apart, are up to 0.15 m deep and 0.30 m in diameter. Manure is placed in the pits to improve plant growth. The pits also concentrate local runoff which is especially useful at the germination and establishment phase.

As in the case of all cropping systems under runoff farming, an improved standard of general husbandry is important to make use of the extra water harvested. Manuring (as described above) is very important in fertility management. Also essential is early weeding: in areas where stone bunds are commonly used, late weeding is often a constraint to production.

11.7 Flood Farming Systems

11.7.1 Permeable Rock Dams

(1) **Background**

Permeable rock dams are a floodwater farming technique where runoff waters are spread in valley bottoms for improved crop production. Developing gullies are healed at the same time. The structures are typically long, low dam walls across valleys. Permeable rock dams can be considered a form of "terraced wadi," though the latter term is normally used for structures within watercourses in more arid areas.

Interest in permeable rock dams has centered on West Africa and has grew substantially in the latter part of the 1980s. This technique is particularly popular where villagers have experienced the gullying of previously productive valley bottoms, resulting in floodwater no longer spreading naturally. The large amount of work involved means that the technique is labor intensive and needs a group approach, as well as some assistance with transport of stone.

(2) **Technical Details**

(i) Suitability

Permeable rock dams for crop production can be used under the following conditions:

- Rainfall: 200–750 mm; from arid to semi-arid areas.
- Soils: all agricultural soils—poorer soils will be improved by treatment.
- Slopes: best below 2 % for most effective water spreading.
- Topography: wide, shallow valley beds.

The main limitation of permeable rock dams is that they are particularly site-specific, and require considerable quantities of loose stone as well as the provision of transport.

(ii) Overall configuration

A permeable rock dam is a long, low structure, made from loose stone (occasionally some gabion baskets may be used) across a valley floor. The central part of the dam is perpendicular to the watercourse, while the extensions of the wall to either side curve back down the valleys approximately following the contour (see Fig. 11.20). The idea is that the runoff which concentrates in the center of the valley, creating a gully, will be spread across the whole valley floor, thus making conditions more favorable for plant growth. Excess water filters through the dam, or overtops during peak flows. Gradually the dam silts up with fertile deposits.

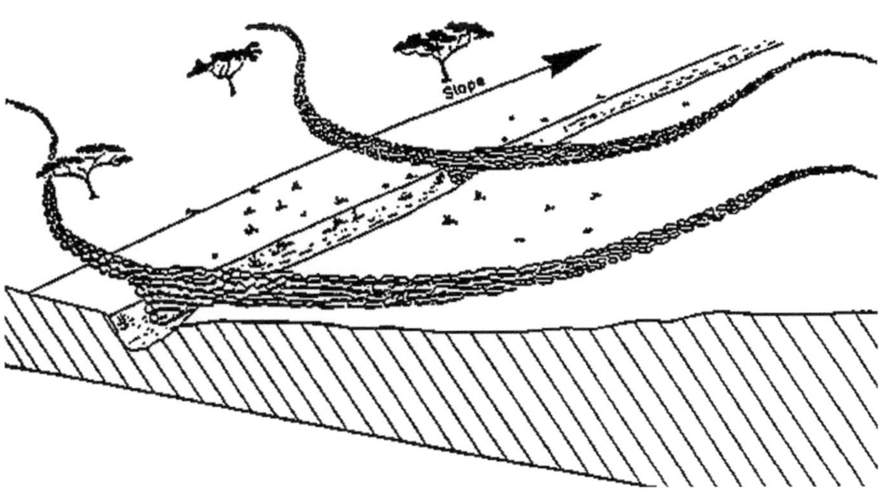

Fig. 11.20 Permeable rock dams: general layout (Courtesy FAO 1991)

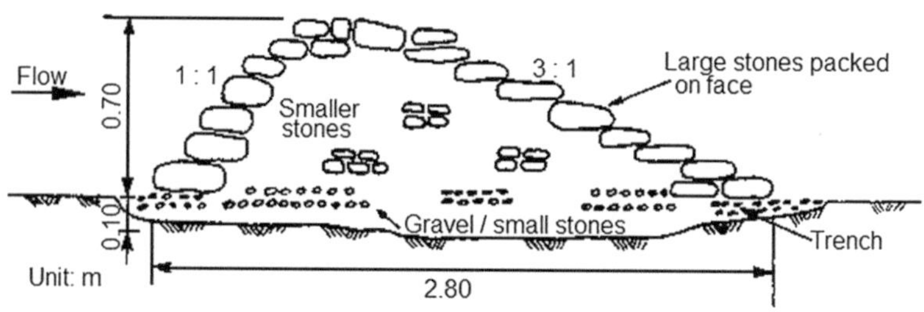

Fig. 11.21 Dam dimensions (Courtesy FAO 1991)

Usually a series of dams is built along the same valley floor, giving stability to the valley system as a whole.

(iii) Catchment:cultivated area ratio

The calculation of the *C*:CA ratio is not necessary as the catchment area and the extent of the cultivated land are predetermined. However, the catchment characteristics will influence the size of structure and whether a spillway is required or not.

(iv) Dam design

The design specifications given below are derived from a number of permeable rock dam projects in West Africa. Each project varies in detail, but the majority conforms to the basic pattern described here.

The main part of the dam wall is usually about 70 cm high although some are as low as 50 cm, see Fig. 11.21. However, the central portion of the dam including the spillway (if required) may reach a maximum height of 2 m above the gully floor. The dam wall or "spreader" can extend up to 1000 m across the widest valley beds, but the lengths normally range from 50 to 300 m. The amount of stone used in the largest structures can be up to 2000 tons.

The dam wall is made from loose stone, carefully positioned, with larger boulders forming the "framework" and smaller stones packed in the middle like a "sandwich." The side slopes are usually 3:1 or 2:1 (horizontal:vertical) on the downstream side, and 1:1 or 1:2 on the upstream side. With shallower side slopes, the structure is more stable, but more expensive.

For all soil types it is recommended to set the dam wall in an excavated trench of about 10 cm depth to prevent undermining by runoff waters. In erodible soils, it is advisable to place a layer of gravel, or at least smaller stones, in the trench.

(v) Quantities and labor

The quantity of stone and the labor requirement for collection, transportation and construction depends on a number of factors and vary widely. Table 11.17 gives the quantity of stone required per cultivated hectare for a series of typical permeable rock dams under different land gradients.

The figures were calculated for a rock dam with an average cross section of 0.98 m^2 (70 cm high, base width of 280 cm) and a length of 100 m. The vertical

Table 11.17 Quantities for permeable rock dams (Courtesy FAO 1991)

Land slope (%)	Spacing between dams[a] (m)	Volume of stone per ha cultivated (m^3)
0.5	140	70
1.0	70	140
1.5	47	208
2.0	35	280

[a]Vertical intervals between adjacent dams = 0.7 m

interval between dams is assumed to be 0.7 m, which defines the necessary spacing between adjacent dams.

Transport of stones by lorries from the collection site to the fields in the valley is the normal method. Considerable labor may be required to collect, and sometimes break, stone. Labor requirements, based on field estimates, are in the range of 0.5 cubic meters of stone per person/day—excluding transport.

(vi) Design variations

Where permeable rock dams are constructed in wide, relatively flat valley floors, they are sometimes made straight across—in contrast to the usual design where the spreader bunds arch back from the center to follow the contour (see Fig. 11.22). With straight dams, the height of the wall decreases from the center toward the sides of the valley to maintain a level crest.

Permeable rock dams are similar in many respects to the "terraced wadis" traditionally used in North Africa and the Middle East. However, the terraced wadi system is used in more arid regions, across clearly defined watercourses. Cross-wadi walls of stone retain runoff to depths of up to 50 cm, with the excess flowing over spillways into successive terraces below. Crops and fruit trees utilize the residual moisture.

The "Liman" system, principally reported from Israel, is used on flood plains or in broad "wadi" beds. Bunds, often of earth, pond water to depths of 40 cm, and excess drains around an excavated spillway. "Limanim" (plural of Liman) may be

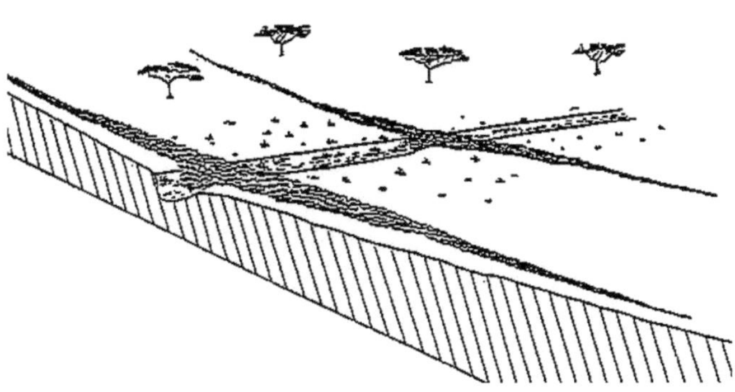

Fig. 11.22 Permeable rock dams: alternative layout (Courtesy FAO 1991)

constructed in series along a wadi bed. This technique is found where rainfall is as low as 100 mm per annum, and is used for crops, fruit trees or forestry.

(3) **Maintenance**

The design given above, with its low side slopes and wide base should not require any significant maintenance work provided the described construction method is carefully observed. It will tolerate some overtopping in heavy floods. Nevertheless there may be some stones washed off, which will require replacing, or tunneling of water beneath the bund which will need packing with small stones. No structure in any runoff farming system is entirely maintenance free and all damage, even small, should be repaired as soon as possible to prevent rapid deterioration.

(4) **Husbandry**

Permeable rock dams improve conditions for plant growth by spreading water, where moisture availability is a limiting factor. In addition, sediment, which will build up behind the bund over the seasons, is rich in nutrients, and this will further improve the crop growth.

This technique is used exclusively for annual crops. In the sandier soils, which do not retain moisture for long, the most common crops are millet and groundnuts. As the soils become heavier, the crops change to sorghum and maize. Where soils are heavy and impermeable, waterlogging would affect most crops, and therefore rice is grown in these zones. Within one series of permeable rock dams, several species of crop may be grown, reflecting the variations in soil and drainage conditions.

11.7.2 Water Spreading Bunds

(1) **Background**

Water spreading bunds are often applied in situations where trapezoidal bunds are not suitable, usually where runoff discharges are high and would damage trapezoidal bunds or where the crops to be grown are susceptible to the temporary waterlogging, which is a characteristic of trapezoidal bunds. The major characteristic of water spreading bunds is that, as their name implies, they are intended to spread water, and not to impound it.

They are usually used to spread floodwater which has either been diverted from a watercourse or has naturally spilled onto the floodplain. The bunds, which are usually made of earth, slow down the flow of floodwater and spread it over the land to be cultivated, thus allowing it to infiltrate.

(2) **Technical Details**

(i) Suitability

Water spreading bunds can be used under the following conditions:

- Rainfall: 100–350 mm; normally hyper-arid/arid areas only.
- Soils: alluvial fans or floodplains with deep fertile soils.
- Slopes: most suitable for slopes of 1 % or below.
- Topography: even.

The technique of floodwater farming using water spreading bunds is very site-specific. The land must be sited close to a wadi or another watercourse, usually on a floodplain with alluvial soils and low slopes. This technique is most appropriate for arid areas where floodwater is the only realistic choice for crop or fodder production.

(ii) Overall configuration

Two design examples are given. The first is for slopes of less than 0.5 %, where the structures are merely straight open ended bunds sited across the slope, which "baffle" (slow and spread) the flow. The second, for slopes greater than 0.5 %, is a series of graded bunds, each with a single short upslope wing, which spread the flow gradually downslope. In each case, crops or fodder are planted between the bunds.

(iii) Catchment:cultivated area ratio

The precise calculation of a catchment:cultivated area ratio is not practicable or necessary in the design of most water spreading bunds (Fig. 11.23). The reasons are that the floodwater to be spread is not impounded—much continues to flow through the system, and furthermore often only part of the wadi flow is diverted to the productive area. Thus the quantity of water actually utilized cannot be easily predicted from the catchment size.

(iv) Bund design

(a) Slopes of less than 0.5 %

Where slopes are less than 0.5 %, straight bunds are used to spread water. Both ends are left open to allow floodwater to pass around the bunds, which are sited at 50 m apart. Bunds should overlap—so that the overflow around one should be intercepted by that below it. The uniform cross section of the bunds is recommended to be 60 cm high, 4.1 m base width, and a top width of 50 cm (Fig. 11.24). This gives stable side slopes of 3:1. A maximum bund length of 100 m is recommended.

(b) Slopes of 0.5–1.0 %

In this slope range, graded bunds can be used. Bunds, of constant cross section, are graded along a ground slope of 0.25 %. Each successive bund in the series downslope is graded from different ends. A short wingwall is constructed at 135° to the upper end of each bund to allow interception of the flow around the bund above. This has the effect of further checking the flow. The spacing between bunds depends on the slope of the land. The bund cross section is the same as that recommended for contour bunds on lower slopes. The maximum length of a base bund is recommended to be 100 m.

Fig. 11.23 Flow diversion system with water spreading bunds in Pakistan (Courtesy FAO 1991)

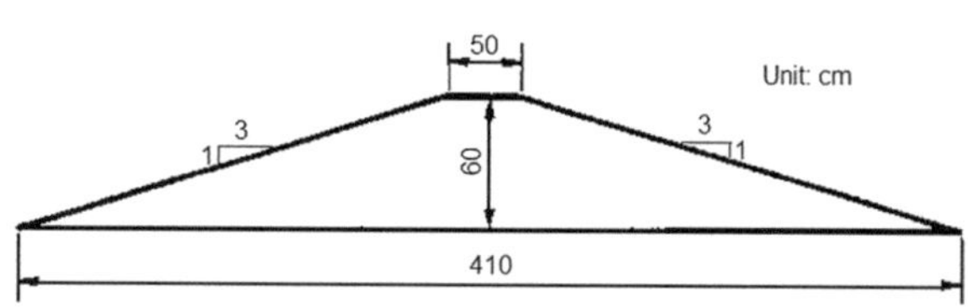

Fig. 11.24 Bund dimensions (Courtesy FAO 1991)

(v) Quantities and labor

Table 11.18 gives details of the quantities and labor involved in construction of water spreading bunds for different slope classes. A bund cross section of 1.38 m^2 is assumed. Labor requirements are relatively high because of the large sized structures requiring soil to be carried.

Table 11.18 Quantities of earthworks for water spreading bunds (Courtesy FAO 1991)

Slope class/technique		No. bunds per ha	Total bund length (m)	Earthworks (m^3/ha)
Level bunds (below 0.5 %)		2	200	275
Graded bund (%)	0.5	2	220	305
	1.0	3	330	455

(vi) Design variations

There are many different designs for water spreading bunds possible, and that given in this material is merely one example. Much depends on the quantity of water to be spread, the slope of the land, the type of soil and the labor available. Existing systems are always worth studying before designing new systems. See Figs. 11.25, 11.26 and 11.27.

(3) **Maintenance**

As is the case in all water harvesting systems based on earth bunds, breaches are possible in the early stages of the first season, before consolidation has taken place. Thus, there must be planning for repair work where necessary and careful inspection after all runoff events. In subsequent seasons, the risk of breaching is diminished, when the bunds have consolidated and been allowed to develop vegetation—which helps bind the soil together, and reduces direct rainfall damage to the structures. Nevertheless with systems which depend on floodwater, damaging floods will inevitably occur from time to time, and repairs may be needed at any stage.

(4) **Husbandry**

Water spreading bunds are traditionally used for annual crops, and particularly cereals. Sorghum and millet are the most common. One particular feature of this

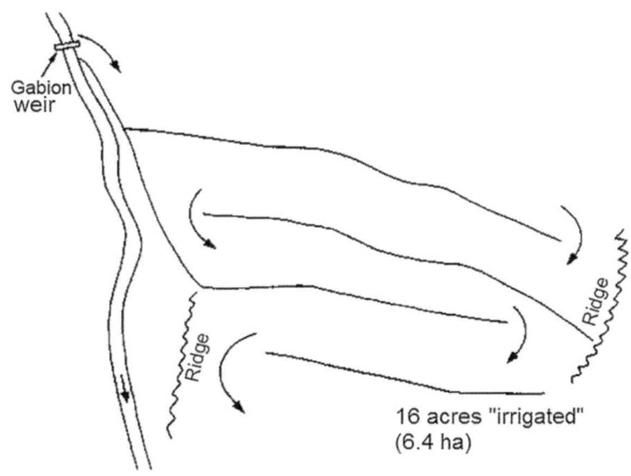

Fig. 11.25 Impala pilot water spreading scheme, Turkana, Kenya (Courtesy FAO 1991)

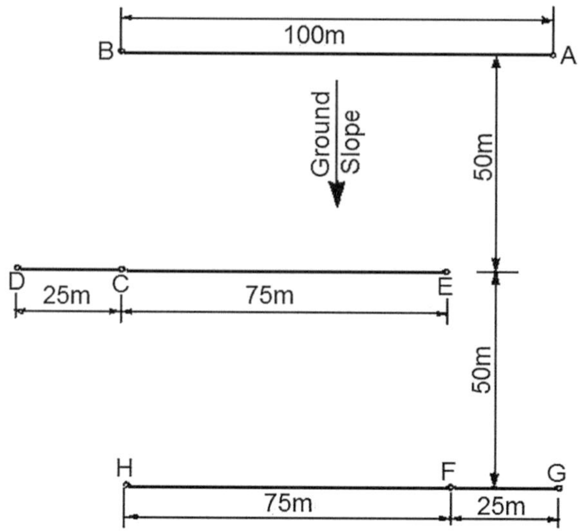

Fig. 11.26 Setting out of level bunds: ground slope <0.5 % (Courtesy FAO 1991)

system, when used in arid areas with erratic rainfall, is that sowing of the crop should be undertaken in response to flooding. The direct contribution by rainfall to growth is often very little. Seeds should be sown into residual moisture after a flood, which gives assurance of germination and early establishment. Further floods will bring the crop to maturity. However, if the crop fails from lack of

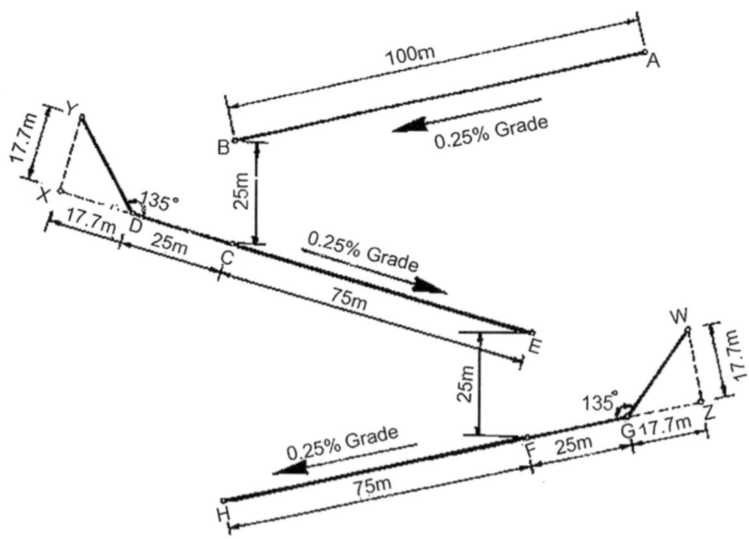

Fig. 11.27 Setting out of graded bunds: ground slope >0.5 % (Courtesy FAO 1991)

subsequent flooding—or if it is buried by silt or sand (as sometimes happens)—the cultivator should be prepared to replant. An opportunistic attitude is required.

Because water spreading usually takes place on alluvial soils, soil fertility is rarely a constraint to crop production. Weed growth however tends to be more vigorous due to the favorable growing conditions, and thus early weeding is particularly important.

11.8 System Planning and Design

11.8.1 Consideration of Major Factors

Factors to be considered in planning and design of runoff farming systems include six major aspects: climate, hydrology, soils, topography, social-economy and agronomy.

(1) **Climate**

Through analysis of data from meteorological services, etc., get clear idea about local climate condition, especially rainfall characteristics, including: total annual rainfall, rainfall patterns (min, max, length of rainy season, etc.), rainfall intensity and evapotranspiration.

(2) **Hydrology**

By means of hydrographical procedure/measurements, local hydrology and water availability to be studied, including intermittent watercourses, rainfall–runoff relationship, soil surface: structure of runoff area, etc.

(3) **Soils**

According to ground truthing, soil cover/land use and satellite imagery, examine the soil characteristics required for runoff farming: medium infiltration rate, high water holding capacity, sufficient depth, nutrients (soil fertility) and low salinity.

(4) **Topography**

Use topographical maps, digital terrain model, satellite data, combine with field measurements to determine land topography.

- >10 % slope: macrocatchments
- <10 % slope: microcatchments

(5) **Social-Economy**

Through interviews, participatory meetings and cost-benefit-analysis, conduct social-economic analysis, including population density, working force, people's priorities, previous experience, land tenure, water laws, accessibility, costs of earthwork/stonework, labor and specialists, machinery, and incentives.

(i) **People's Priorities**

If the objective of runoff farming projects is to assist resource-poor farmers to improve their production systems, it is important that the farmer's priorities are being fulfilled, at least in part. Otherwise success is unlikely.

(ii) **Participation**

It is becoming more widely accepted that unless people are actively involved in the development projects which are aimed to help them, the projects are doomed to failure. It is important that the beneficiaries participate in every stage of the project. When the project is being planned, the people should be consulted, and their priorities and needs assessed. During the construction phase the people again should be involved—not only supplying labor but also helping with field layouts after being trained with simple surveying instruments.

Throughout the course of the season it is helpful to involve people in monitoring, such as rainfall and runoff and recording tree mortality. A further participatory role is in maintenance, which should not be supported by incentives.

After the first season it is the farmers themselves who will often have the best ideas of modifications that could be made to the systems. In this way they are involved in evaluation, and in the evolution of the runoff farming systems.

(iii) **Adoption of Systems**

Widespread adoption of runoff farming techniques by the local population is the only way that significant areas of land can be treated at a reasonable cost on a sustainable basis. It is therefore important that the systems proposed are simple enough for the people to implement and to maintain. To encourage adoption, apart from incentives in the form of tools for example, there is a need for motivational campaigns, demonstrations, training and extension work.

(iv) **Area Differences**

It is tempting to assume that a system which works in one area will also work in another, superficially similar, zone. However, there may be technical dissimilarities such as availability of stone or intensity of rainfall, and distinct socio-economic differences also. For example a system which is best adapted to hand construction may not be attractive to people who normally till with animals. If a system depends on a crop well accepted in one area—sorghum for example—this may be a barrier to acceptance where maize is the preferred food grain.

(v) **Gender and Equity**

If runoff farming is intended to improve the lot of farmers in the poorer, drier areas, it is important to consider the possible effects on gender and equity. In other words, will the introduction of runoff farming be particularly advantageous to one group of people, and exclude others? Perhaps runoff farming will give undue help to one sex, or to the relatively richer landowners in some situations. These are points a projects should bear in mind during the design stage. There is little point in providing assistance which only benefits the relatively wealthier groups.

(vi) **Land Tenure**

Land tenure issues can have a variety of influences on water harvesting projects. On one hand it may be that lack of tenure means that people are reluctant to invest in water harvesting structures on land which they do not formally own. Where land ownership and rights of use are complex it may be difficult to persuade the cultivator to improve land that someone else may use later. On the other hand there are examples of situations where the opposite is the case—in some areas farmers like to construct bunds because it implies a more definite right of ownership.

The most difficult situation is that of common land, particularly where no well defined management tradition exists. Villagers are understandably reluctant to treat areas which are communally grazed.

(vii) **Village Land Use Management**

The whole question of land management by village communities has recently been acknowledged to be extremely important. Degraded land in and around villages can only be improved if land use management issues are faced by the communities themselves. One of the techniques which can assist in rehabilitation of degraded land is runoff farming—but it is only one tool among several others and cannot be effective in isolation. Unless, for example, grazing controls are implemented, there is little point spending money on runoff farming structures for reseeding.

(viii) **Alternate Sources of Water**

The planner must also consider alternate sources of water. These must be compared with runoff farming in cost and in the risk involved. The comparison must take into account the water quality required, operational and maintenance considerations as well as the initial cost. Where alternate water is of better quality, is cheaper to develop, easier to obtain or involves less risk, it should be given priority. An example of this is the development of springs or shallow wells for micro-scale irrigation, prior to runoff farming.

(6) **Agronomy**

(i) Crops

Runoff farming helps crops by providing extra moisture at different stages of growth—although timing cannot be controlled. Periods when the extra moisture can make a significant difference are as follows:

- Around sowing time when germination and establishment can be improved;
- During a mid-season dry spell when a crop can be supported until the next rains;
- While the crop is at the vital stages of flowering and grainfill.

Based on the above understanding, attentions must be paid to following considerations:

(a) Crop choice

The most common cereal crops grown under runoff farming systems are as follows:

- Sorghum (*Sorghum bicolor*) is the most common grain crop under runoff farming systems. It is a crop of the dry areas, and in addition to its drought adaption, it also tolerates temporary waterlogging—which is a common occurrence in some runoff farming systems.
- Pearl Millet (*Pennisetum typhoides*) is grown in the drier areas of West Africa and India, and apart from being drought tolerant, it matures rapidly.
- Maize (*Zea mays*) is occasionally grown under runoff farming but is neither drought adapted nor waterlogging tolerant—but in parts of East and Southern Africa it is the preferred food grain, and farmers are often reluctant to plant millet or sorghum instead.

Legumes are less frequently grown under runoff farming but should be encouraged because of their ability to fix nitrogen and improve the performance of other crops. Suitable legumes are Cowpeas (*Vigna unguiculata*), green grams (*Vigna radiana*), lablab (*Lablab purpureus*), and groundnut (*Arachis hypogea*). All are relatively tolerant of drought and are fast maturing.

(b) Fertility

In dry areas soil fertility is usually the second most limiting production factor after moisture stress. The improvement in the supply of water available to plants under runoff farming can lead to depletion of soil nutrients. Therefore, it is very important to maintain levels of organic matter by adding animal manure or compost to the soil. Inorganic fertilizers are seldom economic for subsistence crops grown under runoff farming.

Crop rotation helps maintain the fertility status. Legumes should be alternated with cereals as often as possible. Intercropping of cereals with legumes—sorghum with cowpeas for example—can also lead to higher overall yields as well as soil fertility maintenance.

Some runoff farming systems actually harvest organic matter from the catchment and therefore build up fertility. This can most clearly be seen with stone bunding techniques which filter out soil and other organic particles, thereby building up fertile deposits.

(c) Other husbandry factors

- Weeds are a problem where runoff farming system is used, due to the favorable growing conditions where water is concentrated. Weeds are especially a problem at the start of the season and therefore early weeding is extremely important.
- Early planting makes the best use of limited rainfall. In some areas it may be best to plant seeds before the rains arrive. This technique is known as "dry planting."
- "Opportunistic" or take-a-chance planting of a quick legume crop like cowpeas can make use of late season or out of season rainstorms.
- Low plant populations in themselves can improve yields in low rainfall zones, and therefore spacing of crops is another important consideration.

(ii) Trees

Runoff farming is used to help tree seedlings become established in dry areas. The microcatchment technique concentrates water around the seedlings, and makes a considerable difference to growing conditions at this vital early stage. In semi-arid areas, tree seedlings in the natural state usually germinate and grow only in years of above average rainfall—runoff farming imitates these conditions.

(a) Choice of species

Table 11.19 summarizes the most important characteristics of the most commonly planted trees in semi-arid areas of Africa and India.

Table 11.19 Characteristics of commonly planted tree species

Species	Characteristics
Acacia albida	"The" agroforestry tree of West Africa. Genuinely multi-purpose. Pods for fodder. Needs water table. Slow at first
Acacia nilotica	Widespread in India and Africa. Likes deep soils and water table. Good fuel/fodder. Quite quick growing
Acacia saligna	Introduced species from Australia. For dune fixation/fodder/windbreaks. Hardy. Fast growing
Acacia Senegal	"Gum arabic" tree producing commercial gum. Good also for fuelwood/fodder. Direct seeding possible. Slow
Acacia seyal	Likes low-lying areas with heavy soils which flood. Good forage/fuelwood. Quite fast early growth
Acacia tortilis	"Umbrella thorn." Good fuelwood and charcoal. Branches for fencing. Pods good fodder. Fast once established
Albizia lebek	From India. Small shade/amenity tree in Sahel. Needs high water table. Foliage for mulch. Rapid growth
Azadirachta indica	Neem tree: from India/Burma. Grown for shade mainly but also good fodder/fuel. Fast growing
Balanites aegyptiaca	"Desert date" widespread and ecologically "flexible." Fodder/edible fruit. Direct seeding possible. Slow
Cassia siamea	Grown for shade, amenity, fuelwood and poles. Better with higher rainfall. Direct seeding possible. Quick
Casuarina equisetifolia	Good on deep sands (also at coasts) so used for dune stabilization. Also fuelwood. Fast growing
Colophosperum mopane	Indigenous to Southern Africa. Poles for construction and leaves for fodder. Firewood. Wood very hard
Eucalyptus camaldulensis	From Australia. Best eucalypt for dry areas. Coppices well. Windbreak/fuelwood. Very quick growing
Prosopis chilensis	Similar to, and often confused with, *P. juliflora*, see below
Prosopis cineraria	Indigenous to NW India where grown as agroforestry tree. Fodder/fuel/building materials. Slow
Prosopis juliflora	Very drought resistant and establishes naturally. May invade better areas. Coppices well. Good fuel supply. Pods for fodder. Quick growth
Ziziphus mauritiana	"Jujube." Produces edible fruit. Can be grafted. Small tree. Branches for fencing. Slow growth

(b) Husbandry

Planting of more than one seedling at a planting station is usually a good idea. The reason is that the additional cost is small compared with the cost of the runoff harvesting structures and there is extra assurance of at least one seedling establishing. It is advisable to plant one seedling at the bottom of the pit/furrow and one by the side. The former may establish better in a dry year, and the second if conditions are wetter.

Direct seeding is a technique which saves all the nursery costs of producing seedlings. However, it does mean that the trees will be slower growing in the early stages. Tree species which can successfully be direct seeded include *Balanites aegyptiaca* and *Prosopis juliflora*.

Weeding is only necessary close to the seedling. The area between the rows of trees should be allowed to grow grass (or even planted with an annual crop) so that some economic benefit can be gained in the first few years, before the trees mature.

(iii) Rangeland and fodder

Although the use of runoff farming for range and fodder is relatively rare, there are some important points to make about husbandry. Often the most important factors are initial fencing and protection, followed by long-term grazing management.

Natural vegetation often gives satisfactory results without reseeding. However, where reseeding is used, it is usually best to collect seeds from local species known to do well in the area.

One advantage of range/grassland systems is that the roots of these perennial species tend to protect structures, and therefore maintenance requirements are reduced.

11.8.2 Identification of Suitable Areas

For selecting suitable runoff farming areas, following methods could be used and combined as needed:

- Field visits
- Experimental plots (e.g., using rainfall simulator)
- Aerial surveys and evaluation of aerial photographs
- Satellite images and their classification and evaluation
- Geographic information systems (GIS)
- Hydrological simulation models (predict runoff from rainfall data)
- Models of plant-soil-water-continuum

11.8.3 Selection of Runoff Farming Techniques

(1) Comparison of runoff farming systems

Flooding farming systems need special condition such as watercourse or river beds, and are easy to distinguish from other type of runoff framing systems. A comparison of microcatchment and macrocatchment systems is given in Table 11.20.

(2) Selection of runoff farming category

A flow chart for identification of runoff farming category is given in Fig. 11.28.

11.9 Project Management

Previous experience has shown that the following factors frequently led a failure of runoff farming projects: technology not appropriate for the country/region; lack of sufficient labor; wrong design; land tenure not properly considered; lack of motivation/involvement of farmers; gender issues; weakness in local capacity; absence of a long-term government policy; not in accordance with culture; too costly; too time-consuming; no monitoring; no cooperation between teams of similar projects. Therefore, successful project implementation needs suitable stage of technology; well organized management activities; dynamic monitoring, reporting and evaluation; proper total project design.

(1) **Suitable Stage of Technology**

Projects should always aim to learn from the people of the target area, in particular about local traditional technology. It is important that the benefits of the new systems should be apparent to the farmer as early as possible. For new techniques

Table 11.20 Comparison of microcatchment and macrocatchment (Courtesy FAO 1991)

Microcatchments	Macrocatchments
+ Simple design + Lower construction costs + Shorter construction period + Design adaptable to specific conditions by varying ratio of catchment/cropping area + High runoff coefficient (only minor water losses due to short conveying distances) + Ratio of catchment/cropping area between 1/1 and 5/1 + Erosion control – Catchment uses potentially arable land – Low crop density, low yield – Higher maintenance demand – Homogenous ground needed for catchment	+ No loss of cropping area for catchment + No change of field borders (property rights are protected) + Large cropped areas possible – Complicated design – Higher construction costs (diversion channels, controlling devices for floods) – Dependent on suitable catchment area near fields – Adaptation to available water amount difficult – Relative small runoff coefficients because of high losses in catchment; therefore need of large catchment areas

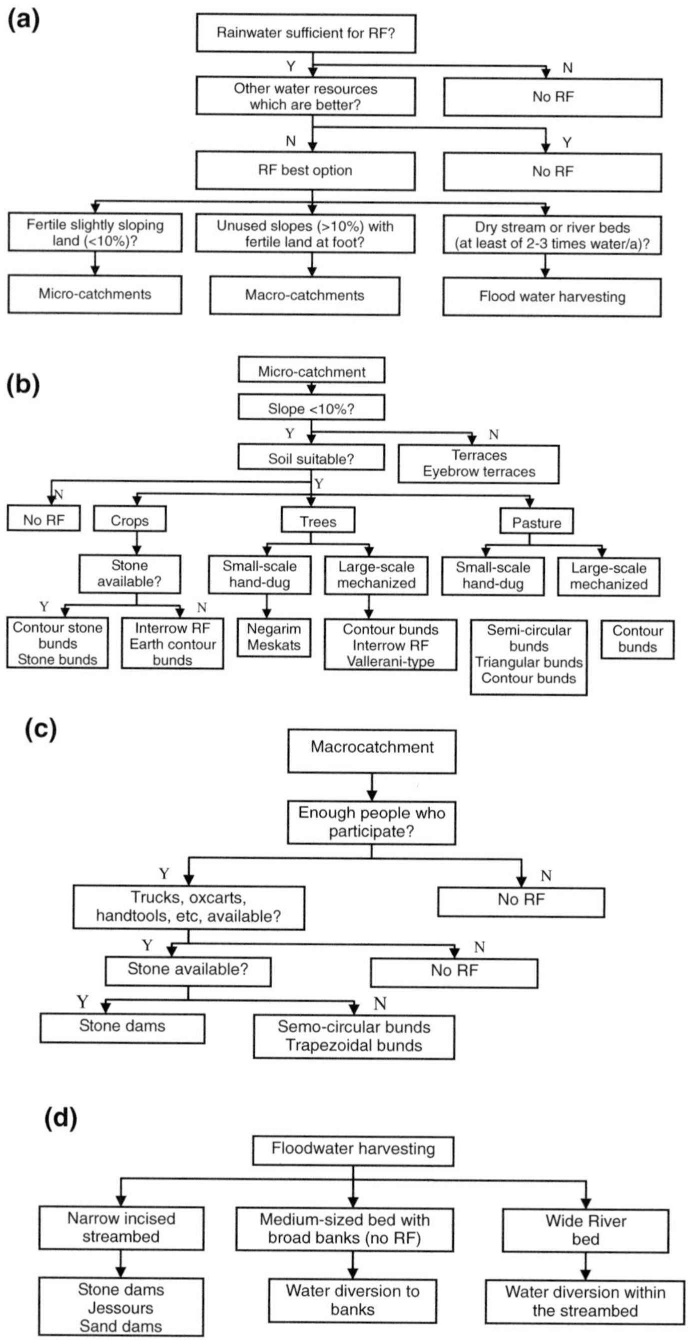

Fig. 11.28 Identification of runoff farming category (Courtesy FAO 1991)

there is often a need for demonstration before people will understand and envisage their effectiveness. Motivation and promotion of awareness among the people with regard to the project objectives and how to achieve them are very important issues.

When quick action is needed, or there is a lack of laborers, or in high rainfall areas, machinery is needed for runoff farming system construction, major types include bulldozers, motor graders, tractors with plough (mainly disc plough) etc. They are normally used for rough shaping of bunds and terraces in combination with "conservation agriculture" practices. While machinery has the advantage of high construction speed, it also faces the problem of repairing, spare parts and fuel, soil compaction and lack of participation of people.

The introduction of inappropriate heavy machinery has been a mistake repeated widely over Africa. Conversely some mechanization—especially where animal traction is a component—can immeasurably speed up work rates and reduce drudgery. The advantage of working by hand is that the people regard the techniques as within their capability. As long as part of the work is voluntary, they will be more willing to carry out maintenance. Nevertheless hand labor is slow, and labor shortages can be a serious constraint in some areas.

(2) **Project Approach**

There are two basically different approaches with regard to runoff farming projects (see Table 11.21).

(i) The Demonstration, Training and Extension Approach:

The technology introduced by the project is relatively simple, and costs per hectare low. The intention is to promote systems which can be taken up and implemented by the people themselves, with a minimum of support. The philosophy behind this approach is that the people themselves must be the prime movers in the development of their own fields and local environment.

(ii) The Implementational Approach:

In this approach the technology may be simple or complex, but it is implemented by the project itself. Machinery is often used, but some projects employ paid (or otherwise rewarded) labor. Costs are often relatively high. The intention is that the project will quickly and efficiently rehabilitate land for the people. The philosophy is that the people are simply unable to undertake the extent of work required using

Table 11.21 Two approaches to runoff farming (Courtesy FAO 1991)

Demonstration, training, extension approach	Implementation approach
= Implementation by people	= Implementation by managers, paid labor
Slow	Quick and efficient rehabilitation of land
Technology simple	Technology simple or complex
Low cost	High cost
More easily adopted	Not easily adopted

their own resources and therefore they require considerable or complete support to implement the project.

Experience shows that it is the first approach which offers the most hope for sustainability once the project has come to an end. Nevertheless there are situations where the introduction of appropriate machinery or support of some labor can be justified.

Runoff farming projects should never have fixed work plans or rigid targets, at least not in the early stages of implementation. The reason quite simply is that it is unrealistic to plan for all contingencies, and arrogant to assume that the techniques and approaches planned from the outset cannot be improved. Learning from experience, and from interaction with the people, is a much better approach. Flexibility should be written into every project document.

(3) **Management Tools**

Many runoff farming projects provide subsidies or incentives for construction. The following points deserve consideration:

- Help and assistance should only be considered as stimulus to the program; too big a subsidy to begin with can cripple future expansion and deter participation.
- It is important that in all cases the beneficiaries should make at least some voluntary contribution toward construction. The level of contribution should rise when incentives are provided.
- Food-for-work is common in projects in drought-prone areas. It is not easy to manage food distribution and development work at the same time. Generally other incentives, such as tools for work, are preferable.
- Incentives/subsidies should not be used for maintenance: this should be the responsibility of the beneficiaries.

A good management tool to identify the responsibilities among donor, government and farmers is an Agreement, which contains at least: project objectives, design, procedures for construction, time-scale, budget and contributions of each party, by-laws and regulations governing the use and maintenance of water harvesting structures and ownership agreement.

(4) **Monitoring, Evaluation and Reporting**

Monitoring, evaluation and reporting are often weak spots in runoff farming projects. Too many projects fail to collect data at even the most basic level. For example crop yields and tree heights are often just estimated. It is also very rare to find any information on the frequency or depth of water harvested. Without a basic monitoring system, projects are starving themselves of data for evaluation. Without clearly written reports, widely circulated, projects are denying to provide others with important information. A suggested monitoring format is presented in Table 11.22.

Table 11.22 Suggested monitoring format for water harvesting projects (Courtesy FAO 1991)

1. *Hydrological data*	
– Rainfall (standard gauges at important sites)	
– Runoff (at least visual recordings of occurrence)	
2. *Inputs*	
–Labor/machinery hours for	(a) Construction
	(b) Maintenance
	(c) Standard agricultural operations
3. *Costs*	
– Labor/machinery use in	(a) Construction
	(b) Maintenance
	(c) Standard agricultural operations
4. *Outputs*	
– Crops: yields of treated plots compared with controls	
– Trees: survival and growth rates	
– Grass/fodder: dry matter of treated plots compared with controls	
5. *Achievements*	
– Area (hectares) covered each season	
– Number farmers/villagers involved/benefiting	
6. *Incentives/support*	
– Quantity and costs	
7. *Training*	
– Number of training sessions	
– Attendance/number of trained personnel	
8. *Extension*	
– Number of farmers visited	
– Number of field days and attendance	

Note Summary Sheets of data are very useful. These could include: (i) labor/ha. (ii) Cost/ha. (iii) Average yield increases over controls. (iv) Total land treated and people benefitted

References

Ben-Asher J. A review of water harvesting in Israel. Working paper for Sub- Saharan Water Harvesting Study. World Bank; 1988.

Critchley W, Reij C. Water harvesting for plant production. World Bank; 1989.

FAO. Water harvesting: a manual for the design and construction of water harvesting schemes for plant production. In: Critchley W, Siegert K, editors. Rome: UN Food and Agriculture Organization; 1991, 133p.

Siegert K. Surface runoff from arid lands (German), Mitteilungen aus dem Leichtweiss Institut, No 58; 1978.

Shanan L, Tadmore N. Micro-catchment systems for arid zone development: a handbook for design and construction. Jerusalem: Hewbrew University; 1979.

Sharma K, Pareek O, Singh H. Microcatchment water harvesting for raising Jujube orchards in an arid climate. Transactions of the ASEA. 1986;29(1):112–8.

Erratum to: Structural Design of the Rainwater Harvesting System

Qiang Zhu

Erratum to: Chapter 3 in: Q. Zhu et al. (eds.), *Rainwater Harvesting for Agriculture and Water Supply*, DOI 10.1007/978-981-287-964-6_3

The original version of this chapter unfortunately contained a mistake. The Figure 3.39 was initially wrong, the correct figure should be the one as given below.

The updated original online version for this chapter can be found at DOI: 10.1007/978-981-287-964-6_3

Q. Zhu (✉)
Gansu Research Institute for Water Conservancy, Lanzhou, China
e-mail: zhuq70@163.com

Q. Zhu et al. (eds.), *Rainwater Harvesting for Agriculture and Water Supply*,
DOI 10.1007/978-981-287-964-6_12

Fig. 3.39 Hand pump for irrigation

Concluding Remarks

Over the past few decades rainwater harvesting has truly come of age and is now widely recognized around the world as a technology that has great potential in tackling the challenges of overcoming water scarcity and providing food security for all in the twenty-first century.

In the 1980s, the spectacular success of the Thai jar program in NE Thailand where several million 2000 L ferrocement water jars were built in less than a decade demonstrated that roof catchment systems could meet domestic household water demand in areas with moderate rainfall. This program was influential and inspirational in encouraging the adoption of RWH for domestic water supply across the developing world.

In the 1990s, the RWH program in Gansu China demonstrated that using a combination of clay tile roofs and cemented courtyard catchments and large excavated 'Shujiao' water cellars rainwater supplies could support household domestic water demand even in semi-arid areas with relatively low and seasonal rainfall.

Since around 2000, the Gansu program has further demonstrated that RWH can also support productive agriculture in a semi-arid region by adopting it in tandem with LORI methods, micro-irrigation systems, and careful water management. In combination with appropriate technologies, such as low cost greenhouses using plastic film, significant increases in crop yields and household incomes can be achieved using quite small quantities of irrigation water.

While the success of the RWH program provides a model with the potential to be replicated in other regions around the world with similar conditions, care needs to be taken to ensure that both the technology and the approach to implementation are adapted and tailored to meet local environmental, economic, and cultural conditions.

Q. Zhu et al. (eds.), *Rainwater Harvesting for Agriculture and Water Supply*,
DOI 10.1007/978-981-287-964-6

Future Challenges and Opportunities

Despite the progress made in rural China a number of key challenges to the continued spread of RWH remain. As a result of the rapid modernization which is now reaching even remote rural locations there are growing calls by government to increase both the quality and quantity of rural domestic water supplies. The Ministry of Water Resources' (MWR) latest policy is aimed at promoting more centralized plants involving treatment and reticulated supplies. However, unless users are charged a much higher price for water, which currently they are unwilling or unable to do, it will prove very difficult to sustain the operation and maintenance costs of centralized water plants in rural areas, especially in those remote, mountainous areas. This focus on centralized, treated, and reticulated supplies will inevitably draw resources and community support away from decentralized RWH systems. Ironically, recent research and demonstration projects around the world are showing that by using first flush and household level treatment of rainwater, high quality potable water supplies can be guaranteed.

Another challenge relates to the use of rainwater for irrigation, some of the tanks built for this purpose are not being used very effectively and in some cases not at all, this is usually because farmers either lack the knowledge or funds to implement drip irrigation systems. This indicates a need for some targeted training of farmers and funding or support mechanisms for improving access to irrigation systems.

Despite these challenges the untapped potential to fully utilize RWH across China to increase agricultural production and rural incomes is very high. A few farmers have become relatively affluent purely through maximizing the opportunities of the RWH and low cost greenhouses and have realized incomes of as much as up to 50 Yuan per cubic meter of water utilized. Some individual families have shown that RWH can be used to fund profitable small-scale horticulture which can support a comfortable rural life and such luxuries as washing machines and hot showers. Such improvements to daily life can help to realize the goal of setting up an overall well-off society in the country.

Around half of China's agriculture is rainfed, and low productivity is common especially in the drier regions. Potentially, if RWH-based irrigation systems could be used to increase agricultural productivity by just 20 % in these areas, this would provide a great boost to the rural economy and help meet the food requirements of China's growing population over the next 50 years. This is certainly feasible since research and demonstrations have shown RWH typically lead to yield increases averaging 40 %. A further opportunity relates to further ecological restoration. Improvements in productivity through supplementary irrigation have already helped many farmers in embracing the state-supported land conversion programs, encouraging them to shift from low-yield crop production on steep hill slopes to planting trees and grass, thus promoting environmental conservation and reducing soil erosion.

The experience from Gansu has clearly demonstrated that RWH provides a decentralized solution for water management that is particularly suitable for populations scattered in mountainous areas. Since the technology is simple and builds on traditional, tried and tested techniques, it is widely accepted by the farmers. Because the implementation is predominantly done by the householders with limited government support, the water users feel ownership of the systems that legally belong to them. Householders therefore have been highly motivated to participate in every stage of project implementation, from planning, design, and construction to operation and maintenance. Unlike large-scale water development schemes that often create significant impacts on ecological systems and can be socially divisive, household-level rainwater harvesting projects are both environmentally friendly and attract significant community support.

Over the past 20 years RWH systems have helped to lift millions of rural people out of poverty. The utilization of rainwater has now become part of an integrated approach for sustainable development in the mountainous areas of Gansu and neighboring semi-arid regions. According to an investigation in 2007, the number of rainwater tanks (and small rainwater storage reservoirs and ponds) constructed in 15 provinces across China amounted to over 10 million units, with a total storage capacity of 4.6 billion m^3. These rainwater systems have provided domestic water supply to 22 million people and supplementary irrigation for 2.8 million ha of rain-fed farmland.

RWH has thus become an important component of water resource development in China. While the approaches to RWH outlined in Part 1 above are tailored specifically to the soil and rainfall conditions of Gansu Province with some adaption, similar techniques could be applied to many arid and semi-arid regions in Africa, South Asia, and Latin America facing similar challenges.

The development of RWH and LORI technologies shows that the world can meet the food and water security needs of a future global population of between 9 and 10 billion people in about 50 years from now. Crucially, using these technologies it will also be possible to meet these future needs in a sustainable way that will not compromise the use of water to provide the numerous environmental services on which we all depend.

Annex

Useful Websites

ASAL Consultants Ltd—Nairobi, asal@wananchi.com Website: www.waterf oraridland.com have produced an excellent series of small, well-illustrated handbooks that will be of great value to water technicians, engineers, planners, and builders implementing a range of appropriate water supply options in rural Africa by E. Nissen-Petersen and others. In 2006/2007, these include: *Water for Rural Communities*, 52p; *Water Supply by Rural Builders*, 60p; *Water Surveys and Designs*, 58p; *Water from Rock Outcrops*, 55p; *Water from dry river beds*, 60p; *Water from Roads*, 57p; *Water from Small Dams* 58p; *Water from Roofs* 78p, CDs, PowerPoints, and Videos on these topics are also available for purchase. The handbooks can be accessed at the website and can be downloaded for free at http://www.waterforaridland.com/Books/Water%20from%20roads.pdf

Asian Institute of Technology (AIT) International Ferrocement Information Center (IFIC)—E-mail: geoferro@ait.ac.th Website: http://www.ait.ac.th/clair/centers/ific

Centre for Science and the Environment (CSE)—http://oneworld.org/cse/html/cmp/cmp43.htm—Rainwater harvesting page—a very active Indian Group, Website: www.rainwaterharvesting.org 41 Tughlakabad Institutional Area, New Delhi 110062, India E-mail: cse@cseindia.org

Development Technology Unit, School of Engineering, University of Warwick—Coventry CV4 7AL, UK. E-mail: dtu@eng.warwick.ac.uk. Website www.eng.warwick.ac.uk/DTU/rainwaterharvesting/index.htm—a number of case studies from around the world, with good descriptions. Roofwater Harvesting: A Handbook for Practictioners, Thomas T. and Martinson D 2007 IRC

FAKT—Consult for Management, Training, and Technologies Email: fakt@fakt-consult.de Website: www.fakt-consult.de, FAKT have developed a web-based Toolkit on Rainwater Harvesting: http://rainwater-toolkit.net and have various

Q. Zhu et al. (eds.), *Rainwater Harvesting for Agriculture and Water Supply*,
DOI 10.1007/978-981-287-964-6

resources including a video "Mvua ni Maji—Rain is Water" (language versions in English, French, German, and Swahili) about Ugandan women who visited Kenya to learn about rainwater harvesting in 1996

Global Applied Research Network (GARNET)—info.lut.ac.uk/departments/cv/wedc/garnet/tncrain.html Site of the Global Applied Research Network (GARNET) Rainwater Harvesting Page.

GRIWAC (China)—Rainwater Harvesting in the Loess Plateau of Gansu, China—a paper presented at the 9th **IRCSA** Conference in Brazil, Email: gssk@163.com

International Rainwater Catchment Systems Association (IRCSA)—Archive of conference papers: www.eng.warwick.ac.uk/ircsa/

IRC (The International Water and Sanitation Centre)—This site has some great freely downloadable resources on Rainwater Harvesting and Water and Sanitation in general—including the excellent book ***Roofwater Harvesting*** by Thomas and Martinson 2007. PO Box 93190, 2509 AD The Hague, Netherlands Email: www.ircwash.org

The **International Rainwater Harvesting Alliance** (IRHA)—was created in Geneva in November 2002 following recommendations formulated during the World Summit for Sustainable Development in Johannesburg two months earlier. The mandate called for federation and unification of the disparate rainwater harvesting (RWH) movement around the world—http://www.irha-h2o.org/

Lanka Rainwater Harvesting Forum (LRWF)—c/o Practical Action South Asia, 5 Lionel Eridisinghe Mawatha, Colombo 5, Sri Lanka E-mail: rwhf@itdg.lanka.net Website: http://www.rainwaterharvesting.com/

People for promoting Rainwater Utilisation—1-8-1 Higashi-Mukojima, Sumida City, Tokyo, Japan E-mail: murase-m@tc4.so-net.ne.jp

Practical Action—www.practicalaction.org an excellent source of information on WATSAN appropriate technologies and innovations and publishers of Waterlines.

Rain Centre (India)—http://www.rainwaterharvesting.org/raincentre.htm: Raincenters are a network of permanent exhibitions that seek to spread water literacy among urban Indians. They define the role played by every Indian citizen in harvesting rainwater and using it to combat the menace of water scarcity. The raincenters provide people the know-how to harvest rain.

RAIN Foundation—www.rainfoundation.org The Netherlands-based RAIN foundation has been working since 2003 to spread, develop, and implement rainwater harvesting systems around the world and currently supports partner projects in Africa, Asia, and Europe.

Rainwater Harvesting (Australia)—This is a commercial site offering a range of products all designed to collect, store, and distribute higher volumes of cleaner water for all homes, specializing in water catchment, storage, insect/pest control, and leaf and debris removal systems used in both rainwater harvesting and home protection. http://rainharvesting.com.au/knowledge-center/rainwater-harvesting-intro/

Rainwater Tank Performance Calculator—This is a free online tool to assist users in designing roof catchment systems providing guidance on the performance of rainwater tanks of specified volumes from a given catchment area. To use this tool, 10 years of mean monthly rainfall data input is needed from the user. www.warwick.ac.uk/fac/sci/eng/research/civil/crg/dtu-old/rwh/model

Rainwater Wiki—www.rain4food.net/wiki/#akvopedia:Rainwater Harvesting: A knowledge library on Tools, Technologies and Innovation Pilots, all specifically about rainwater harvesting. It has been developed by the Rainwater for Food Security program, in collaboration with Akvo Foundation.

Roofwater harvesting discussion forum—www.jiscmail.ac.uk/lists/rwh.html

SA WATER (South Australian Water Corp.)—www.sacentral.sa.gov.au/agencies/saw

SEARNET—The mission of SearNet is to network among its member associations within the region for the promotion of rainwater harvesting and utilization. There are a total of 18 countries covered in the joint venture between CSE and GWP-Associated program in Eastern and Southern Africa, India, Pakistan, Sri Lanka, Nepal, Bangladesh, and Bhutan. www.Searnet.net

SimTanka—Software for sizing reliable rainwater harvesting systems with covered storage tanks—SimTanka, is freely available. http://sourceforge.net/projects/simtanka/

SODIS—The SODIS Reference Center based at Eawag/Sandec is engaged in providing information, technical support, and advice to local institutions in developing countries for the worldwide promotion and dissemination of the Solar Water Disinfection Process www.sodis.ch

SOPAC South Pacific Applied Geoscience Commission—excellent source of information on domestic rainwater harvesting on small and remote islands www.sopac.org

The Pelican Tank Rainwater Collection System—a packaged RWH collection system developed in Australia for developing countries www.trade.altconcepts.net

UNEP publications—www.unep.or.jp/ietc/Publications/TechPublications/TechPub-8e/index.html. Sourcebook of Alternative Technologies for Freshwater Augmentation in Some Countries in Asia—another in this series of **UNEP** publications www.unep.or.jp/ietc/Publications/TechPublications/TechPub-8d/index.html#1—'Sourcebook of Alternative Technologies for Freshwater Augmentation in Small Island Developing States' that includes some useful information on RWH

World Bank—Save the Rain—This tool allows you to identify your own home on Google Earth and estimate how much rain you could potentially collect from your roof each year—http://save-the-rain.com/SR2/

World Meteorological Organisation (WMO)—www.wmo.ch

WELL—www.lboro.ac.uk/well/resources/technical-briefs/36-ferrocement-water-tanks.pdf A technical brief on how to make a ferrocement water tank

Note: *The above websites which were current at the time of publication are a good starting point for searching for information online but new website sites and online resources are appearing all the time—so undertaking your own search using Google or other reputable search engine would be worthwhile.*

MIX
Papier aus verantwortungsvollen Quellen
Paper from responsible sources
FSC® C105338

If you have any concerns about our products,
you can contact us on
ProductSafety@springernature.com

In case Publisher is established outside the EU,
the EU authorized representative is:
Springer Nature Customer Service Center GmbH
Europaplatz 3, 69115 Heidelberg, Germany

Printed by Libri Plureos GmbH
in Hamburg, Germany